Advances in

ECOLOGICAL RESEARCH

VOLUME 16

Advances in

ECOLOGICAL RESEARCH

Edited by

A. MACFADYEN

School of Biological and Environmental Studies, New University of Ulster, Coleraine, County Londonderry, Northern Ireland

E. D. FORD

Center for Quantitative Science, University of Washington, 3737 15th Avenue Seattle, Washington 98195, USA

VOLUME 16

1987

ACADEMIC PRESS

Harcourt Brace Jovanovich, Publishers

London Orlando
San Diego New York Austin Boston
Sydney Tokyo Toronto

ACADEMIC PRESS INC. (LONDON) LTD.
24/28 Oval Road
London NW1

United States Edition published by
ACADEMIC PRESS INC.
Orlando, Florida 32887

British Library Cataloguing in Publication Data

Advances in ecological research.
Vol. 16
1. Ecology
I. Macfadyen, A. II. Ford, E. D.
574.5 QH541

ISBN 0-12-013916-2

Typeset by Paston Press, Loddon, Norfolk
and printed in Great Britain
by St Edmundsbury Press
Bury St Edmunds, Suffolk

Contributors to Volume 16

D. BINKLEY, *School of Forestry and Environmental Studies, Duke University, Durham, North Carolina 27706, USA.*

R. BUCKLEY, *Amdel, PO Box 114, Eastwood, South Australia 5063, Australia.*

C. H. GIMINGHAM, *Department of Plant Sciences, University of Aberdeen, St Machar Drive, Aberdeen, AB9 2UD.*

R. J. HOBBS, *CSIRO Division of Wildlife and Rangelands Research, LMB 4, PO Midland, Western Australia 6056, AUSTRALIA.*

E. N. G. JOOSSE, *Department of Animal Ecology, Vrijeuniversiteit, Postbus 7161, 1007 M.C., Amsterdam, THE NETHERLANDS.*

E. KUNO, *Laboratory of Entomology, Faculty of Agriculture, Kyoto University, Kyoto 606, JAPAN.*

D. RICHTER, *School of Natural Resources, University of Michigan, Ann Arbor, Michigan 48109, USA.*

H. A. VERHOEF, *Department of Animal Ecology, Vrijeuniversiteit, Postbus 7161, 1007 M.C., Amsterdam, THE NETHERLANDS.*

Contents

Contributors to Volume 16 v

Nutrient Cycles and H^+ Budgets of Forest Ecosystems

D. BINKLEY and D. RICHTER

I. Introduction 2
II. The nature of Soil Acidity 3
III. Composing H^+ Budgets 7
A. Cellular Level 7
B. Root/Soil Interface 8
IV. Whole Ecosystem Nutrient Cycling 9
A. The Carbon Cycle 9
B. The Nitrogen Cycle 12
C. The Phosphorus Cycle 15
D. The Sulfur Cycle 19
E. The Calcium Cycle 21
V. Ecosystem and Environment 22
A. Nitrogen Inputs and Outputs 24
B. Sulfur Inputs and Outputs 26
C. Cation Inputs and Outputs 27
VI. Time Scales 27
A. Seasonal Time Scales 28
B. Annual Time Scales 28
C. Secondary Plant Successional Time Scales 29
D. Pedogenic Time Scales 30
VII. Effects of Forest Management 31
VIII. Effects of Species on H^+ Budgets and Soil Acidity 32
A. Afforestation 32
B. Comparisons of Tree Species within Plantations 34
C. Importance of Below Ground Litter, Microflora and Soil Animals . 36
IX. Sensitivity of Forest Ecosystems to Acidification from Atmospheric Deposition 37
A. Buffering Processes in Forest Soils 38
X. A Chestnut Oak Forest Case Study 40
XI. Conclusions 44
Acknowledgements 45
References 45

Ant–Plant–Homopteran Interactions

R. BUCKLEY

I. Summary . . . 53
II. Introduction . . . 54
III. The Twofold Components – Direct or Pairwise Interactions . . . 56
A. Ant–Plant . . . 56
B. Homopteran–Plant . . . 57
C. Ant–Homopteran . . . 60
IV. The Indirect Components . . . 66
A. Plants Affecting Ants via Homoptera . . . 66
B. Plants Affecting Homoptera via Ants . . . 66
C. Ants Affecting Plants via Homoptera . . . 67
D. Ants Affecting Homoptera via Plants . . . 68
E. Homoptera Affecting Ants via Plants . . . 68
F. Homoptera Affecting Plants via Ants . . . 68
V. Discussion and Conclusions . . . 70
A. Effect on Host Plant Populations, and Long-Term Persistence of APHI . . . 70
B. Abiotic Factors, and Embedding in Higher-Order Interactions . . . 72
C. Directions for Research . . . 73
Acknowledgements . . . 73
References . . . 74

Vegetation, Fire and Herbivore Interactions in Heathland

R. J. HOBBS and C. H. GIMINGHAM

I. Introduction . . . 87
II. Vegetation Dynamics and Productivity . . . 90
A. The Dynamics of *Calluna vulgaris* . . . 90
B. Production and Organic Matter Accumulation . . . 95
C. Community Dynamics . . . 103
III. Effects of Fire on Habitat and Vegetation . . . 112
A. Fire Characteristics . . . 112
B. Ecosystem Effects . . . 117
C. Post-Fire Vegetation Development . . . 120
IV. Herbivore Dynamics and Interactions with Vegetation . . . 129
A. Red Grouse . . . 129
B. Red Deer . . . 137
C. Mountain Hares . . . 140
D. Sheep . . . 142
E. Invertebrates . . . 144
V. Management and Conservation of Heathlands . . . 147
A. Grazing and Burning . . . 149
B. Recreation . . . 152
C. Conservation . . . 152
VI. Conclusions . . . 155
VII. Summary . . . 157
Acknowledgements . . . 158
References . . . 158

Developments in Ecophysiological Research on Soil Invertebrates

E. N. G. JOOSSE and H. A. VERHOEF

I. Introduction 175
II. Coastal Sand Dunes and Heathlands 177
A. Drought 177
B. Cold and Heat 190
C. Starvation 200
D. Mineral Shortage 206
E. Toxic Compounds 209
III. Polluted Environments 213
A. Air Pollution 214
B. Heavy Metals 216
IV. Summary and Conclusions 233
References 234

Principles of Predator–Prey Interaction in Theoretical, Experimental and Natural Population Systems

E. KUNO

I. Introduction 250
II. Mathematical Models for Predator–Prey Interaction 252
A. Framework of Predator–Prey Interaction: Classic Models . . . 252
B. Basic Models 253
C. Extended Models 256
III. Patterns of Predator–Prey Dynamics in Theoretical Population Systems . 262
A. Classic Models 262
B. Basic Models 262
C. Extended Models 274
IV. Regulation of Prey Population by Predation: Its Possibility in Theoretical Population Systems 281
A. Prey Population Regulation by Predation – its Definition and Detection 282
B. Prey Population Regulation in Classic Models 282
C. Prey Population Regulation in Basic Models 283
D. Prey Population Regulation in Extended Models 283
E. Robustness of Prey Population Regulation in a Varying Environment 288
F. Non-Regulatory Suppression of Prey Density by Predation . . . 296
V. Conflict of Interests between Individual and Population: Analysis of an Evolutionary Paradox in Theoretical Predator–Prey Systems 299
A. Selection Among Prey Individuals 299
B. Selection Among Predator Individuals 301
C. Selection Between "Specialists" and "Generalists" in Either Predator or Prey Species 302
D. Predator–Prey Conflict in their Coevolution 305
VI. Predator–Prey Dynamics in Experimental Population Systems . . . 311
A. Protozoan Predator–Prey Systems in the Laboratory 312
B. Arthropod Predator–Prey Systems in the Laboratory 313
C. Cases of Biological Pest Control in the Field 316

VII. Predator–Prey Dynamics in Natural Population Systems 318
A. Insect Populations – Epidemic Species 319
B. Insect Populations – Endemic Species 320
C. Populations of Birds and Mammals 324
VIII. Discussion and Conclusions 326
Acknowledgements 331
References 331

Nutrient Cycles and H^+ Budgets of Forest Ecosystems

D. BINKLEY and D. RICHTER

I. Introduction . . . 2
II. The nature of Soil Acidity . . . 3
III. Composing H^+ Budgets . . . 7
A. Cellular Level . . . 7
B. Root/Soil Interface . . . 8
IV. Whole Ecosystem Nutrient Cycling . . . 9
A. The Carbon Cycle . . . 9
B. The Nitrogen Cycle . . . 12
C. The Phosphorus Cycle . . . 15
D. The Sulfur Cycle . . . 19
E. The Calcium Cycle . . . 21
V. Ecosystem and Environment . . . 22
A. Nitrogen Inputs and Outputs . . . 24
B. Sulfur Inputs and Outputs . . . 26
C. Cation Inputs and Outputs . . . 27
VI. Time Scales . . . 27
A. Seasonal Time Scales . . . 28
B. Annual Time Scales . . . 28
C. Secondary Plant Successional Time Scales . . . 29
D. Pedogenic Time Scales . . . 30
VII. Effects of Forest Management . . . 31
VIII. Effects of Species on H^+ Budgets and Soil Acidity . . . 32
A. Afforestation . . . 32
B. Comparisons of Tree Species within Plantations . . . 34
C. Importance of Below Ground Litter, Microflora and Soil Animals . . . 36
IX. Sensitivity of Forest Ecosystems to Acidification from Atmospheric Deposition . . . 37
A. Buffering Processes in Forest Soils . . . 38
X. A Chestnut Oak Forest Case Study . . . 40
XI. Conclusions . . . 44
Acknowledgements . . . 45
References . . . 45

ADVANCES IN ECOLOGICAL RESEARCH Vol. 16
ISBN 0-12-013916-2

I. INTRODUCTION

The acidity of a forest soil affects a wide range of ecological processes, including the solubility and exchange reactions of inorganic nutrients and toxic metals, the activities of soil animals and microorganisms, and the weathering of soil minerals. Changes in soil acidity result from an array of interacting processes that produce and consume hydrogen ions (H^+). One source of acid is deposition from the atmosphere, in precipitation, fog-drip and impaction of dry aerosols. Increases in the acidity of precipitation in industrialized regions have led to concern over increased H^+ input to aquatic and terrestrial ecosystems. In some regions, aquatic ecosystems have been acidified by acid deposition (Drablos and Tollan, 1980; Bubenick, 1984; Dillon and LaZerte, 1985), but the case is unclear for terrestrial ecosystems (Johnson and Siccama, 1983; Burgess, 1984; Postel, 1984). Recent forest declines in Europe ("Waldsterben" in Germany) are associated with a suite of air pollutants, and acid deposition is one of several candidate causes. Because acid input from the atmosphere is only one component of the H^+ budget of an ecosystem (Table 1), assessments of the probable impacts of acid deposition must include an accounting of both the buffering capacity of ecosystems and the natural production of H^+ within the ecosystems.

Table 1

Magnitude of some H^+ fluxes (kmol H^+ ha^{-1} annually). The H^+ in rain may not have a proportional effect on ecosystem acidity, as internal ecosystem processes may consume H^+. Fertilizers may acidify soils, whether added in acid form or salt form. Mineral weathering is a major process consuming H^+ in soils.

Component	Source	Sink
1. 100 cm of rain at pH 5	0·1	
2. 100 cm of rain at pH 4	1·0	
3. Nutrient accumulation in biomass of 450-year-old Douglas-fir forest[a]	0·2	
4. Nutrient accumulation in biomass of 4-year-old poplar plantation[b]	5·0	
5. 200 kg ha^{-1} N fertilizer	0–8	
6. 14 kg N ha^{-1} as nitric acid rain	0–1	
7. 14 kg N ha^{-1} of ammonium salt in rain	0–2	
8. Weathering of 1500 kg ha^{-1} of glacial till[c]		2·2

[a] Calculated from Sollins *et al.* (1980).
[b] Calculated from Wittwer and Immel (1980).
[c] From Likens *et al.* (1977) and Driscoll and Likens (1982).

The components of H^+ budgets have been examined separately for several decades, including the acidity associated with N fertilizers (Volk and Tidmore, 1946; Abruna *et al.*, 1958) and the maintenance of electroneutrality during ion uptake by the secretion of H^+ or OH^- (Ulrich, 1941; review by Mengel and Kirkby, 1979). Studies of the H^+ budget of entire ecosystems began in the 1970s (Pierre and Banwart, 1973; Pierre *et al.*, 1971; Helyar, 1976; Reuss, 1977), and a H^+ budget for a forest ecosystem was first estimated as a part of the Swedish Coniferous Forest Project (Sollins, 1976). By 1985, more than two dozen H^+ budgets have been calculated for forest ecosystems (Andersson *et al.*, 1980; Sollins *et al.*, 1980; Ulrich, 1980; Driscoll and Likens, 1982; Nilsson *et al.*, 1982; Ulrich and Pankrath, 1983; Johnson *et al.*, 1985), and the number will probably double by the end of the decade. These budgets differ in structure and compartments, but all indicate that the generation and consumption of H^+ within ecosystems is very substantial relative to atmospheric deposition.

An understanding of the potential impacts of acid deposition, and forest management in general on ecosystem acidity requires a framework for evaluating the interacting effects of H^+ sources, H^+ pools and H^+ sinks. In this paper we review the nature of soil acidity, and explore the role of nutrient cycles in the H^+ dynamics of forest ecosystems. We focus on the importance of spatial and temporal scales in determining the effects of nutrient cycles on H^+ dynamics, and discuss the effects of forest management and vegetation types. We continue our discussion by considering the factors that regulate the sensitivity of forest ecosystems to acidification, and conclude by presenting an integrated H^+ budget for a chestnut oak forest. This H^+ budgeting provides the necessary framework for the evaluation of ecosystem sensitivity to internal and external sources of H^+.

We used a wide variety of sources to develop this synthesis. Some of the major references are Handley (1954), Pearson and Adams (1967), Helyar (1976), Reuss (1977), Mengel and Kirkby (1979), Clarkson and Hanson (1980), Leninger (1982), Bielski and Ferguson (1983), Bohn *et al.* (1979), Adams (1984).

II. THE NATURE OF SOIL ACIDITY

Of the large amount of H^+ present in acidic forest soils, only a very small fraction exists as free H^+ in soil solution. For example, an extremely acid soil might have a total acidity of 100 to 1000 kmol of H^+ per hectare to a depth of 60 cm, yet contain only 0·1 kmol ha^{-1} of free H^+ in soil solution (Fig. 1). Unravelling H^+ budgets requires some definitions about soil acidity.

Three types of soil acidity are recognized, based on conventional and convenient analytical techniques. The first is free H^+ in soil solution,

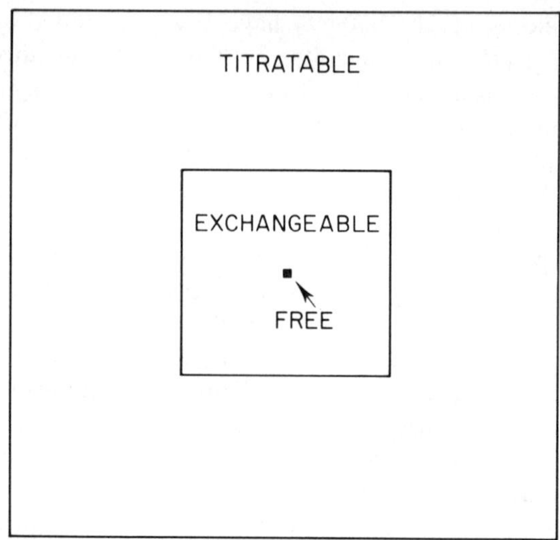

Fig. 1. Soil pH measurements account only for H^+ free in soil solution. Exchangeable acidity includes free H^+ plus H^+ (and aluminium) on cation exchange sites. Titratable acidity includes all H^+ pools that must be neutralized to raise pH to 7 or 8. Size of the boxes corresponds to the relative pool sizes in the chestnut oak forest in Table 7.

empirically approximated by the pH (negative logarithm of the H^+ activity, or concentration) of soil-water suspension. In many cases, a soil-dilute salt (such as 0·02 *M* $CaCl_2$) suspension is used, commonly giving a pH about 0·5 units lower than that of a soil-water suspension.

A second type of acidity is exchangeable (or extractable) by concentrated salt solutions. Exchangeable acidity is typically estimated by extracting soil samples with 1 *M* KCl, filtering and titrating the extract to a neutral endpoint. Much of the exchangeable acidity derives from exchangeable aluminium (Al); displacement of Al into the soil solution is often accompanied by a pH-dependent release of H^+ from the aluminium ion's (Al^{3+}) hydration sphere (described later).

The third type of soil acidity is not readily released by concentrated salt solutions, but can be removed (or neutralized) by titration with an alkaline solution. This titratable acidity is composed of hydrous oxides of Al and iron (Fe), Al cations strongly adsorbed to organic matter, and undissociated, weak organic acids. The titratable (or variable-charge) pool of H^+ reacts slowly but forms the major, long-term buffering mechanism of pH in forest soils. Typically, total acidity is determined with $BaCl_3$-triethanolamine (TEA) buffered at pH 8·2.

Although the free H^+ in soil solution is often said to be measured by taking the pH of a suspension of soil in water or in dilute salt, soil pH should generally be considered only an empirical index of free H^+ in solution. The pH reading of such a suspension is the product of complex electrode behavior. Soil pH has major method-caused variations attributable to both the soil : solution ratio and to the eletrolyte concentration of the suspension; soil pH values are usually lower with increasing soil-to-water ratios and as the concentration of the extracting salt increases. The salt effect occurs because cations displace H^+ and Al^{3+} from exchange sites, lowering the pH. Even the spatial placement of the electrode within the soil suspension or supernatant (known as the suspension effect) can affect the pH reading; the pH of the suspension is typically lower than that of the supernatant.

Because of these complexities, soil pH measurements cannot provide precise interpretations regarding chemical solubilities. However, soil pH is a valuable empirical measure of the chemical status of the soil. Forest soils typically range in pH from 3·5 to 7·0. Values below 3·5 are found mainly in acid-sulfate cat clays and mine spoils (where concentrations of sulfuric acid are high), and also in some extremely acid forest humus and peats. Forest soils with pH above 7·0 occur most often on base-rich parent materials, such as limestones that contain very weatherable carbonate minerals. Values above 7·0 indicate alkaline conditions, although many soil acids are weak enough to be dissociated only at a pH greater than 7·0. Therefore, some soil scientists prefer the term soil reaction to acidity or alkalinity.

In addition to pH, soil acidity can be defined by the relative distribution of acid cations (H^+ and Al^{3+}) and so-called base cations (Ca^{2+}, Mg^{2+}, K^+, Na^+) that are electrostatically attracted to the negative charges of the soil. As described later, Al^{3+} is considered an acidic cation because it causes water molecules to dissociate. The other major cations are not involved in water hydrolysis, and have unfortunately been referred to historically as base cations. They are not bases in the sense used by chemists, and this terminology can muddle discussions of acids and bases in soils. The proportion of the soil's negative charges (the cation exchange capacity, or CEC) that is balanced by acid and "base" cations is termed acid saturation or base saturation.

Acid or base saturation would be a simple and specific soil property except that CEC determinations are very method-dependent. In fact, nearly all of the variation in CEC obtained by different procedures results from the amount of acidity measured by each method. Two common procedures used for CEC determinations involve extraction of soil with unbuffered salt solutions at the pH of the salt-soil suspension (termed the effective CEC or CEC_e) or with $BaCl_2$-TEA solution buffered at pH 8·2 (total CEC or CEC_t). Effective CEC is obtained by summing the titratable acidity in the salt-solution extract with the exchangeable base cations in the extract. In contrast, total CEC is the sum of the salt-exchangeable base cations and the soil acidity neutralized during the extraction buffered at pH 8·2. A pH of 8·2 is often chosen because complete neutralization of acid cations adsorbed to clay minerals, organic matter and hydrous oxides occurs between pH 8 and 9 (Coleman and Thomas, 1964). This is also the theoretical pH of a soil saturated with calcium carbonate, and represents a common maximum pH in soils of humid climates.

Total acidity (acidity$_t$) neutralized by $BaCl_2$-TEA extraction is often much larger than that exchanged by unbuffered salt solutions (acidity$_e$). The difference between acidity$_e$ and acidity$_t$ depends on the type of cation exchange complex and its base saturation. Moreover, the ratio of acidity$_e$ to acidity$_t$ indicates the strength of soil acidity. For example, organic matter typically has a high affinity for H^+ and Al^{3+}, especially in relation to the replacing power of base cations in unbuffered salt solutions. Therefore organic matter is a relatively weak acid with a low acidity$_e$/acidity$_t$ ratio. Kaolinite and trioctohedral vermiculite also have low acidity$_e$/acidity$_t$ ratios, whereas dioctohedral vermiculite and illite have intermediate ratios. Montmorillonite is a strongly acidic clay, with a high ratio (Thomas and Harward, 1984). The relationship between pH and acid saturation thus depends markedly on the type of exchange complex in the soil. Mehlich (1941, 1942) demonstrated that montmorillonite and illite clays have a pH of about 7 at about 20% acid saturation, whereas organic matter and kaolinite clay have a pH of 7 at 50–70% acid saturation.

Hydrolytic reactions with soil aluminum are responsible for much of the variation among soils in pH, base and acid saturation, and buffering capacity. Hydrolysis with a cation occurs when the charge/size ratio of a hydrated cation is large enough to rupture O–H bonds of the hydration water molecules. Cations that are highly charged and relatively small hydrolize readily. In soils, Al^{3+} and Fe^{3+} are the primary cations that contribute to acidity via hydrolysis. Some simplified hydrolysis reactions for aluminium are (Bohn *et al.*, 1979):

$$
\begin{aligned}
Al^{3+} + 6\,H_2O &\leftrightarrow [Al(H_2O)_6]^{3+} && (\text{pH} < 4{\cdot}5)\\
&\leftrightarrow [Al(H_2O)_5OH]^{2+} + H^+ && (\text{pH } 4{\cdot}5\text{–}5{\cdot}0)\\
&\leftrightarrow [Al(H_2O)_4(OH)_2]^{+} + H^+ && (\text{pH } 5{\cdot}0\text{–}6{\cdot}5)\\
&\leftrightarrow [Al(H_2O)_3(OH)_3]^{0} + H^+ && (\text{pH } 6{\cdot}5\text{–}8{\cdot}0)
\end{aligned}
$$

Complex reactions result from hydrolysis of Al^{3+} in soils (cf. Johnson *et al.*, 1981; Driscoll *et al.*, 1985). Displacement of Al^{3+} from exchange sites, for example, can be followed by varying degrees of hydrolysis, readsorption of hydrolysis products to the exchange sites, and reaction of H^+ with soil minerals releasing new Al^{3+} by mineral weathering. Hydrolysis reactions dominate the H^+ status of acid soils, and Al^{3+} export from soils poses a threat to aquatic ecosystems. The complexity of these reactions presents major challenges in the prediction of changes in soil pH caused by increased H^+ input or H^+ consumption in forest ecosystems.

Which of the three pools of acidity, free H^+, salt-exchangeable H^+, or titratable H^+, is most important to ecosystem processes? Obviously, free H^+ in solution has important effects on nutrient availability and the solubility of toxic metals such as Al^{3+}. Salt-exchangeable acidity also has immediate importance because cation exchange reactions are rapid and the H^+ and Al^{3+} content of soil solution is controlled in large part by displacement and adsorption of H^+ and Al^{3+}. In addition, the availability of nutrient cations held on exchange sites is strongly affected by the acid and base saturation of the exchange complex. The importance of titratable acidity arises in buffering free H^+, Al^{3+}, and exchangeable acidity. An evaluation of changes in soil acidity (acidification or alkalinization, van Breemen *et al.*, 1983, Reuss and Johnson, 1986) requires accounting of the sizes and flux rates for all three pools. Indeed, the empirical classification of acidity into these pools is based upon convenient analytical methods, and any division of the acidity continuum should be recognized as at least partly artificial.

III. COMPOSING H^+ BUDGETS

Because H^+ concentrations and fluxes are important from the level of sub-cellular organelles to whole ecosystems, the pools and fluxes of H^+ that must be measured in a forest ecosystem depend on the perspective of the investigator. Many key components at a cellular perspective usually can be ignored at an ecosystem level. Our focus in this paper is on the role of nutrient cycles in H^+ budgets at the level of the ecosystem and its environment. To set the stage, we briefly discuss some processes affecting H^+ flux at two finer levels of resolution.

A. Cellular Level

Fluxes of H^+ within cells are important for a wide range of processes. The formation of high-energy adenosine triphosphate (ATP) from adenosine diphosphate (ADP) and inorganic phosphate relies in part on gradients of H^+ concentration across the membranes of chloroplasts and mitochondria (cf. Goodwin and Mercer, 1983). The opening and closing of leaf stomata is driven by changes in the osmotic potential of guard cells caused by changes in K^+ concentrations; electrogenic pumping of H^+ out of guard cells induces K^+ uptake into the cell (cf. Assmann *et al.*, 1985). Cells also produce a variety of organic acids that serve to buffer cellular pH and to provide H^+ for export from roots to balance the charge of cation nutrients taken up from the soil (cf. Mengel and Kirkby, 1979).

B. Root/Soil Interface

At a higher level, cellular H^+ cycles can be ignored as they represent no net flux beyond cell walls. Physiologists and soil scientists focusing on the root/soil interface often attempt to identify the balance of positively and negatively charged nutrient ions taken up by plants (Fig. 2). Uptake of cations and anions must be equal to maintain electroneutrality within both the plant and soil. The uptake of positively and negatively charged nutrients is rarely balanced, however, so plants must release or absorb a balancing ion. Excess cation uptake can be balanced by release of H^+, and excess anion uptake by absorption of H^+ or release of OH^- (note that at typical plant and soil pH and CO_2 levels, OH^- is actually HCO_3^-). For H^+ accounting, the uptake of each nutrient can be accounted for as the absorption or release of H^+ equivalents and then the net H^+ flux can be calculated by summing all ions.

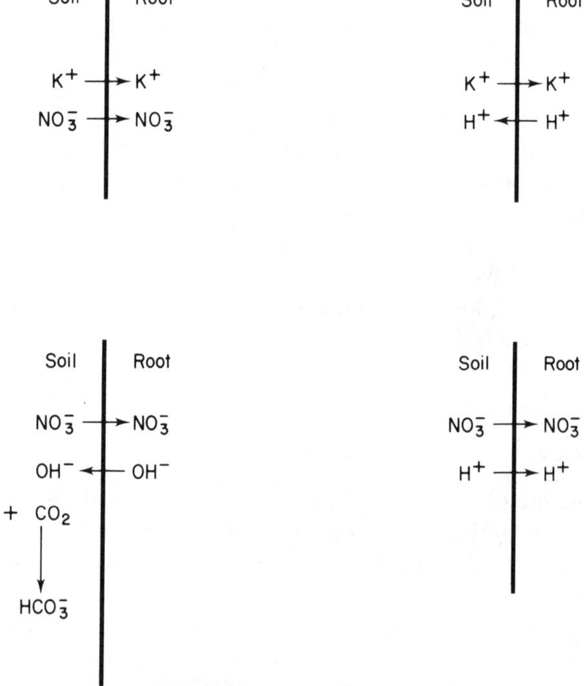

Fig. 2. At the root/soil interface, the uptake of cations and anions must balance to maintain electroneutrality. Absorption or release of H^+ or OH^- may balance any charge deficit among nutrients. For accounting purposes, the uptake of each cation and anion can be considered equivalent to H^+ absorption or release, and the resultant H^+ fluxes summed to obtain the net flux.

Nitrogen is the nutrient required in the greatest quantity by plants, and in agricultural soils, the nitrate anion is the primary form used by plants. Therefore, the total charge of anion nutrients exceeds that of cation nutrients. The excess anion uptake must be counterbalanced by absorption of H^+ (or release of OH^-) which increases the pH of the soil around the root (the rhizosphere or rhizoplane). Forest soils are typically less fertile than agricultural soils, and trees may take up both nitrate and ammonium. Ammonium is a cation (NH_4^+), and its uptake is associated with the release of H^+.

IV. WHOLE ECOSYSTEM NUTRIENT CYCLING

These fine-resolution examples illustrate an important feature of H^+ budgets. At the root/soil level, the only cellular level H^+ fluxes of importance are net fluxes of H^+ into or out of cells. The fluxes within cells do not affect the H^+ of the cells' environment. At the whole-ecosystem scale, H^+ fluxes at the root/soil level may be balanced by fluxes elsewhere in the ecosystem, allowing the closed cycle to be left out of the whole-ecosystem H^+ budget. This section summarizes the H^+ budget for the major nutrient cycles within forest ecosystems, and the next section considers inputs and outputs from ecosystems.

A. The Carbon Cycle

Carbon compounds constitute the majority of the biomass of forests, and C dynamics have a major influence on ecosystem H^+ budgets. In the simplest case, CO_2 can be fixed by photosynthesis into reduced C compounds, which are then respired back to CO_2 with no net generation or consumption of H^+ (Fig. 3, Table 2). Respired CO_2 may return to the atmosphere, or react with water in the soil to form carbonic acid (H_2CO_3). Carbonic acid dissociates to H^+ and HCO_3^- between pH 5 and 8.

The picture becomes more complex as the fate of C compounds is followed through the ecosystem. As noted in the previous section, the regulation of pH and maintenance of anion/cation balance within plants are achieved by synthesis of organic acids (Table 2, Eqn. 3). The degree of dissociation of these acids varies with cellular pH; most leaves fall within the range of pH from 4·0 to 7·0. The metabolism of both P and S will generate one additional H^+ in tissues with near-neutral pH relative to more acidic levels.

Two sources of organic acids are important within soils. First, low-molecular-weight organic acids are added to the soil in litterfall and are synthesized within the soil by microbes. At any point in time, this pool is fairly small, but the turnover rate of this short-term pool may be rapid. The second pool,

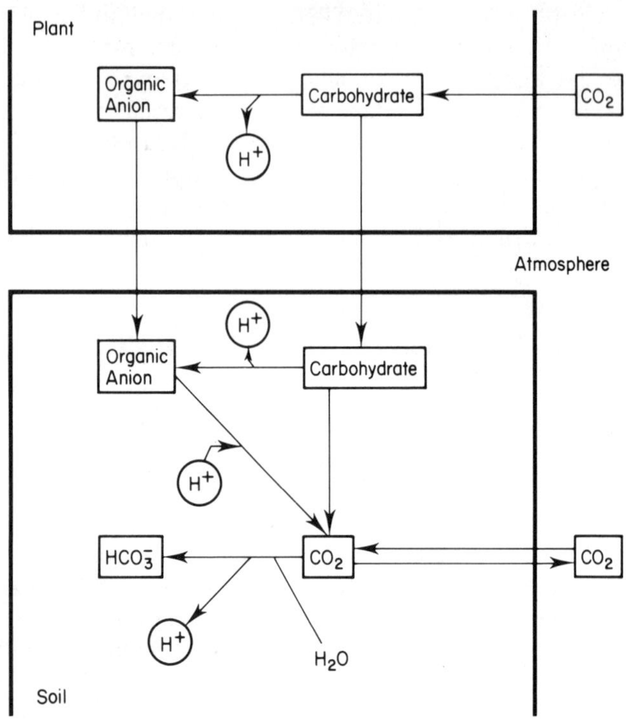

Fig. 3. The carbon cycle generates H^+ primarily by two mechanisms. Conversion of neutral carbohydrates to organic acids is a fundamental part of plant metabolism. Carbon dioxide reacts with water to form carbonic acid, which is especially important at high partial pressures of CO_2 in soils.

complex organic acids, is more stable and turns over on a much longer time period. These complex organic molecules are largely synthesized by microbes and their acidity derives from carboxyl groups (R-COOH) and phenol groups (R-OH). Carboxyl groups are stronger acids than phenolic groups, and dominate the organic acid pool in forest soils.

Soil organic matter is generally divided into three groups, based on solubility in acid and alkaline solutions. Fulvic acids are soluble in both acid and alkaline solutions; humic acids dissolve only in alkaline solutions. Residual organic matter not soluble in alkaline solutions is termed humin. Fulvic acids are generally more acidic (per unit weight) than humic acids, and are more mobile due to greater solubility in acid forest soils. For acid soils in temperate climates, Schnitzer (1980) reported that fulvic acids contain about 9 to 14 mmol g^{-1} of titratable acidity, compared to 6 to 9 mmol g^{-1} for humic acids.

Table 2
Carbon transformations and H^+ flux.

Reaction	H^+ flux		
	Plant		Soil
1. Photosynthesis			
$6CO_2 + 6H_2O \rightarrow C_6H_{12}O_6 + 6O_2$	0		—
2. Respiration and combustion			
a. Carbohydrates			
$C_6H_{12}O_6 + 6O_2 \rightarrow 6CO_2 + 6H_2O$	0	or	0
b. Organic acid anions			
$CH_3COOK + 2O_2 + H^+ \rightarrow 2CO_2 + 2H_2O + K^+$	$-1H^+$	or	$-1H^+$
3. Organic acid formation			
$Pyruvate^{1-} + CO_2 + ATP^{4-} + H_2O \rightarrow Oxaloacetate^{2-} + ADP^{3-} + HPO_4^{2-} + 2H^+$	$+2H^+$		—
4. Partial decomposition of phenolic ring			
catechol (ring with OH, OH) $\xrightarrow{+O_2}$ *cis*-muconic acid (COOH, COOH) $\longrightarrow$ (COO⁻, COO⁻) $+2H^+$	—		$+2H^+$
5. Carbonic acid–bicarbonate equilibrium			
$CO_2 + H_2O \leftrightarrow H_2CO_3 \leftrightarrow HCO_3^- + H^+$ (pH 4·5–8)	—		$+1H^+$

Substantial quantities of organic acids are deposited on the forest floor in litterfall. Several surveys have reported that fresh litter from most tree species ranges in pH from 3·5 to 6·7 (Hesselman, 1925; Plice, 1934; Melin, 1930; Broadfoot and Pierre, 1939; Mattson and Koutler-Andersson, 1941; Lutz and Chandler, 1946) with the pH of most litter falling within the range of pH 4·5 to 5·5. Mattson and Koutler-Andersson (1941) characterized the total acidity (OH^- consumed in titration to pH 7·0) of litter from 8 European tree species. Birch and ash had the least-acid litter (about 160 micromoles H^+ per gram of litter). Pines had intermediate levels of acidity (about 340 micromoles H^+ per gram), and oak and alder had the most acid (about 500 micromoles per gram). Acidity was not well related to the content of base cations in the litter, indicating that litter acidity is not simply a function of cation content. The types of acids present in fresh litter vary, but most are low-molecular-weight acids. For example, Scots pine (*Pinus sylvestris* L.) litter in one study had a pH of 4·5 and a total acidity of 160 micromoles per gram of needles (Muir *et al.*, 1964). About 25% of the acid was accounted for by shikimic acid, another 50% was divided evenly among citric acid, phosphoric acid and quinic acid, 10% was malic acid, and about 15% remained undetermined. We are unaware of any similar data on below-ground litter in forests.

These low-molecular-weight acids are easily degraded by decomposer microbes, and typically last only days or weeks after litterfall (Nykvist, 1959, 1963). Therefore the titratable acidity in fresh litter probably contributes relatively little to the long-term acidity of forest floors. At an ecosystem scale, the amount of tritratable acidity deposited annually on the forest floor in 2,000 kg ha^{-1} of litterfall would range from about 0·3 to 1·0 kmol H^+ ha^{-1}.

Another component of the carbon cycle's H^+ budget involves the equilibrium between CO_2, H_2O and carbonic acid (H_2CO_3). Respiration from roots and from decomposer microorganisms elevates the CO_2 partial pressure of the soil atmosphere, often to 100 times atmospheric levels (Reuss and Johnson, 1985). The CO_2/carbonic acid equilibrium (Table 2, Eqn. 5) results in the formation of substantial carbonic acid, much of which dissociates if soil pH is above 5. Although some of the H^+ from carbonic acid may remain in the soil solution, most adsorbs onto cation exchange sites, protonates organic anions or variable-charge surfaces, or is consumed in mineral weathering. These reactions promote further dissociation of carbonic acid. Although some bicarbonate may be retained on anion exchange sites, bicarbonate leaching roughly matches the H^+ addition from the dissociation of carbonic acid. The negative charge of bicarbonate anions leaching from the soil must be matched by cations, resulting in leaching losses of nutrients such as calcium, magnesium and potassium.

When bicarbonate reaches aquatic systems with low CO_2 partial pressure, degassing of the water releases CO_2 and H_2O. One H^+ is consumed from the aquatic system to form H_2O, and so the generation of bicarbonate by high CO_2 partial pressures in the soils of terrestrial ecosystems can provide a major source of alkalinity (acid neutralization capacity) to aquatic ecosystems. As the carbonic acid/bicarbonate equilibrium is sensitive to soil pH, a fairly small change in the pH of a forest soil may result in carbonic acid (with no acid neutralizing capacity) export to lakes rather than bicarbonate. This mechanism may allow small changes in soil pH to be amplified to large changes in the pH of poorly buffered lakes. If this untested scenario is realistic, small increases in forest soil acidity may reduce the export of alkalinity to lakes, reducing the lake's ability to neutralize acids from other sources (Reuss and Johnson, 1985).

B. The Nitrogen Cycle

The various transformations of the N cycle generate and consume large quantities of H^+. The complete cycle is balanced with no net change in H^+

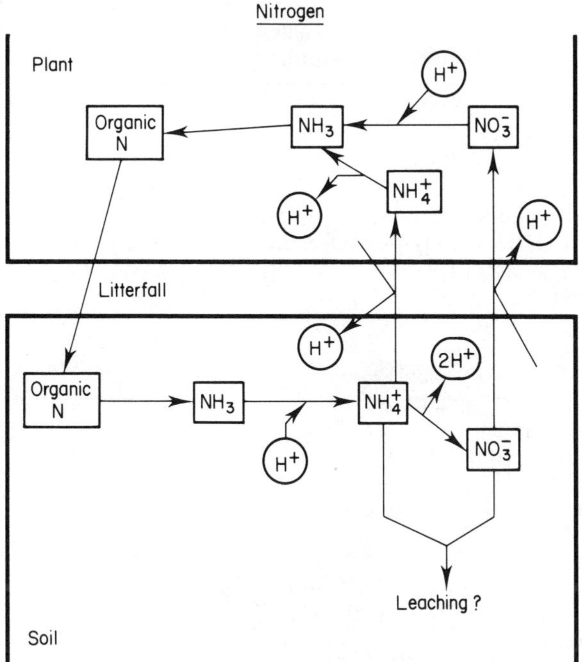

Fig. 4. Oxidation and reduction reactions in the N cycle generate and consume H^+, but transfer of N from soil organic pools to plant organic pools has no net H^+ flux. Oxidation of organic N produces ammonia (NH_3), which protonates and forms ammonium (NH_4). Uptake of ammonium is balanced by equivalent H^+ release from the root, balancing H^+ consumed in protonating ammonia. Oxidation of ammonium (nitrification) forms nitrate plus 2 H^+; one H^+ balances that consumed to protonate ammonia, the other balances nitrate's charge in uptake, and is later consumed as nitrate is reduced to organic N.

(Fig. 4, Table 3). That is, the release of inorganic N from decomposing organic matter (N mineralization) followed by plant uptake and assimilation into organic forms is a pathway that begins and ends with organic N (typically in the amide form, R-NH_2). Despite the importance of ammonium versus nitrate at the root/soil level, the complete N cycle can be ignored if the focus is on the whole ecosystem. Only N inputs, outputs, or accumulations of inorganic forms affect the net ecosystem H^+ budget.

The first step in N mineralization (Fig. 4) is the conversion of organic N to ammonia (Table 3, Eqn. 1). Although the N atom is transformed from R-NH_2 to NH_3, there is no net change in H^+. This is because the step that severs the carbon–nitrogen bond adds oxygen to the C and allows one of the carbon's H atoms (a proton *plus* electron) to remain with the N. As the H brings along its own electron, there is no flow of hydrogen ions.

Table 3
Nitrogen transformations and H^+ flux.

Reaction	H^+ flux	
	Plant	Soil
1. Ammonification + hydrolysis		
$CH_2NH_2COOH + 1\frac{1}{2}O_2 \rightarrow 2CO_2 + H_2O + NH_3$	—	0
$NH_3 + H^+ \rightarrow NH_4^+$ (pH < 9)	—	$-1H^+$
2. Nitrification		
$NH_4^+ + 2O_2 \rightarrow NO_3^- + H_2O + 2H^+$	—	$+2H^+$
3. Ammonium absorption		
NH_4^+ (soil) $\rightarrow$ NH_4^+ (plant)	$-1H^+$	$+1H^+$
H^+ (soil) $\swarrow\!\nwarrow$ H^+ (plant)		
4. Nitrate absorption		
NO_3^- (soil) $\rightarrow$ NO_3^- (plant)	$+1H^+$	$-1H^+$
H^+ (soil) $\nearrow\!\searrow$ H^+ (plant)		
5. Ammonium uptake and dissociation		
$NH_4^+ \rightarrow NH_3 + H^+$	$+1H^+$	—
6. Ammonia assimilation		
$NH_3 + (CH_2)_2(COOH)_2CHNH_2 \rightarrow$ $(CH_2)_2COOHNH_2CHNH_2 + H_2O$	0	—
7. Nitrate reduction		
$NO_3^- + 9H^+ + 8e^- \rightarrow NH_3 + 3H_2O$	$-1H^+$	—
8. Ammonium salt input + assimilation		
Net reaction: $NH_4^+ \rightarrow$ Organic-N $+ H^+$	0	$+1H^+$
9. Nitrate salt input + assimilation		
Net reaction: $NO_3^- \rightarrow$ Organic-N	0	$-1H^+$
10. Nitric acid input + assimilation		
Net reaction: $HNO_3 \rightarrow$ Organic-N	0	0
11. Nitrogen fixation		
$N_2 + 6H^+ + 6e^- \rightarrow 2NH_3$	0	—
12. Denitrification		
$2NO_3^- + 10H^+ + 8e^- \rightarrow N_2O + 5H_2O$	—	$-2H^+$
$N_2O + 2H^- + 2e^- \rightarrow N_2 + H_2O$	—	0
13. Ammonium leaching		
Organic-N $\rightarrow$ NH_4^+ leached	—	$-1H^+$
NH_4 input $\rightarrow$ NH_4^+ leached	—	0
14. Nitrate leaching		
Organic-N $\rightarrow$ NO_3^- leached	—	$+1H^+$
NH_4^+ input $\rightarrow$ NO_3^- leached	—	$+1$ or $+2H^+$
NO_3^- input $\rightarrow$ NO_3^- leached	—	0 or $-1H^+$
15. Urea fertilization + hydrolysis		
$(NH_2)_2CO + 2H_2O \rightarrow CO_2 + NH_3$	—	0
$CO_2 + H_2O \rightarrow H_2CO_3$ (<pH 4) $\leftrightarrow$ $HCO_3^- + H^+$ (pH 5–8)	—	0 or $+1H^+$
$NH_3 + H^+ \rightarrow NH_4^+$ (pH<9)	—	$-1H^+$

In acid soils, ammonia readily protonates to form ammonium (NH_4^+) which temporarily reduces the soil pool of H^+. If the ammonium cation is absorbed by a microbe or plant, an equivalent H^+ must be released into the soil for charge balance to be maintained (Table 3, Eqn. 3), balancing the H^+ consumed in forming ammonium from ammonia. The assimilation of ammonium within the plant releases one H^+, again balancing the H^+ lost from the plant in the uptake of ammonium.

The N cycle may also involve oxidation of soil ammonium to nitrate (nitrification) by autotrophic bacteria, generating 2 H^+ for every nitrate formed (Table 3, Eqn. 2). One of these H^+ may be considered to balance the H^+ consumed in protonating ammonia; therefore nitrification yields a net +1 H^+ for each nitrate ion formed. If a plant or microbe takes up the nitrate, one balancing H^+ is also taken up (Table 3, Eqn. 4), again resulting in a balanced soil H^+ cycle. Assimilation of nitrate involves reduction to ammonia, which consumes 1 H^+ (Table 3, Eqn. 7) and leaves the H^+ budget of both the plant and soil balanced. The cycle from soil organic-N to plant or microbial organic-N can be ignored from the standpoint of ecosystem H^+ accounting.

Fluxes of H^+ may also have reciprocal effects on N cycling. Nitrogen leaching from ecosystems can generate substantial quantities of H^+ (see later sections), and changes in soil pH may alter N mineralization rates which could alter leaching rates. For example, net mineralization of ammonium from organic-N often increases when forest soil pH is raised (e.g., Montes and Christensen, 1979), but occasionally decreases (e.g., Adams and Cornforth, 1973). Nitrification is also sensitive to soil pH, generally increasing as pH is raised from extremely acid to moderately acid levels in forest soils (Zottl, 1960; Focht and Verstraete, 1977; Montes and Christensen, 1979; Robertson, 1982). If nitrate leaching lowered soil pH, nitrification may be reduced, providing negative feedback on further soil acidification. However, increased nitrate leaching and soil acidification do occur simultaneously in some ecosystems (cf. Van Miegroet and Cole, 1984, 1985), so this feedback may not be very strong.

C. The Phosphorus Cycle

Unlike N, P undergoes no oxidation or reduction reactions in forest ecosystems, but chemical reactions of the trivalent phosphate anion (PO_4^{3-}) can influence H^+ fluxes. In weathering reactions, complete dissolution of phosphate minerals (congruent reactions) consumes 1 to 2 H^+ from the soil solution for each phosphate group released (Fig. 5; Table 4, Eqn. 1a and 1b). In many cases, weathering is not congruent and H^+ may be generated as phosphate minerals weather to form other secondary minerals (Table 4, Eqn. 1c). This contrary behavior is associated with the degree of protonation

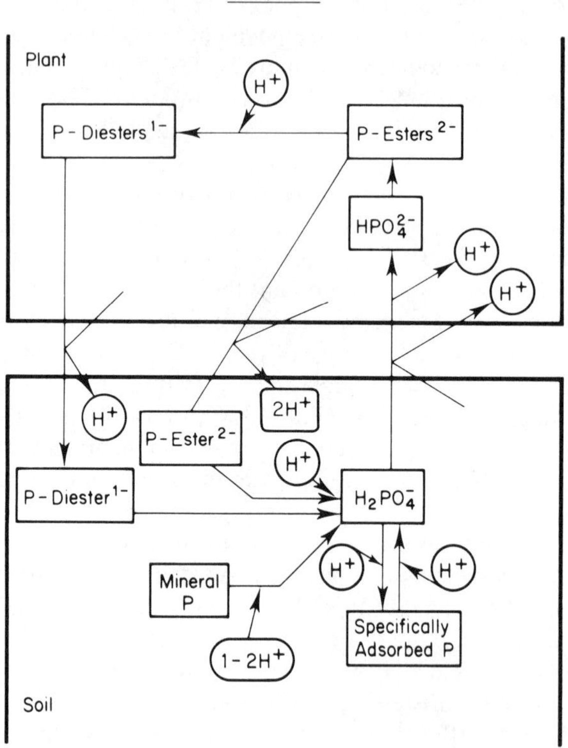

Fig. 5. The P cycle includes no reduction or oxidation steps, but is complicated by the variable number of H^+ that associate with phosphate (PO_4). Phosphate weathering consumes H^+ if minerals dissolve completely, but may generate H^+ at intermediate steps. Phosphate may be adsorbed to soil sesquioxides, with associated H^+ flux. Phosphate also enter into single (ester) or double (diester) bonds, with different H^+ fluxes. A complete P cycle, from inorganic phosphate in the soil into the plant and back again, has no net effect on the H^+ budget. A partial cycle, such as from the soil into the plant, may have a variable effect depending on pH and types of organic compounds formed.

of phosphoric acid; at pH 4–7, the $H_2PO_4^-$ form predominates. Incongruent weathering generates H^+ by releasing H_3PO_4 that then dissociates. The congruent weathering reactions release PO_4^{3-} groups that then consume H^+ from the soil solution.

Phosphate may be held on soil surfaces by two mechanisms. The first is electrostatic attraction, termed anion exchange or non-specific adsorption, to positively charged surfaces on variable-charge surfaces and broken edges of clays. Phosphate also enters into ligand exchange reactions (specific anion

Table 4
Phosphorus transformations and H^+ flux.

Reaction	H^+ flux	
	Plant	Soil
1. Mineral weathering		
a. Fluoroapatite		
$Ca_5F(PO_4)_3 \rightarrow 5Ca^{2+} + F^- + 3PO_4^{3-}$	—	0
$3PO_4^{3-} + 6H^+ \rightarrow 3H_2PO_4^-$ (pH 4–7)	—	$-6H^+$
b. Variscite (or strengite with Fe^{III})		
$AlPO_4 \cdot 2H_2O \rightarrow Al^{3+} + 3PO_4^{3-} + 2H_2O$	—	0
$Al^{3+} + 6H_2O \rightarrow Al(OH)_2^+ \cdot 4H_2O + 2H^+$ (pH 5–6)	—	$+2H^+$
$Al(OH)_3^{2+} \cdot 3H_2O + 3H^+$ (pH 3–5)	—	$+3H^+$
$3PO_4^{3-} + 6H^+ \rightarrow 3H_2PO_4^-$ (pH 4–7)	—	$-6H^+$
c. Monocalcium phosphate		
$Ca(H_2PO_4)_2 \cdot 2H_2O + 2H_2O \rightarrow CaHPO_4 + H_3PO_4$	—	0
$H_3PO_4 \rightarrow H_2PO_4^- + H^+$ (pH 4–7)	—	$+1H^+$
2. Decomposition		
a. Phosphoester bonds		
$R\text{-}PO_4^{2-} + H_2O + H^+ \rightarrow R\text{-}OH + H_2PO_4^-$ (pH 4–7)	—	$-1H^+$
b. Phosphodiester bonds		
$R\text{-}PO_4^- - R + 2H_2O \rightarrow 2(R\text{-}OH) + H_2PO_4^-$ (pH 4–7)	—	0
c. Inorganic phosphate		
$HPO_4^{2-} + H^+ \rightarrow H_2PO_4^-$ (plant pH > 5·5, soil pH < 5·5)	—	$-1H^+$
3. Specific anion adsorption		
$2(FeOOH) + H_2PO_4^- + H^+ \rightarrow (FeO)_2HPO_4 + 2H_2O$	—	$-1H^+$
4. Plant uptake		
$H_2PO_4^-$ (soil) $\rightarrow$ $H_2PO_4^-$ (plant) if pH < 5·5; $HPO_4^{2-} + H^+$ if pH > 5·5; H^+ (soil) $\rightarrow$ H^+ (plant)	$+2H^+$	$-1H^+$
5. Plant assimilation		
a. Phosphoester bonds		
$R\text{-}OH + HPO_4^{2-} \rightarrow R\text{-}PO_4^{2-} + H_2O$ (plant pH > 5·5)	0	—
$R\text{-}OH + H_2PO_4^- \rightarrow R\text{-}PO_4^{2-} + H_2O + H^+$ (plant pH < 5·5)	$+1H^+$	—
b. Phosphodiester bonds		
$HPO_4^{2-} + 2(R\text{-}OH) + H^+ \rightarrow R\text{-}PO_4^- - R + 2H_2O$ (plant pH > 5·5)	$-1H^+$	—
$H_2PO_4^- + 2(R\text{-}OH) \rightarrow R\text{-}PO_4^- - R + 2H_2O$ (plant pH < 5·5)	0	—
6. Superphosphate fertilization		
$Ca(H_2PO_4)_2 \cdot H_2O \rightarrow Ca^{2+} + 2H_2PO_4^- + H_2O$ (pH 4–7)	—	0

adsorption) with Al and Fe hydroxides, to surfaces of soil particles and to the oxide minerals found in many tropical soils. Specific adsorption releases OH^- to the soil solution that consumes H^+ to form water (Table 4, Eqn. 3). At low soil pH, the OH^- ligand already may be protonated, and the adsorption of P would release H_2O. In this case, a negatively charged molecule (phosphate) would replace a neutral molecule (water), leaving a net negative charge on the surface that would serve as a cation exchange site. At very high pH, the OH^- ligand may be dissociated, and adsorption of P would result in the consumption of 2 H^+. This pool of specifically adsorbed phosphate is relatively slowly available to plants. Desorption releases H^+ as water is split to supply OH^- which replaces desorbed phosphate.

Plant uptake of $H_2PO_4^-$ must be accompanied by an equivalent H^+ uptake. Once inside the plants a second H^+ is released (producing HPO_4^{2-}) if pH is near 7 (Table 4, Eqn. 4). In plant tissues below pH 5·5, however, most inorganic phosphate remains as $H_2PO_4^-$.

The next step is the incorporation of phosphate into organic molecules, with monoester (C–O–P) or diester (C–O–P–O–C) bonds. However, much of the P in plants remains as inorganic phosphate, with a reported range of about 25%–75% (Bielski and Ferguson, 1983). Formation of a phosphoester bond releases a water molecule, but no H^+ is consumed (Table 4, Eqn. 5a). A second water molecule is released if the phosphate group forms a second ester bond, and in this case 1 H^+ is consumed from within the plant (Table 4, Eqn. 5b). With the death and decay of plant tissues, no H^+ is released in breaking diester bonds, but one is released in severing monester bonds (Table 4, Eqns 2a and 2b).

We found little information on the proportions of plant P represented by inorganic phosphate, monoesters and diesters. Bielski (1973) reported that inorganic phosphate comprised 68% of the total P content of young pea leaves; diesters (such as DNA, RNA and phospholipids) accounted for another 25%. Monoesters accounted for the remaining 7%. Working through the equations in Table 4, the P remaining as inorganic phosphate would represent a gain of 2 H^+ (Eqn. 4) inside the plant (if plant pH were about 7) and the diesters would represent the consumption of 1 H^+ (Eqn. 6a). Monoesters would have no net effect (Eqn. 6b). The net balance would be an increase in the plant of about 1·1 H^+ for every $H_2PO_4^-$ taken up. If the pH of plant tissues were below 6, then the net balance would be reduced to an increase of about 0·1 H^+ for each phosphate group taken up. This ratio would vary with the composition of plant P pools, but on average would range from 0 to 1 H^+ produced in the plant per phosphate.

The N cycle produced a net 0 flux of H^+ when half complete (N from soil organic-N to plant organic-N) or when complete (from soil organic-N to soil organic-N). Although a complete P cycle also has a 0 net H^+ flux, any inorganic P from the soil that accumulates in plant biomass represents a net flux of -1 H^+ from the soil, and about 0 to $+1$ H^+ in the plant. The decomposition of organic matter should result in a consumption of about 0 to 1 H^+ for each phosphate released.

D. The Sulfur Cycle

As with the N cycle, the sulfur cycle includes oxidation and reduction transformations. Similar to the P cycle, sulfur is also present in rock minerals and its major anion (sulfate, SO_4^{2-}) can be specifically adsorbed on soil iron and aluminium hydroxides.

The weathering of sulfur minerals such as pyrite is a potent source of H^+; the precise number of H^+ released per sulfate group depends on the mineral being weathered and the endproduct minerals (Fig. 6; Table 5, Eqn. 1).

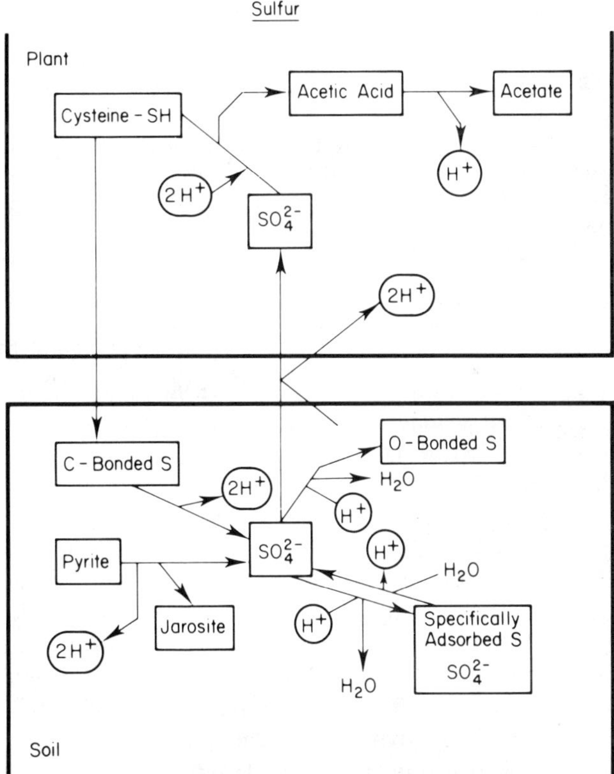

Fig. 6. The sulfur cycle combines the complexities of the N and P cycles. As with the N cycle, S transfer from soil organic matter into plant organic matter appears balanced; however, production of acetic acid may affect the H^+ of the carbon cycle. Weathering of reduced sulfur minerals generates sulfuric acid.

Table 5
Sulfur transformations and H^+ flux.

Reaction	H^+ flux	
	Plant	Soil
1. Pyrite weathering (to jarosite)		
$12FeS_2 + 45O_2 + 30H_2O + 4K^+ \rightarrow 4KFe_3(OH)_6(SO_4)_2 + 16SO_4^{2-} + 36H^+$	—	$+36H^+$
2. Specific anion adsorption		
$2FeOOH + SO_4^{2-} + 2H^+ \rightarrow (FeO)_2SO_4 + H_2O$	—	$-2H^+$
3. Sulfate reduction		
$SO_4^{2-} + 10H^+ + 8e^- \rightarrow H_2S + 4H_2O$	—	$-2H^+$
4. Plant uptake		
SO_4^{2-} (soil) $\rightarrow SO_4^{2-}$ (plant) $2H^+$ (soil) $\rightarrow 2H^+$ (plant)	$+2H^+$	$-2H^+$
5. Plant assimilation		
$SO_4^{2-} + 2H^+ +$ acetyle serine $\rightarrow$ Cysteine-SH $+ CH_3COOH$	$-2H^+$	—
$CH_3COOH \rightarrow CH_3COO^- + H^+$ (pH > 5)	$+1H^+$	—
6. Decomposition		
Cysteine–SH $+ H_2O \rightarrow H_2S + NH_3 + CH_3$–CO-COOH	—	0
$H_2S + 4H_2O \rightarrow SO_4^{2-} + 2H^+ + 4H_2$	—	$+2H^+$
$NH_3 + H^+ \rightarrow NH_4^+$ (pH < 9)	—	$-1H^+$
7. Sulfate esters (largely formed by microbes)		
$SO_4^{2-} + HO–CH_2 \rightarrow R + H^+ \rightarrow SO_4^- - CH_2–R + H_2O$	$-1H^+$ (microbe)	—

Sulfuric acid is much stronger than phosphoric acid, and dissociates totally in soil solutions. Sulfate may also be adsorbed and desorbed non-specifically, or specifically on hydroxide coatings, consuming or generating 1–2 H^+ in the soil solution (Table 5, Eqn. 2).

Plant uptake of sulfate is accompanied by uptake of 2 H^+ to maintain charge balance (Table 5, Eqn. 4). The form of sulfur compounds in plants is variable. Most appears in reduced C–S bonds (amino acids such as cysteine), but a variable amount remains as inorganic sulfate. The portion of sulfate remaining as a free anion contributes no further to the plant H^+ budget. Assimilatory reduction to form C–S bonds consumes 2 H^+ (Table 5, Eqn. 5), leaving the plant balanced in relation to the 2 H^+ taken up with the SO_4^{2-}. Acetic acid, however, is a byproduct of the reduction of sulfate, and acetic acid will dissociate to acetate and H^+ if plant pH is above 5. Alternatively, acetic acid may be metabolized to water and carbon dioxide and have no net effect on the H^+ budget of the plant.

In some microbes and some plants, such as the mustard family (Brassicaceae) which has a high oil content, sulfate esters (C–O–S) may also represent a sizable fraction of plant S (Roy and Trudinger, 1970). Sulfate esters also appear to comprise a substantial portion of the organic S in soils. The production of esters from sulfate consumes only 1 H^+ (Table 5, Eqn 6) and the net effect is +1 H^+ in the microbe or plant (2 H^+ entered with SO_4^{2-}, and only one was consumed).

After tissue death, decomposition of reduced organic S compounds is generally accompanied by oxidation to sulfate (Table 5, Eqn. 7), which releases 2 H^+ to the soil solution.

The general pattern of H^+ flux in the uptake, assimilation and decomposition portions of the S cycle resembles that of the N cycle, with the exception that acetic acid may be formed when sulfate is reduced to C–S compounds. If the acetic acid dissociates, the uptake and assimilation of sulfate leaves the plant with a net increase of 1 H^+. Subsequent decomposition and release of organic S compounds completes the S cycle with no net flux of H^+. However, the production of acetic acid has altered the H^+ of the C cycle, and should be considered as outlined in the C cycle discussion.

E. The Calcium Cycle

In general, cycles of nutrient cations are simpler than the cycles of N, P and S. Although important differences exist among cycles of cations nutrients (cf. Johnson *et al.*, 1985), we discuss calcium cycling as broadly representative of cation nutrients. The weathering of calcium minerals typically consumes H^+, but the number of H^+ per Ca^{2+} released varies among minerals (Fig. 7; Table 6, Eqns 1a–1c). A pool of readily available Ca^{2+} is held on cation exchange sites. The adsorption of Ca^{2+} on cation exchange sites removes either an equivalent charge of other base cations, or H^+ (or Al^{3+}). For H^+ accounting purposes, the adsorption and desorption of base cations can be tabulated relative to H^+ flux, and the net change in H^+ pools calculated by summing the flux for each base cation.

Plant uptake of Ca^{2+} is matched by excretion of 2 H^+ from the plant into the soil (Table 6, Eqn. 4), and bonding of Ca^{2+} with organic molecules releases 2 H^+ within the plant (Table 6, Eqn. 5). Subsequent decomposition of plant tissues consumes 2 H^+ (Table 6, Eqn. 6), balancing the overall cycle.

Once Ca^{2+} enters the ecosystem through weathering, subsequent cycling of the element does not alter the H^+ status of the site; a complete cycle of Ca^{2+} from soil organic matter back to soil organic matter is neutral in the overall H^+ budget. However, the Ca^{2+} cycle resembles the P cycle in that half a cycle, from Ca^{2+} on cation exchange sites to plant Ca^{2+}, is not balanced with regard to H^+ flux. As biomass accumulates in a forest ecosystem, Ca^{2+} storage results in a net flux of H^+ into the soil. When the forest degenerates or is consumed by fire, H^+ is consumed in the soil, thus completing the cycle.

It is important to note that this pattern applies only to the net accumulation

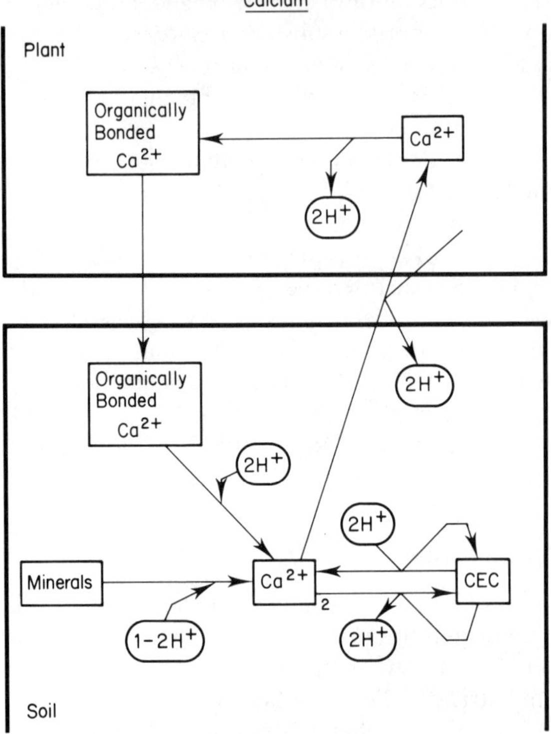

Fig. 7. The calcium cycle is fairly simple and represents nutrient cations. Weathering consumes H^+; ionic Ca^{2+} may displace H^+ from exchange sites or be taken up (in exchange for H^+) by plants. Plant uptake of Ca^{2+} released from soil organic matter has no net effect on the H^+, but uptake from exchange sites represents H^+ release into the soil.

of biomass in the entire ecosystem. If Ca^{2+} taken up by plants originated from the decomposition of organic matter in the soil, then the half-cycle would balance as in the N cycle. Cation uptake by plants represents a net flux of H^+ into the soil only if it represents a net transfer of cations from inorganic to organic pools.

V. ECOSYSTEM AND ENVIRONMENT

The addition and loss of nutrients in forest ecosystems is an important component of the annual H^+ budget, but the balance is not always straightforward. As discussed above, the weathering of minerals typically consumes H^+ from the soil, but in some cases (such as weathering of S and N minerals) large quantities of H^+ can be generated.

Table 6
Calcium transformations and H^+ flux.

Reaction	H^+ flux	
	Plant	Soil
1. Mineral weathering		
a. Carbonates		
$CaCO_3 \rightarrow Ca^{2+} + CO_3^{2-}$	—	0
$CO_3^{2-} + H^+ \rightarrow HCO_3^-$ (pH 5–8)	—	$-2H^+$
$CO_3^{2-} + 2H^+ \rightarrow H_2CO_3$ (pH 2–5)	—	$-4H^+$
b. Feldspars		
$CaAl_2SiO_8 + 8H^+ \rightarrow Ca^{2+} + 2Al^{3+} + 2H_4SiO_4$	—	$-8H^+$
$2Al^{3+} + 4H_2O \rightarrow 2Al(OH)_2^+ + 4H^+$ (pH 4·5–6·5)	—	$+4H^+$
c. Phosphate minerals		
$Ca_5F(PO_4)_3 \rightarrow 5Ca^{2+} + 3PO_4^{3-} + F^-$	—	0
$3PO_4^{3-} + 3H^+ \rightarrow 3HPO_4^{2-}$ (pH 7·5–11·5)		$-3H^+$
$3PO_4^{3-} + 6H^+ \rightarrow 3H_2PO_4^-$ (pH 2·5–7.5)	—	$-6H^+$
2. Cation exchange adsorption		
Clay micelle–2H + $Ca^{2+} \rightarrow$ Clay micelle–Ca + $2H^+$	—	$+2H^+$
3. Cation exchange desorption		
Clay micelle–Ca + $2H^+ \rightarrow$ Clay micelle–2H + Ca^{2+}	—	$-2H^+$
4. Plant uptake		
Ca^{2+} (soil) $\rightarrow$ Ca^{2+} (plant)	$-2H^+$	$+2H^+$
$2H^+$ (soil) $\leftarrow$ $2H^+$ (plant)		
5. Incorporation into organic matter		
Pectin–COOH + Ca^{2+} + HPO_4–Phospholipid $\rightarrow$ Pectin–COO–Ca–PO_4–Phospholipid + $2H^+$	$+2H^+$	—
6. Organic matter decomposition or combustion		
R–COO–Ca–OOC–R + $2H^+ \rightarrow Ca^{2+}$ + 2R–COOH	—	$-2H^+$

Increased acidity of precipitation and total atmospheric deposition in industrialized regions is due to high levels of sulfuric acid (H_2SO_4) and nitric acid (HNO_3). Unpolluted rain should have a pH of about 5·6, due to the natural production of carbonic acid from carbon dioxide and water. About 20% of the carbonic acid would dissociate under unpolluted conditions, giving a H^+ concentration of $10^{-5 \cdot 6}$ mol/L. As with soils, the pH of rain characterizes the intensity of H^+, but does not index the capacity of the water to resist changes in acidity. It might appear that rain pH would be reduced to 4·6 by addition of enough sulfuric or nitric acid to provide a 10-fold increase in H^+. The actual amount of strong acid needed is somewhat greater, because most of the bicarbonate produced from carbonic acid would reprotonate, acting as a buffer against acidification. Similarly, it might appear that pH 4·6 rain would be 10-times more acidic than pH 5·6 rain, but that applies only to the intensity factor of acidity and not to the capacity factor. If a strong base were added to the water, the large pool of undissociated carbonic acid would begin to dissociate and buffer any increase in pH. From a titratable acidity standpoint, rain with a pH of 4·6 can neutralize only about 2 to 3 times as much strong base as could be neutralized by rain with a pH of 5·6 at equilibrium with atmospheric carbon dioxide.

The addition of sulfuric and nitric acids might be expected to acidify an ecosystem, but biologic utilization of the sulfate and nitrate anions alters the net effect on the H^+ budget. The next section focuses on inputs and outputs for the N cycle, followed by comments on the S cycle.

A. Nitrogen Inputs and Outputs

Nitrogen may enter the nutrient cycles of a forest ecosystem via the atmosphere or fertilization in a variety of forms, including: nitrate as a salt or acid, ammonium as a salt, nitrous oxides deposited on leaves, organic-N, N_2 via biological N fixation, and ammonia (a base) and urea fertilizers.

The H^+ effect of nitrate input varies with the fate of the nitrate within the forest (Fig. 8). Nitrate deposited as a salt (e.g., potassium nitrate) that is used by microbes or plants will consume 1 H^+, reducing the overall acidity of the ecosystem (Table 3, Eqn. 9). However, use of the accompanying base cation would include an equivalent amount of H^+ generation, giving an overall net balance. If the nitrate is not used in the system, it may leach out

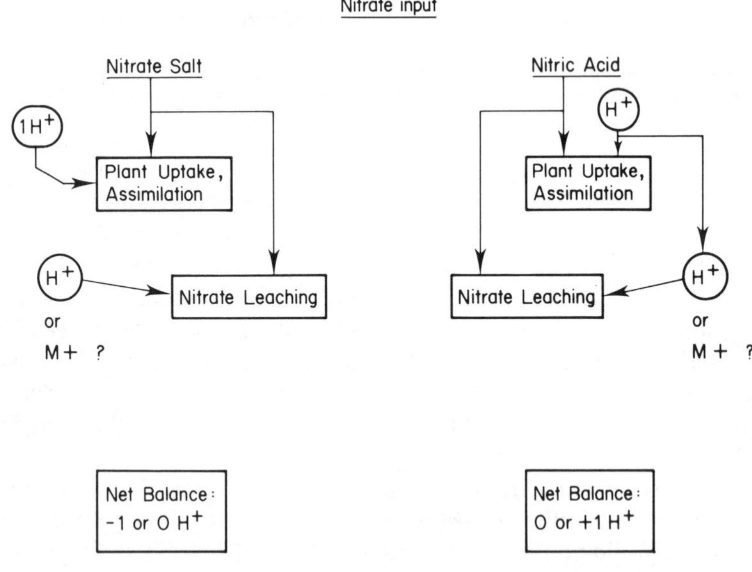

Fig. 8. Input of nitrate salt results in consumption of 1 H^+ per nitrate if assimilated by plants or microbes consumes H^+, preventing any acidification.

in company with a nutrient cation (no H^+ net effect) or an acidic cation (net $-1\ H^+$ for the ecosystem) (Table 3, Eqn. 14). If the incoming nitrate anion is accompanied by H^+ (nitric acid), plant utilization results in no net H flux, and nitrate leaching would yield a zero or $+1\ H^+$ net balance for the ecosystem (Figure 8, Table 3, Eqns 10 and 14). Surprisingly, if the nitrate in nitric acid is utilized by plants or microbes, there is no acidifying effect in the ecosystem.

Although ammonium enters an ecosystem as a salt, subsequent processing within the ecosystem can generate substantial quantities of H^+ (Fig. 9). If ammonium is taken up by plants or microbes, 1 H^+ is released to the soil (Table 3, Eqn. 8). Because the ammonium did not originate within the ecosystem through ammonia hydrolysis (Table 3, Eqn. 1), this H^+ represents a net input to the ecosystem.

When the added ammonium is oxidized to nitrate (nitrified), 2 H^+ are added to the ecosystem (Table 3, Eqn. 2). If the nitrate is then used by the vegetation, 1 H^+ is consumed, leaving a net increase of 1 H^+ (Table 3, Eqns 4 and 7). Leaching of the nitrate accompanied by 1 H^+ (or equivalent Al^{3+}) would also yield a net increase of 1 H^+; if accompanied by a base cation the net increase would be 2 H^+ for the ecosystem for each ammonium added (Table 3, Eqn. 14). It may seem counterintuitive, but the input of nitric acid

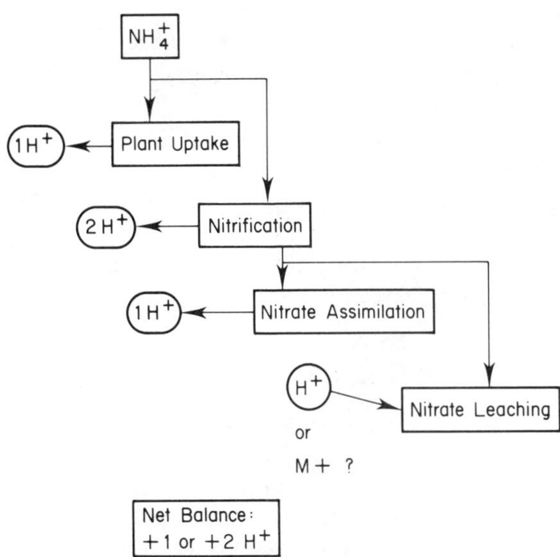

Fig. 9. Input of ammonium salt may involve the production of 1 or 2 H^+, depending on the fate of the ammonium ion.

may have no net effect on the ecosystem H^+ budget, whereas ammonium salt input can generate up to 2 H^+ for each ammonium added.

Although rarely estimated, the dry deposition of nitric acid vapour and nitrogen dioxide (NO_2) gas (followed rapidly by hydrolysis to nitric acid) onto canopies exceeds the wet deposition input over much of eastern North America (cf. Husar, 1985; Lindberg *et al.*, 1986). These forms have the same net H^+ effect as that nitric acid input (Table 3, Eqn. 10).

Biological nitrogen fixation converts N_2 to ammonia (NH_3). The hydrogen atoms added to the N_2 are accompanied by electrons, so there is no net effect on the H^+ budget. However, nitrogen fixation may indirectly alter the H^+ budget through subsequent effects on N cycling. Plants with an internal source of N may take up less from the soil, leaving more in the soil that may leach as nitrate. Increases in ecosystem N content due to N fixation may also accelerate N mineralization, and if mineralization exceeds plant uptake, nitrate leaching again may increase.

The deposition of organic-N compounds from the atmosphere has no net effect, as cycling of this N resembles soil organic-N or plant organic-N.

Fertilization can have substantial effects on the H^+ budget of a forest ecosystem. Consider the addition of 200 kg of ammonium-N per hectare. If the ammonium were utilized by the microbes and vegetation, 14 kmol H^+ ha^{-1} would be generated. If some of the added ammonium were nitrified and the nitrate leached, the net H^+ increase would be even greater. If the 200 kg of N were added as nitrate salt (e.g., potassium nitrate), utilization of the nitrogen would neutralize 14 kmol H^+ ha^{-1}. Note, however, that increased uptake of K^+ might partly offset this neutralization. Finally, if the same amount of N were added as ammonium nitrate salt and all of it were utilized, there would be no net H^+ effect.

Fertilization of forests with urea $[(NH_2)_2CO]$ may increase soil pH in the short run as hydrolysis of urea forms ammonia which protonates (consumes H^+) to form ammonium (Table 3, Eqn. 15). However, subsequent nitrification or uptake of the ammonium by plants or microbes balances this by generating an equivalent quantity of H^+. In agriculture, liquid ammonia is a common form of N fertilizer, with H^+ budget effects similar to those of urea.

B. Sulfur Inputs and Outputs

Accounting for the H^+ effects of sulfate inputs must include the accompanying cations (whether H^+ or other cation) and subsequent fate of the sulfate. Both wet and dry deposition are important vectors of sulfur inputs in polluted regions. In eastern North America, dry deposition probably exceeds wet deposition (cf. Husar, 1985; Lindberg *et al.*, 1986). Sulfuric acid (H_2SO_4) or sulfur dioxide (SO_2) gas (oxidized to sulfuric acid after deposition) are the two major forms of S deposited in forests. Uptake and reduction of SO_4^{2-} by plants or microbes consume H^+ and may balance any H^+ which accompanied the sulfate. If sulfate entered as a salt (e.g. calcium sulfate), subsequent uptake and reduction can consume H^+ from the ecosystem. Similarly, specific adsorption may consume H^+, offsetting the acidification effect.

In general, ecosystems contain much more plant-available S relative to annual vegetation requirements than N. Many forest ecosystems are unable to retain all of the S deposited from the atmosphere (Johnson, 1984). Therefore, inputs of sulfuric acid typically result in greater increases in leaching losses of cations than found with nitric acid, and may play a larger role in soil acidification.

C. Cation Inputs and Outputs

Precipitation in different regions contains variable quantities of base cations. Annual input rates of 2–10 kg ha^{-1} are common for K^+, and Ca inputs are often 5–20 kg ha^{-1} annually (for a literature compilation, see Kimmins *et al.*, 1985). If these cations are taken up by plants or microbes, a net release of H^+ into the soil will occur. Similarly, adsorption of these cations on exchange sites will displace H^+ (the salt effect) into the soil solution. This would increase base saturation, but tend to decrease soil solution pH. The displaced H^+ may then leach from the system, be consumed in mineral weathering, or remain in soil solution. The addition of base cations to an ecosystem may decrease soil pH almost as much as the direct addition of H^+, but would reduce exchangeable acidity rather than increase it.

The actual effect of cation deposition depends in part on the anion composition of precipitation. If cations are balanced by an anion, such as chloride (Cl^-), that is very mobile in soils, the cations may quickly leach from the system with no net effect. If the cations are accompanied by nitrate, utilization of the nitrate will not allow leaching of the cations and a net generation of H^+ would be expected if plants used the nutrient cation. However, this H^+ generation would balance the H^+ consumed in nitrate reduction, giving no net change in the soil system.

Cations accompanied by bicarbonate, carbonate, or hydroxide represent a special case. These alkaline anions directly consume H^+ and increase soil pH. Uptake of added cations would partially counter the H^+ consumption.

VI. TIME SCALES

Just as the differing perspectives of plant physiologists and ecosystem biogeochemists require different H^+ budget components, the appropriate time scales for H^+ budgets also vary with perspectives. Processes that are important at a time scale of several seconds can often be ignored in analyses dealing with days or weeks. Similarly, some seasonal patterns in nutrient cycling may be unimportant in annual H^+ budgets. The geologic time scale of soil formation entails a simpler accounting framework than a single sequence of secondary plant succession. Four time scales are of primary interest at the ecosystem level: seasonal, annual, successional, and pedogenic.

A. Seasonal Time Scales

The soil acidity measured by a pH electrode may vary several tenths of a unit on a weekly basis, corresponding to a several-fold change in the concentration of free H^+ in the soil solution. A variety of processes are involved, such as periodic changes in the rates of decomposition of organic matter, root and microbial respiration, nitrification, concentrations of inorganic ions and organic acids, and nutrient uptake by plants and microbes. For example, Ulrich (1983a) speculated that nitrification might occur during periods when soils are too dry to permit nitrate uptake by plants, resulting in a depression in soil pH for part of the year. Nitrate production is also inhibited at low soil moisture levels, but some field data have supported this proposed pattern (cf. Rehfeuss, 1981; Ulrich and Matzner, 1986). Seasonal changes in pH may be driven by changes in the ionic strength of soil solutions (see Reuss and Johnson, 1986).

Another seasonal flux of H^+ that might be important in some forest soils occurs during the melting of snowpacks in polluted regions. Intensive studies in Norway have shown that 50–80% of the pollutants are rinsed from the snowpack by the first 30% of the meltwater (Seip, 1980). This process may strongly depress pH of streams and lakes, largely due to high concentrations of nitric acid. The impact of the acid pulse in the meltwater on soil pH is uncertain, because much of the meltwater travels through soil macropores and does not interact with the soil matrix. The importance of this seasonal pulse of acidity in both streams and soils will become clearer as current research projects are completed (cf. Dillon *et al.*, 1984).

B. Annual Time Scales

Many of the processes included in the within-season time scale can be amalgamated and ignored when the level of resolution is expanded to an annual cycle. For example, the N cycle will largely balance with respect to H^+ generation and consumption over an annual period. Changes in the pools of inorganic ammonium and nitrate that were important on short time scales become minor between years. A substantial shift of soil organic N into plant organic N may occur, but as noted earlier this half-cycle is balanced with respect to H^+ flux. One of the most important components is the accumulation of cation nutrients from inorganic soil pools into plant tissues, which represents a net H^+ flux into the soil on an annual time scale. The accumulation of organic acids may also be of great importance in some ecosystems.

C. Secondary Plant Successional Time Scales

As an ecosystem develops on soil laid bare by disturbance, secondary succession may involve substantial changes in nutrient cycling and biomass production (see Gorham *et al.*, 1979). Until the nutrient uptake rates of regenerating vegetation match mineralization rates, nutrient outputs may be relatively high. Indeed, increased temperature and moisture in disturbed areas may increase nutrient mineralization when plant uptake is at a minimum. An increase in decomposition rate might have two opposing effects. Mineralization of ammonium might increase and be accompanied by an increase in nitrate production and loss, which would generate H^+ in the soil. Conversely, the decomposition release of nutrient cations would entail consumption of H^+.

Few studies have examined changes in soil pH following disturbance and regeneration, but some insights are available from two studies. Nykvist and Rosen (1985) reported soil pH increased (about 0·5 to 1·0 units) after harvest in several Norway spruce (*Picea abies* L) stands, and that H^+ consumption seemed to be greater than H^+ production. Horne and Binkley (1986) measured H^+ pools across an age sequence wave induced by a root-rot fungus. They found that destruction of the old-growth mountain hemlock [*Tsuga mertensiana* (Bong.) Carr.] stand increased forest floor pH by 0·5 units and decreased titratable acidity by 80%. Forest floor acidity had not recovered to predisturbance levels within 70 years.

As vegetation develops and organic matter accumulates in the soil, high rates of plant production increase nutrient demands, with two effects. Any nutrient leaching losses would be diminished, and nitrate production might be largely coupled with plant uptake with no net effect on the H^+ budget. Cation nutrients should also be sequestered in plant biomass at a more rapid rate, increasing inputs of H to the soil.

As succession proceeds, the respiration requirement of accumulated biomass gradually approaches or exceeds the rate of photosynthesis, and biomass accumulation ceases or becomes negative. At this point, H^+ generation by cation accumulation in biomass ceases, and nutrient leaching may increase. Finally, most unmanaged forest ecosystems are subject to devastating disturbances such as fire, wind storms, and insect or pathogen epidemics, which will restart the secondary succession sequence. Associated with such disturbances are increased rates of decomposition and nutrient release.

In the case of both decomposition and fire, the oxidation of organic matter consumes H^+. The oxidation of organic anions (dissociated acids) consumes H^+, and the liberation of cation nutrients (such as calcium) from organic bonds requires replacement by H^+. For these reasons, the ash residue remaining after a fire is often alkaline, in the range of pH 7–9. Nitrogen in organic matter may also be volatilized in the fire, but this does not consume or generate H^+. The generation and consumption of H^+ by vegetation growth can typically be ignored in long-term H^+ assessments. An exception arises when forest biomass is removed from a site rather than decomposed or burned in place. Forest harvesting interrupts this natural cycle, resulting in a net generation of H^+ over the rotation due to cation uptake and removal in harvests.

D. Pedogenic Time Scales

A complete secondary successional sequence, from pioneer communities through devastation of mature communities, results in little net change in ecosystem organic matter. Most of the biomass produced is consumed either by saprophytes or fire, with a relatively small (and rarely measured!) fraction becoming incorporated into long-term soil organic matter pools.

Just as a complete N cycle yields no net H^+ flux, the complete biomass cycle of a secondary successional sequence has no net H^+ flux. Carbon dioxide is taken from the atmosphere, bound with cation nutrients (generating H^+), and later oxidized, releasing the cations and consuming H^+. Therefore, successional accumulations and depletions of ecosystem biomass can be considered neutral in composing H^+ budgets on soil-genesis time scales. Indeed, Bockheim (1980) summarized the available literature on pedogenic chronosequences and concluded that soil pH related better to the age of the soil than to any other factor. At this gross level of resolution, the key H^+ budget processes are those dealing with inputs and outputs from the ecosystem, coupled with any long-term increases in pools such as soil organic matter.

Vegetation does have a major effect on rates of nutrient input and output on pedogenic time scales, so the role of vegetation in long-term H^+ budgets is not trivial. For example, Crocker and Major (1955) examined soil development on recessional moraines and found soil pH declined from 8·0 to 5·5 in 32 years under the influence of nitrogen-fixing green alder. The pH of uncolonized soil remained unchanged over the same period. The processes responsible for the alder effects were not examined, but could include nitrate leaching, cation uptake, and carbonic acid production.

Leaching losses of base cations are typically small on an annual basis, but become very significant over pedogenic time.

VII. EFFECTS OF FOREST MANAGEMENT

Forest management affects H^+ budgets primarily through alteration of nutrient cycles. Harvesting the boles of an old-growth forest in the Pacific Northwest United States (Sollins *et al.*, 1980) might remove 15–20 kmol of cation charge per hectare (0·04–0·05 kmol (+) ha^{-1} annually over 400 years). This is substantial compared to current H^+ input in precipitation of 0·03 kmol ha^{-1}, but small compared to H^+ inputs in many industrialized regions. If branches and leaves were also removed, the net H^+ increase would be doubled, and removal of large woody debris on the forest floor would further increase the total. If the vegetation decayed naturally or burned in a wildfire, the cation release would consume equivalent quantities of H^+, leaving the ecosystem H^+ budget in balance. However, directional (noncyclic) pedogenic processes such as erosion, weathering, and nutrient uptake from subsoil can provide a net change in the H^+ budget on a successional time scale.

Intensively managed plantations accumulate biomass and nutrient cations rapidly, causing a net flux of H^+ into the soil. This acidification is an acceleration of a natural process, but harvesting the plantation prevents the natural consumption of H^+ that would accompany decomposition. Wittwer and Immel (1980) reported that a 4-year-old poplar plantation contained about 20 kmol(+) ha^{-1} of base cations; a complete biomass harvest would result in a very large net H^+ production of about 5 kmol(+) ha^{-1} annually.

Fertilization may have a variety of net effects on the H^+ budget, as described earlier, but the acceleration of cation accumulation in forest biomass always increases the net H^+ generation. Fertilization with lime materials is common for controlling soil pH in forest nurseries. As the $CaCO_3$ dissolves in water to Ca^{2+} and CO_3^{2-}, the CO_3^{2-} consumes 1 or 2 H^+, and raises the soil pH. Liming is also used occasionally in forest stands to reclaim acidic soils (Baule and Fricker, 1970), and is used operationally in Sweden to counter the acidifying effect of natural acid production and acid deposition. If the input of H^+ plus internal H^+ generation equalled 1 kmol of H^+ ha^{-1} annually, about 2,000 kg ha^{-1} of lime (as $CaCO_3$) would be needed once a rotation to neutralize 40 years of H^+.

Some other management impacts are less direct than those of fertilization. For example, removing vegetation from a site may decrease nutrient uptake for one to several years and result in higher nitrate losses. This effect is often small, however, in comparison with the removal of cations in the harvested biomass. Fires are a common practice in the management of some forest types, and burning may substantially decrease soil acidity. For example, about 1 million ha of pine forests are burned annually in the Southeastern United States (Richter *et al.*, 1982) to meet a variety of management objectives. Such low intensity fires reduce the pool of soil H^+ by consuming organic acids and releasing base cations, with a concommitant consumption of H^+ in combustion of the organic matter. This reduction in "natural" acidity might allow acid deposition to be buffered with little net change in soil pH. In one study, fires at 2-year intervals for 24 years in a loblolly pine (*Pinus taeda* L.) plantation reduced total acidity of the forest floor plus 0–10 cm mineral soil by about 95 kmol ha^{-1} (Binkley, 1986). The pH of the upper mineral soil in burned plots was 4·1, compared with 3·7 in unburned stands. Titration with HCl showed that lowering the pH of the burned soil to the pH of the unburned soil might require about 25 kmol H^+ ha^{-1}. This experiment would indicate that such a burning regime might offset the effects of acid deposition by several decades. However, the total picture is more complicated and the entire H^+ budget would need to be followed to assess the long-term impacts on both biogeochemical cycles and site productivity.

VIII. EFFECTS OF SPECIES ON H^+ BUDGETS AND SOIL ACIDITY

A. Afforestation

Soils beneath forests typically differ from those beneath grasslands and other types of vegetation, but on a broad scale these differences may reflect the climatic patterns that also determine the type of vegetation. In some cases, such as the boundary between forests and grasslands, climatic factors are constant and differences in soils may primarily reflect the effects of vegetation.

The invasion of grassland by trees might substantially alter H^+ budgets and soil acidity. Dormaar and Lutwick (1966) characterized soil profiles in a grassland and beneath invading stands of 45-year-old balsam poplar (*Populus balsamifera* L.), 85-year-old aspen (*Populus tremuloides* Michx.) and 150-year-old Douglas-fir. The number of profiles sampled ranged from 1 to 3 for each vegetation type, so the comparisons may reflect limited sampling in addition to any vegetation effects. Jenny (1980) summarized these profiles (to 80 cm) and showed the grassland and Douglas-fir soils both had about 170 kmol ha^{-1} of exchangeable H^+ (assuming a bulk density of 1). The balsam poplar soil had about 250 kmol H^+ ha^{-1}, and the aspen soil contained only 100 kmol H^+ ha^{-1}. The profiles differed substantially in organic matter and N contents; the grassland soils appeared to have about 150 Mg ha^{-1} more organic C and 12·5 Mg ha^{-1} more N than the Douglas-fir soil. The changes in soil organic matter and N content were probably too large to have occurred within the timespan of the current forests, so the differences among stands in exchangeable H^+ may not be fully attributed to the current vegetation. These data are intriguing, but we know of no systematic study of H^+ budgets of grasslands and invading forests that quantify the key processes responsible for large chains in soil chemistry and H^+ budgets.

Afforestation of moorlands and bogs should alter ecosystem production, soil aeration and nutrient cycling rates, and therefore H^+ budgets and soil acidity. Miles and Young (1980) characterized the differences in soil properties associated with first-generation stands of birch (*Betula pendula* Roth and *B. pubescens* Ehrh.) invading moorlands in Scotland and England. After 90 years, soil pH (10–15 cm) increased from 3·8 in the *Calluna* heathland to 4·9 under birch. The increase in pH was associated with a decrease of about 40 kmol exchangeable H^+ ha^{-1}, or roughly 0·5 kmol H^+ ha^{-1} annually. The authors attributed the increase in pH to the increase in Ca^{2+}, but the increase in exchangeable Ca^{2+} (about 4 kmol (+) ha^{-1} over 80 years) was much too small to account for the decrease in exchangeable H^+. The increase in base saturation was due primarily to a decrease in exchangeable acidity rather than to an increase in exchangeable bases. The authors noted that the interaction of improved substrate quality (birch litter rather than heather), increased pH, and increased Ca^{2+}, on soil fauna and microflora might generate further improvements in soil conditions.

Pine plantations have been established on large areas of peat bogs in Great Britain, and in many cases soil pH has declined. Williams and Cooper (1978) examined 6 plantations of lodgepole pine (*Pinus contorta* Dougl.), ranging in age from 12 to 47 years. Across all sites, pH declined from an average of 3·2 in unplanted bogs to 3·0 under pines. Exchangeable acidity increased by about 15%, or roughly 50 kmol H^+ ha^{-1} (about 0·5 kmol H^+ ha^{-1} annually). Extractable base cations declined by only about 7 kmol ha^{-1}, so the increase in exchangeable acidity and decrease in base saturation derived from an increase in cation exchange capacity rather than from a depletion of exchangeable cations. The authors concluded that the plantations had greater rates of evapotranspiration, which improved aeration in the peat and allowed for an increase in decomposition and production of humic and fulvic acids.

Afforestation might also affect rates of soil acidification by altering deposition from the atmosphere (Harriman and Morrison, 1981). Tall canopies with large leaf areas would receive greater inputs than low canopies of grasses or shrubs, and this difference in H^+ deposition could easily match the effect of differences among vegetation types in nutrient uptake, decomposition, and other internal-ecosystem processes.

B. Comparisons of Tree Species Within Plantations

The belief that various tree species have different effects on soils dates back at least to the nineteenth century (see Handley, 1954). Growth rates, nutrient use, litter quality, and soil microenvironments differ among species, so it is evident that species might differ in effects on H^+ and soil acidity (Miles, 1985). Changes in soil chemistry around single plants have been documented in some cases (cf. Zinke, 1962; Grubb *et al.*, 1969), but not others (Jenny, 1980). At a whole-ecosystem scale, the early evidence supporting this belief was weak (Stone, 1975b). Baule and Fricker (1970) noted that much of the soil degradation attributed to spruce monocultures relative to native beech forests in Europe arose from litter raking (and nutrient removal) in the spruce forests. Concerns about soil degradation by monocultures of radiata pine in New Zealand (*Pinus radiata* D. Don) also have not been supported by intensive studies; any declines in second rotation production appear related to nutrient availability and plant competition rather than to soil degradation (cf. Will, 1984). Most early studies finding major changes in mineral soils within one rotation were often unreplicated and not quantitative (cf. Fischer, 1928). A few early studies were replicated and provided stronger evidence that species can alter soil acidity. For example, Read and Walker (1950) found substantially higher soil pH and exchangeable cations beneath scattered red cedar (*Juniperus virginiana* L.) in a red pine (*Pinus resinosa* Ait.) plantation than in the rest of the plantation. However, even this well-replicated study could not separate cause and effect; the distribution of the red cedar may well have resulted from pre-existing soil conditions rather than have caused the differences.

Species might differ in effects on H^+ budgets and soil acidity by several mechanisms: (1) canopy effects on deposition rates; (2) biological nitrogen fixation; (3) nutrient uptake and litter quality; and (4) indirect effects on soil animals and microflora.

Canopy characteristics substantially influence the deposition rate of atmospheric chemicals, and any differences among species in canopy structure would be especially important in polluted regions. For example, Ulrich (1983b) reported S deposition in a beech forest was 25 kg ha^{-1} annually, compared to about 60 kg ha^{-1} annually in a nearby spruce forest. In such cases, any species effect on soil acidity would be dominated by effects on atmospheric deposition rather than by any within-system process.

Nitrogen-fixing trees may indirectly alter H^+ budgets even though the N fixation process does not generate or consume H^+. For example, nitrogen-fixing red alder (*Alnus rubra* Bong.) can enhance soil fertility to levels where over 50 kg of nitrate-N (3·5 kmol H^+ ha^{-1}) leaches from the forest each year (Van Miegroet and Cole, 1984), and pH can be reduced by 0·5 to 1·0 units in the presence of alder (cf. Franklin *et al.*, 1968; DeBell *et al.*, 1983). Less fertile sites with alder show no decrease in pH, probably due to less nitrate leaching (Binkley, 1983).

Aside from extreme cases of pollution or high rates of N fixation, the differences among species on H^+ budgets and soil acidity are much less clear. Replicated, experimental plantations of various species offer the best opportunity to evaluate species effects on H^+ budgets, and the few available experiments do not encourage broad generalizations. For example, conifers are generally considered to acidify soils more than hardwoods (see an excellent review by Handley, 1954). However, Alban (1982) examined soils beneath 40-year-old plantations of jack pine (*Pinus banksiana* Lamb.) red pine (*P. resinosa* Ait.), white spruce [*Picea glauca* (Moench) Voss] and aspen, and found that forest floor pH was the same under spruce and aspen (5·5), but lower under the pines (5·0). The pattern was reversed in the upper mineral soil (5·6 and 6·0, respectively), which would be expected if nutrient cation uptake (and H^+ release) were greater in aspen and spruce stands than in the pine stands. In fact, calculations from Perala and Alban (1982) of the nutrient content of biomass showed cation accumulation in the aspen and spruce stands [about 120 kmol (+) ha^{-1} at age 40] exceeded that of the pine stands [about 65 to 70 kmol (+) ha^{-1}] by 50 to 55 kmol (+) ha^{-1}. No other experiment on species effects on H^+ budgets has been this thorough, but this study (and see Challinor, 1968) does not support broad generalizations about effects of conifers, or the relative tendency of conifers and hardwoods to acidify soils.

Species also affect H^+ budgets through the acid content of litter, and the production of organic acids in the decomposition of litter. As noted earlier, the acid content of fresh litter varies with species, but these low-molecular-weight acids are readily oxidized by microbes, and have no long-term effect on soil acidity. Incomplete decomposition of litter can generate more stable organic acids; high lignin content of conifer needles appears to retard decomposition (Swift *et al.*, 1979) resulting in large accumulations of organic acids.

Silvicultural prescriptions occasionally call for interplanting hardwood species in conifer plantations to speed decomposition and reduce surface soil acidity (cf. Lutz and Chandler, 1946; Gosz, 1985), but the supporting data are unconvincing (Miles and Young, 1980). Some well-designed experiments again do not support broad generalizations. For example, Thomas (1968) tested the ability of dogwood (*Cornus florida* L.) leaves (with a calcium content of 2·8%) to increase decomposition of loblolly pine (*Pinus taeda* L.) needles. After one year, weight loss from needles in litter bags was identical with (ratio of needles: dogwood leaves was 4:1) and without dogwood leaves. Another well-designed study of the effects of mix species plantations involves 0·2 ha replicated plots in the Gisburn Forest in England of pure and mixed stands of Scots pine (*Pinus sylvestris* L.), Norway spruce (*Picea abies* L.) and black alder [*Alnus glutinosa* (L.) Gaertn.]. Although the N-fixing hardwood did increase nitrification, the greatest nitrate production occurred in the spruce/pine stand (Brown and Harrison, 1983). Spruce growth was also improved in association with alder, but greatest spruce growth occurred in the spruce/pine stand. Earthworm biomass followed the same pattern.

All of these studies underscore the limitations of broad generalizations about effects of species on soil acidity, and the importance of direct evaluation of processes affecting H^+ flux. Mixtures of hardwood species in conifer plantations might increase decomposition and pH in the upper soil if several conditions were met: if cation uptake by the hardwood occurred from deep in the soil, if hardwood litter had high concentrations of base cations, low concentrations of aluminium, and if hardwood litter were low in lignin and decomposed rapidly. Few experiments have characterized all of these key variables.

C. Importance of Below Ground Litter, Microflora and Soil Animals

Another limitation in understanding the potential effects of litter quality and decomposition on H^+ budgets derives from a lack of knowledge on the decomposition of below-ground litter inputs. The annual turnover of biomass belowground exceeds aboveground litterfall in most forests (cf. Vogt *et al.*, 1986). In addition, most aboveground litter is respired rather than mixed into the mineral soil, so the largest effect of plant litter on acidity in the mineral soil probably derives from the death and decomposition of fine roots, mycorrhizae and associated microflora.

The activities of animals and microbes may affect H^+ budgets, and tree species clearly differ in effects on soil animals and microbes. These processes may be so interactive that cause and effect are almost impossible to separate. For example, species producing litter of high palatability for earthworms may increase earthworm activity which might increase decomposition and pH, further improving litter quality and earthworm activity. Tree species also differ in their mycorrhizal symbioses. For example, Read *et al.* (1985) noted that mycorrhizal associates of beech developed sheaths around roots but did not extend many hyphae into the soil. In coniferous soils, mycorrhizae commonly form both sheaths and an extensive network of hyphae in the soil. These authors speculated that the seasonality of nutrient release might vary between beech and conifers, causing a different pattern of mycorrhizal association. In turn, the distribution of the fungi would affect patterns in nutrient uptake and H^+ generation.

The activities of bacteria and fungi generate and consume H^+, and may vary with soil pH. As noted earlier, nitrification generates H^+ but nitrifying bacteria are generally depressed by low soil pH. Fungi commonly respire carbohydrates incompletely, producing small-molecular-weight acids such as oxalic acid. Oxalic acid precipitates with calcium to form a very insoluble salt, and accumulations of oxalate may be large. For example, a Douglas-fir soil in Oregon contained about 20 kmol (−) ha^{-1} of oxalate (Sollins *et al.*, 1981). Oxalate production may aid in P uptake by mycorrhizal fungi, which could increase growth of trees, introducing a variety of other changes in H^+ fluxes.

It may not be possible to separate the direct effects of tree species from their indirect effects on soil animals and microbes, but it is important to realize the comparative effects of each species may differ among sites because of differences in effects on non-tree biota.

IX. SENSITIVITY OF FOREST SOILS TO ACIDIFICATION FROM ATMOSPHERE DEPOSITION

Perhaps the most significant questions of nutrient cycles and H^+ flux relate to the net effects on the acid-base status of the soil. How large a change in soil pH and cation nutrient availability will result from a given increase in H^+ inputs?

The internal generation of H^+ within ecosystems has been shown to increase soil acidity. For example, Ugolini (1968) characterized the first 50 years of soil development following the retreat of a glacier in Alaska, and found a drop in pH similar to that reported by Crocker and Major (1955). Green alder was the major colonizer, and within 50 years the pH of the upper mineral soil dropped from 8·3 to 6·0. Over the same period, exchangeable acidity increased from 0 to about 25 kmol ha^{-1} (0–10 cm soil depth), and exchangeable base cations decreased from 40 kmol ha^{-1} to 30 kmol ha^{-1}. Cation exchange capacity thus increased from 40 kmol ha^{-1} to 55 kmol ha^{-1}, reflecting an accumulation of organic matter despite decreasing pH. These trends continued at least to age 250, when the vegetation was dominated by black spruce. By then, soil pH had declined to 4·5, exchangeable acidity had risen to 65 kmol ha^{-1}, exchangeable bases had dropped to 20 kmol ha^{-1}, and cation exchange capacity had increased to 85 kmol ha^{-1} in the upper 10 cm of soil. Base saturation declined over the entire sequence from near 100% to about 25%, due to roughly equal contributions of a decrease in base cations and an increase in cation exchange capacity.

Similar acidification of soils through acid deposition have occurred in soils located near major industrial sources of sulfur dioxide. For example, the pH of soil near a large copper ore smelter in Tennessee, USA, was reduced to the range of 3·2 to 3·8, compared with 4·5 to 5·0 in similar soils farther from the smelter (Wolt and Lietzke, 1982). Regional pollution levels are much lower than near such point sources, but long-term data on soil acidity in a heavily-polluted part of Czechoslovakia showed an acceleration of acidification between 1940 and 1980 (Pelisek, 1983). The pH of the Al horizon declined from 3·6 to 3·1 over this period, with smaller but significant declines in lower horizons. Part of this decrease in pH may have stemmed from natural acidification associated with stand development, but pollution is so extreme in this region [S deposition averages 100 kg ha^{-1} (up to about 6 kmol H^+ ha) annually across Czechoslovakia] that deposition dominates H^+ inputs.

Increased acidification of soils due to atmospheric deposition across broad regions has not been well documented. Much of Northern Europe and eastern North America receive about 10 to 40 kg ha^{-1} of S deposition annually. Any effects of these lower levels of H^+ inputs would become apparent much more slowly than has been the case in Czechoslovakia, and would be very difficult to demonstrate without any unpolluted control sites.

One approach to this problem has been the experimental addition of large quantities of H^+ as concentrated acids over short time periods (cf. Lodhi, 1982). These experiments should result in overestimates of the effects of atmospheric deposition for two reasons. First, applications of concentrated acids provide much greater concentration gradients in the soil than would arise from many years of input at lower concentrations. High concentration gradients may displace base cations from exchange sites which would not be affected by lower concentrations. Second, a variety of processes in forest soils consume H^+, and their ability to buffer the increase in H^+ input depends upon the solution concentration and duration of the experiment. The interactions between soil solution, exchange sites and mineral weathering make it difficult to simulate the effects of long-term acid deposition in a short-term experiment.

Despite these factors that should overestimate the sensitivity of soils to acidification under abnormally-high inputs, most short-term experiments have found fairly small effects of large additions of acids (Abrahamsen, 1980; Tamm, 1976; Bjor and Teigen, 1980; Stuanes, 1980). For example, Federer and Hornbeck (1985) found that 30 to 50 keq ha^{-1} of H^+ was needed to lower the pH of soil samples from Northeastern U.S. by 1 unit. Such rapid experiments allow for no gradual adjustment by soil processes (see later) that can substantially buffer pH. If these inputs were spread out over 30 to 50 years, the effects on soil acid-base status might be barely detectable.

A. Buffering Processes in Forest Soils

Four processes play major roles in buffering the acid-base status of forest soils: cation exchange, mineral weathering, anion adsorption, and nutrient uptake from deep in the soil. The retention and release of cations from soils are broadly predictable, based on distributions of acid and base cations in the soil solution and on soil exchange sites. Ions that dominate the exchange site are easily displaced by other ions in the soil solution. Thus, naturally acid soils should respond to increased H^+ input by releasing H^+ or Al^{3+} from exchange sites. Moderately acid soils retain H^+ and release base cations to the soil solution. Although these generalizations are common, relatively few experimental data have verified the patterns. In one of the best demonstrations, Wiklander (1973) found that dilute acid solutions were markedly inefficient in displacing base cations from exchange sites that were dominated by acid cations, and that the ability of H^+ to displace base cations decreased as the concentration of base cations in soil solution increased. He also found that the type of cation exchange complex (i.e., clay minerals and organic exchange sites) strongly influenced the net effect of dilute mineral acids on ion exchange reactions. These principles have not been demonstrated with many soils, but they suggest that acidification would occur most rapidly in soils with relatively low CEC and high pH and moderate-to-high base saturation (see Reuss and Johnson, 1986). Ugolini's (1968) example of soil development after glacial retreat is an excellent example of this type of soil. However, this combination is rare in humid climates of North America and Europe, mainly because such soils would already have acidified under natural conditions over the past several thousand years. Naturally acidified soils tend to be characterized by acid-dominated exchange sites, and thus tend to be resistant to further acidification.

As discussed earlier, many nutrient cycling steps consume H^+. The weathering of soil minerals is one of the most important. The mineral fraction of soils has an enormous capacity to consume H^+ as evidenced by the fact that intense acidification of soils is usually limited to the surface layer of soil profiles (soil pH generally increases with depth). This pattern also indicates limited mixing of upper and lower soil horizons on a short-term scale. However, tree falls mix surface and subsurface soil layers, and about 15% to 50% of the soil surface of natural forests consists of either pits or mounds. This microtopography can persist for two or three centuries, during which time the acid-base status of the upturned soil comes to match that of undisturbed soil (Stone, 1975a). Mixing by soil animals has a similar effect.

The removal of exchangeable bases from soils, through plant uptake or leaching, is also a primary factor driving mineral weathering, which resupplies base cations to the exchange complex. Thus, saturating soils with H^+ can increase the input of base cations. Wiklander and Koutler-Andersson (1963) found that H^+ saturation (the removal of almost all of the exchangeable base cations) of several Scandinavian soils caused a release of base cations in amounts equivalent to 10% to 20% of the CEC. Similarly, repeated cropping of unfertilized soils can remove more base cations than are present on exchange sites at any one time. The resupply of base cations to exchange sites remains an unresolved problem in the prediction of plant-available nutrients, and these poorly quantified processes are very important to assessments of the effects of increased H^+ inputs on forest soils.

A third process that buffers rapid changes in soil acidity is sulfate adsorption. Cation leaching requires an accompanying anion for electroneutrality, so adsorption of sulfate reduces cation leaching. Sulfate adsorption can consume H^+ by releasing OH^-, depending on the degree of protonation of OH^- and H_2O ligand groups, to the soil solution. This process is especially important in soils with large accumulations of aluminium and iron hydrous oxides, such as Ultisols in the Southeastern United States (Johnson and Todd, 1983). Spodosols are generally regarded as having low sulfate adsorption capacity, but some types may show substantial capacity (Wiklander, 1975; Morrison, 1983).

Although the cycling of cation nutrients through biomass and back to the soil is a neutral process with respect to H^+ flux, this cycle can result in spatial redistribution of acid and base cations within an ecosystem. Nutrient cations taken up in the subsoil would be associated with excretion of H^+ into the relatively well-buffered subsoil. Subsequent decomposition of the plant tissue near the soil surface would consume H^+ and tend to moderate the acidity of the upper soil. Uptake of nutrient cations from subsoils is probably an important component of nutrient cycling, but it has been poorly quantified and probably underestimated in many cases (Comerford *et al.*, 1984; Johnson *et al.*, 1985).

In sum, short-term acceleration of soil acidification cannot be expected except under rather unusual and specific circumstances. The most susceptible soils are characterized by: (1) low cation exchange capacity (low in clay and organic matter); (2) moderate or high pH; (3) low in weatherable minerals and rates of renewal by soil mixing or erosion; (4) low sulfate adsorption capacity; (5) shallow profiles, and (6) atmospheric inputs of concentrated mineral acids. Because few soils exhibit these features, any effects of region-wide increases in H^+ deposition rates should become detectable only over time periods of many decades or centuries. On this time scale, the production and consumption of H^+ by nutrient cycling processes will largely mediate the response of forest ecosystems. For these reasons, evaluations of the effects of acid deposition must include the H^+ budget of the ecosystem and its environment.

X. A CHESTNUT OAK FOREST CASE STUDY

Although a great variety of ecosystem analyses provide data on components of H^+ budgets, non have provided a complete list of the H^+ fluxes associated with the key processes outlined in this review. Many studies have estimated

H^+ input in wet deposition and the accumulation of base cations in biomass, but very few have measured the three components of soil acidity. A lack of information on rates of mineral weathering is probably the most critical gap.

To integrate the discussion of nutrient cycles and H^+ flux, we chose a 30 to 80 year-old deciduous forest dominated by chestnut oak (*Quercus prinus* L.) in Tennessee (Cole and Rapp, 1980; Johnson *et al.*, 1982; Richter *et al.*, 1983: Johnson *et al.*, 1985; Lindberg *et al.*, 1986; Richter, 1986). The data set for this ecosystem is one of the best available, especially regarding atmospheric deposition.

Inputs of H^+ from the atmosphere are dominated by dry deposition (Table 7). Bulk precipitation pH averages about 4·2, and with an annual rainfall of 150 cm, for an input of 0·7 kmol H^+ ha^{-1}. Dry deposition of H^+ associated with sulfate and nitrate add another 0·9 kmol ha^{-1}, for a total of 1·6 kmol H^+ ha^{-1} of deposition annually. The inputs of both nitrate and sulfate are dominated by dry deposition. The added nitrate is retained in the ecosystem through assimilation by microbes and plants, resulting in the consumption of 0·5 kmol H^+ ha^{-1} annually. About 0.6 kmol (−) ha^{-1} of sulfate is retained in this ecosystem, mostly through soil adsorption, which should consume 0·5 to 1 H^+ for each equivalent of sulfate. For simplicity, Table 7 lists a H^+ consumption equivalent to the sulfate retained. Deposition of calcium, potassium and ammonium (Lindberg *et al.*, 1986) sums to 0·7 kmol (+) ha^{-1}, but leaching losses are about 0·3 kmol (+) ha^{-1} (Johnson *et al.*, 1985). If plant uptake accounts for the difference, then 0·4 kmol H^+ would be produced from the cation nutrient component of deposition.

The production of carbonic acid is important in this forest, generating about 0·6 kmol H^+ ha^{-1} annually. This figure was derived from measurements of bicarbonate in soil solution, based on 1 H^+ produced for each HCO_3^-. The majority of the carbonic acid is not dissociated at the pH of this soil, so it may be left out of the H^+ budget calculation for the undisturbed forest. In an evaluation of the effect of disturbance, it would be important to account for any change in this pool of titratable acidity.

The production and accumulation of organic acids have not been studied in this forest. As noted in earlier examples, these acids can either accumulate or be consumed at rates on the order of 0·1 to 0·5 kmol H^+ ha^{-1} annually, and such a rate would be significant relative to other portions of the H^+ budget of this forest.

Although the forest cycles large quantities of N, P and S from organic pools in the soil into biomass, these cycles are balanced with no net H^+ flux. The annual uptake of nutrient cations for aboveground biomass is about 6·4 kmol ha^{-1}; if all nutrient cations were obtained from inorganic pools, this would represent a production of 6·4 kmol H^+ ha^{-1} to maintain charge balance in nutrient uptake. Much of the cation supply comes from decomposing organic matter (which consumes H^+), so the net H^+ generation

Table 7
The H^+ budget of a chestnut oak forest in Tennessee, USA (data sources cited in text).

Process	Pool size or annual rate (kmol (+) ha^{-1})	Net ecosystem (kmol H^+ ha^{-1})
Atmospheric deposition	*annually*	
H^+		
Wet	0·7	+0·7
Dry	0·9	+0·9
Nitrate		
Wet	0·2	
Dry	0·3	
Assimilation	0·5	−0·5
Sulfate		
Wet	0·7	
Dry	0·9	
Assimilation + adsorption	0·6	−0·6
Cations (Ca + Mg + NH_4)		
Wet	0·3	
Dry	0·4	
Assimilation	0·4	+0·4
Bicarbonate production	0·6	+0·6
Organic acid accumulation	unknown, perhaps important	
Biomass accumulation of internal-ecosystem nutrients		
Nitrogen, sulfur	no net effect on H^+ budget	
Phosphate	negligible effect on H^+ budget	
Base cations	0·9	+0·9
Weathering, subsoil uptake of base cations	unknown, probably large	
Pool	*pool size*	
Free H^+	<0·1	
Exchangeable H^+	100	
Titratable H^+	750	
Base cation content of boles	30	
Whole-tree base cation content	52	

would be considerably less. Most of the cations taken up by the forest are returned annually to the forest floor in litterfall and throughfall, leaving about 1·1 kmol ha^{-1} to accumulate in aboveground biomass annually. Accumulation in belowground tissues has not been measured, but probably equals about 20% of the aboveground value, giving a total of 1·3 kmol (+) ha^{-1} accumulated in biomass annually. This net accumulation probably comes from inorganic pools in the soil, but if a net transfer of cations occurred from soil organic matter into plant biomass, the H^+ would be somewhat less. In Table 7, we assumed that 0·4 kmol (+) ha^{-1} of cations deposited from the atmosphere were taken up by the vegetation, leaving a requirement of only 0·7 kmol (+) ha^{-1} from soil pools. The nutrient content of the forest floor may have reached a steady state, and if so, the uptake of nutrient cations would result in the production of equivalent quantities of H^+.

At this point in the H^+ budget, annual H^+ generation sums to 3·6 kmol ha^{-1}, compared with H^+ consumption of only 1·3 kmol ha^{-1}, with the contribution of organic acid accumulation unknown. Some H^+ (or Al^{3+}) may leach from the soil, but as with the vast majority of forest ecosystems, this flux is negligible (<0·1 kmol H^+ ha^{-1} annually). Most of the remaining 2·3 kmol H^+ ha^{-1} must accumulate on exchange sites or be consumed in weathering. These two major processes (changes in H^+ pools and weathering) have not been measured for this forest. Changes in the size of the exchangeable or total H^+ pools would be difficult to detect, even if the rate matched that of other fluxes. Weathering rates are especially difficult to measure. These two components of H^+ are very important and have a high priority for research in many ecosystems.

Moving on to a time scale of a rotation or successional sequence, destruction of the forest by a hurricane or fire would release the cations in plant biomass and balance the H^+ generated in cation uptake. Harvesting only the boles would result in a net production of about 30 kmol ha^{-1} of H^+ (or 0·5 kmol ha^{-1} annually if the stand were 60 years old). Whole-tree harvest without leaves in winter would give a net H^+ production of about 47 kmol ha^{-1} (0·8 kmol ha^{-1} annually) or with leaves in summer of 52 kmol ha^{-1} (0·9 kmol ha^{-1} annually). As with the annual time scale, the rate of mineral weathering and changes in exchangeable and total H^+ pools over a rotation are unknown for this forest. Rates of weathering and changes in acid and base saturation may both be significant relative to other H^+ budget components, so the H^+ budget for even this well-characterized forest is far from complete.

This case study illustrates several key points. Classical measurements of H^+ in wet deposition would account for less than half of the actual rate of deposition. However, internal ecosystem processing of the nitrate and sulfate anions of the strong acids neutralizes about two-thirds of the H^+ in acid deposition, giving a new H^+ loading 0·5 kmol ha^{-1} annually rather than 1·6 kmol ha^{-1} annually. Next, the production of bicarbonate produces a large proportion of the H^+ added to this ecosystem, and this production would be very sensitive to any alteration in soil pH. A decrease in soil pH would reduce H^+ production by this process, and an increase in pH would increase H^+ production, so the bicarbonate pathway of producing H^+ should buffer H^+ budgets relative to changes in other components. Further, H^+ generation due to accumulation of cations in biomass is also large relative to other H^+ sources, and forest harvest prevents later recycling that would neutralize this production. Finally, it would be difficult to evaluate the likely effects of changes in annual H^+ production on soil pH even in this well-studied ecosystem for two major reasons. The rates of organic acid accumulation and mineral weathering are unknown, and the dynamic response of soil H^+ pools and extractable base cation pools to changes in H^+ inputs have not been characterized.

XI. CONCLUSIONS

Both atmospheric deposition and natural nutrient cycling processes contribute to the production and consumption of H^+ in forest ecosystems. The effects of these fluxes on soil pH depends upon net changes in the large, long-term pools of H^+ measured as exchangeable and titratable acidity, and upon the dynamic equilibrium between these large buffers and the H^+ free in soil solution. Assessments of probable changes in soil acidity under the influence of various pollution regimes or management schemes requires the construction of H^+ budgets with appropriate spatial and temporal scales. For example, uptake of nitrate may be accompanied by uptake of H^+, increasing the pH of the rhizosphere. At a larger scale, however, the earlier production of nitrate produced an equivalent amount of H^+, so the H^+ balance in the whole soil would remain unchanged. The H^+ flux associated with nitrate uptake would be very important for studies of the rhizosphere environment, but could be ignored at a whole-ecosystem level.

Although internal ecosystem cycling of nutrients generate and consume very large quantities of H^+, these cycles generally balance each other and have no effect on soil acidity in the long term. Trends in soil acidification over long periods are dominated by non-cyclic transfers of H^+ and other elements across ecosystem boundaries, including atmospheric deposition, bicarbonate production, leaching, and forest harvest.

Frameworks for H^+ accounting such as the one used in Table 7 are easy to develop, but it is exceedingly difficult to obtain precise information for many key components. The chemical composition of rainfall can be measured easily, but dry deposition onto forest canopies is very problematic. Similarly, exchangeable acidity and base cations can be measured at one point in time, but accurate documentation of changes over time are rare. Most estimates of mineral weathering rates are crude, and few studies have attempted to monitor changes in soil organic matter over long periods. Any of these poorly quantified processes could dominate the H^+ budget in some cases, so precise characterization of H^+ budgets may be an elusive goal.

Within these limitations, a H^+ budget approach can advance ecosystem studies in three directions. First, assessments of the probable impacts of acid deposition must account for the processing of the deposited chemicals within the ecosystem, as nitric acid is neutralized if plants assimilate the nitrate. Further, comparisons of the magnitudes of H^+ input from deposition and from within ecosystem generation can also allow a rough evaluation of the need for concern. Deposition is of most concern when it rivals or exceeds the rate of H^+ production by internal sources. Second, a H^+ budget approach can improve empirical evaluations of changes in soil pH. Many studies have measured changes in forest soil pH after treatments such as fire, or beneath contrasting vegetation types, but few have quantified the changes in exchangeable and titratable acidity, or the change in equilibrium between free acidity and these buffers. Simple, empirical H^+ budgets could improve predictions of changes in soil acidity. Finally, a H^+ budget approach to computer simulation of ecosystem dynamics would provide a stimulating direction for integrating ecosystem biogeochemistry, and would allow gaming with various portions of such models to determine the likely importance of poorly quantified processes.

ACKNOWLEDGEMENTS

This synthesis was funded in part by NSF grant #BSR 841678, and by the Electric Power Research Institute's Integrated Forest Effects Study. We gratefully acknowledge the assistance of those who carefully reviewed earlier drafts, including W. Reiners, J. Pye, D. Valentine, W. Schlesinger, R. Church, J. Siedow, I. Burke, and four anonymous reviewers.

REFERENCES

Abrahamsen, G. (1980). Impact of atmospheric sulfur deposition on forest ecosystems. *In* "Atmospheric Sulfur: Environmental Impact and Health Effects" (Eds D. S. Shriner, C. R. Richmond, and S. E. Lindberg), pp. 397–416. Ann Arbor Science Publishers, Ann Arbor, Michigan.

Abruna, R., Pearson, R. W. and Elkins, C. B. (1958). Quantitative evaluation of soil reaction and base status changes resulting from field applications of residually acid-forming nitrogen fertilizers. *Soil Sci. Soc. Am. Proc.* **22**, 539–542.

Adams, F. (Ed.) (1984). "Soil Acidity and Liming" Agronomy Society of America, Madison, Wisconsin.

Adams, S. N. and Cornforth, I. S. (1973). Some short-term effects of lime and fertilizers on a Sitka spruce plantation II. Laboratory studies on litter decomposition and nitrogen mineralization. *Forestry* **46**, 39–47.

Alban, D. H. (1982). Effects of nutrient accumulation by aspen, spruce and pine on soil properties. *Soil Sci. Soc. Am. J.* **46**, 853–861.

Andersson, F. T., Fagerstrom, T. and Nilsson, S. I. (1980). Forest ecosystem responses to acid deposition – hydrogen ion budget and nitrogen/tree growth model approaches. *In* "Effects of Acid Precipitation on Terrestrial Ecosystems" (Eds T. C. Hutchinson and M. Havas), pp. 319–334. Plenum Press, New York.

Assman, S. M., Simoncini, L. and Schroeder, J. I. (1985). Blue light activates electrogenic ion pumping in guard cell protoplasts of *Vicia faba*. *Nature* **318**, 285–287.

Baule, H. and Fricker, C. (1970). "The Fertilizer Treatment of Forest Trees" BLV Verlagsgesellschaft, Munich.

Bielski, R. L. (1973). Phosphate pools, phosphate transport, and phosphate availability. *Ann. Rev. Plant Physiol.* **24**, 225–252.

Bielski, R. L. and Ferguson, I. B. (1983). Physiology and metabolism of phosphate and its compounds. *In* "Inorganic Plant Nutrition" (Eds A. Lauchli and R. L. Bielski), pp. 422–449. Springer-Verlag, New York.

Binkley, D. (1983). Interaction of site fertility and red alder on ecosystem production in Douglas-fir plantations. *For. Ecol. Managem.* **5**, 215–227.

Binkley, D. (1986). Prescribed fire and soil acidity in loblolly pine. *Soil Sci. Soc. Am. J.* (in press).

Bockheim, J. G. (1980). Solution and use of chronofunctions in studying soil development. *Geoderma* **24**, 71–85.

Bjor, K. and Teigen, O. (1980). Effects of acid precipitation on soil and forest: 6. *In* "Ecological Effects of Acid Precipitation" (Eds D. Drablos and A. Tollen) pp. 200–201. SNSF Project, Oslo-As, Norway.

Bohn, H. L., McNeal, B. L. and O'Connor, G. A. (1979). "Soil chemistry" John Wiley, New York.

Broadfoot, W. M. and Pierre, W. H. (1939). Forest soil studies: I. Relation of rate of decomposition of tree leaves to their acid-base balance and other chemical properties. *Soil Sci.* **48**, 329–347.

Brown, A. H. F. and Harrison, A. F. (1983). Effects of tree mixtures on earthworm populations and nitrogen and phosphorus status in Norway spruce (*Picea abies*) stands. *In* "New Trends in Soil Biology", pp. 101–108. Dieu-Brichart, Ottignies – Louvain-la-Neuve.

Bubenick, D. V. (1984). "Acid Rain Information Book" Noyes Publications, Park Ridge, New Jersey.

Burgess, R. L. (Ed.) (1984). "Effects of Acidic Deposition on Forest Ecosystems in the Northeastern United States: An Evaluation of Current Evidence" Institute of Environmental Program Affairs, State University of New York, Syracuse.

Challinor, D. (1968). Alteration of soil characteristics by four tree species. *Ecology* **49**, 286–290.

Cole, D. W. and Rapp, M. (1980). Elemental cycling in forest ecosystems – a synthesis of the IBP. *In* "Dynamic Properties of Forest Ecosystems" (Ed. D. E. Reichle), pp. 341–409. Cambridge University Press, Cambridge.

Coleman, N. T. and Thomas, G. W. (1964). Buffer curves of acid clays as affected by the presence of ferric iron and aluminum. *Soil Sci. Soc. Amer. Proc.* **28**, 187–190.

Comerford, N. B., Kidder, G. and Mollitor, A. V. (1984). Importance of subsoil fertility to forest and nonforest plant nutrition. *In* "Forest Soils and Treatment Impacts" (Ed. E. L. Stone), pp. 381–404. University of Tennessee, Knoxvile.

Clarkson, D. T. and Hanson, J. B. (1980). The mineral nutrition of higher plants. *Ann. Rev. Plant Phys.* **31**, 239–298.

Crocker, R. L. and Major, J. (1955). Soil development in relation to vegetation and surface age at Glacier Bay, Alaska. *J. Ecol.* **43**, 427–448.

DeBell, D. S., Radwan, M. A. and Kraft, J. M. (1983). "Influence of Red Alder on Chemical Properties of a Clay Loam Soil in Western Washington" USDA Forest Service Res. Pap. PNW-313. Pac. Northwest For. Range Exp. Stn., Portland, Oregon.

Dillon, P. J., Yan, N. D. and Harvey, H. H. (1984). Acidic deposition—effects on aquatic ecosystems. *RC Crit. Rev. Env. Cont.* **13**, 167–195.

Dormaar, J. F. and Lutwick, L. E. (1966). A biosequence of soils of the rough fescue prairie-poplar transition in southwestern Alberta. *Can. J. Earth Sci.* **3**, 457–471.

Driscoll, C. T. and Likens, G. E. (1982). Hydrogen ion budget of an aggrading forested watershed. *Tellus* **34**, 283–292.

Driscoll, C. T., van Breemen, N. and Mulder, J. (1985). Aluminum chemistry in a forested spodosol. *Soil Sci. Soc. Amer. J.* **49**, 437–444.

Drablos, D. and Tollan, A. (Eds.) (1980). "Ecological Impact of Acid Precipitation" SNSF Project, Oslo-As, Norway.

Federer, C. A. and Hornbeck, J. W. (1985). The buffer capacity of forest soils in New England. *Water, Air, Soil Pollut.* **26**, 163–173.

Fisher, R. T. (1928). Soil changes and silviculture on the Harvard Forest. *Ecology* **9**, 6–11.

Focht, D. D. and Verstraete, W. (1977). Biochemical ecology of nitrification and denitrification. *Adv. Microb. Ecol.* **1**, 135–214.

Franklin, J. F., Dyrness, C. T., Moore, D. G. and Tarrant, R. F. (1968). Chemical soil properties under coastal Oregon stands of alder and conifers. *In* "Biology of Alder" (Eds J. M. Trappe, J. F. Franklin, R. F. Tarrant, and G. M. Hanson), pp. 157–172. USDA Forest Service, Portland, Oregon.

Goodwin, T. W. and Mercer, E. I. (1983). "Introduction to Plant Biochemistry" Pergamon Press, Oxford.

Gorham, E., Vitousek, P. M. and Reiners, W. A. (1979). The regulation of chemical budgets over the course of terrestrial ecosystem succession. *Ann. Rev. Ecol. Syst.* **10**, 53–84.

Gosz, J. (1985). Biological factors influencing nutrient supply in forest soils. *In* "Nutrition of Plantation Forests" (Eds G. D. Bowen and E. K. S. Nambiar), pp. 119–146. Academic Press, London.

Grubb, P. J., Green, H. E. and Merrifield, R. C. J. (1969). The ecology of chalk heath: its relevance to the calcicole-calcifuge and soil acidification problems. *J. Ecol.* **57**, 175–213.

Handley, W. R. C. (1954). "Mull and Mor Formation in Relation to Forest Soils" Forestry Comm. Bull. #23, Her Majesty's Stationery Off., London.

Harriman, R. and Morrison, B. R. s. (1981). Forestry, fisheries, and acid rain in Scotland. Scottish Forestry **35**, 89–95.

Helyar, K. R. (1976). Nitrogen cycling and soil acidification. *J. Austr. Inst. Agric. Sci.* 1976, 217–221.

Horne, A. L. and Binkley, D. (1986). Acidity of forest floor and mineral soil: changes in response to stand development in mountain hemlock. *Bull. Ecol. Soc. Amer.* **67**, *in press.*

Husar, R. B. (1985). Chemical climate of North America: sources and deposition with special emphasis on high elevation locations. *In* "Air Pollutants Effects on Forest Ecosystems" pp. 5–38. The Acid Rain Foundation, St. Paul, Minnesota.

Jenny, H. (1980). "The Soil Resource: Origin and Behavior" Springer Verlag, New York.

Johnson, A. H. and Siccama, T. G. (1983). Acid deposition and forest decline. *Environ. Sci. Tech.* **17**, 294–306.

Johnson, D. W. 1984. Sulfur cycling in forests. *Biogeochem.* **1**, 29–44.

Johnson, D. W. and Todd, D. E. (1983). Relationships among iron, aluminum, carbon, and sulfate in a variety of forest soils. *Soil Sci. Soc. Am. J.* **47**, 792–800.

Johnson, D. W., Henderson, G. S., Huff, D. D., Lindberg, S. E., Richter, D. D., Shriner, D. D., Todd, D. E. and Turner, J. (1982). Cycling of organic and inorganic sulphur in a chestnut oak forest. *Oecologia* **54**, 141–148.

Johnson, D. W., Richter, D. D., Lovett, G. M. and Lindberg, S. E. (1985). The effects of atmospheric deposition on potassium, calcium, and magnesium cycling in two deciduous forests. *Can. J. For. Res.* **15**, 773–782.

Johnson, N. M., Driscoll, C. T., Eaton, J. S., Likens, G. E. and McDowell, W. H. (1981). 'Acid rain,' dissolved aluminum and chemical weathering at the Hubbard Brook Experimental Forest, New Hampshire. *Geoch. Cosmoch. Acta* **45**, 1421–1437.

Kimmins, J. P., Binkley, D., Chatarpaul, L. and Decatanzaro, J. (1985). "Biogeochemistry of Temperate Forest Ecosystems: Literature on Inventories and Dynamics of Biomass and Nutrients" Canadian Forestry Service Infor. Rep. PI-X-47 E/F, Chalk River, Ontario.

Leninger, A. L. (1982). "The Principles of Biochemistry" Worth Publishers, New York.

Likens, G. E., Bormann, F. H., Pierce, R. S. and Eaton, J. S. (1977). "Biogeochemistry of a Forested Ecosystem" Springer-Verlag, New York.

Lindberg, S. E. Lovett, G. M., Richter, D. D. and Johnson, D. W. (1986). Atmospheric deposition and canopy interactions of major ions in a forest. *Science* **231**, 141–145.

Lodhi, M. A. K. (1982). Effects of H ion on ecological systems: effects on herbaceous biomass, mineralization, nitrifiers and nitrification in a forest community. *Amer. J. Bot.* **69**, 474–478.

Lutz, H. J. and Chandler, R. F. Jr. (1946). "Forest Soils" Chapman and Hall, London.

Mattson, S. and Koutler-Andersson, E. (1941). The acid-base condition in vegetation, litter and humus: I. Acids, acidoids and bases in relation to decomposition. *Lantbr-Hoqsk. Ann* **9**, 1–26.

Mehlich, A. (1941). Base unsaturation and pH in relation to soil type. *Soil Sci. Soc. Am. Proc.* **6**, 150–156.

Mehlich, A. (1942). The significance of percentage base saturation and pH in relation to soil differences. *Soil Sci. Soc. Am. Proc.* **7**, 167–174.

Mengel, K. and Kirkby, E. A. (1979). "Principles of Plant Nutrition" International Potash Institute, Berne, Switzerland.

Melin, E. (1930). Biological decomposition of some types of litter from North American Forests. *Ecology* **11**, 72–81.

Miles, J. (1985). The pedogenic effects of different species and vegetation types and the implications of succession. *J. Soil Sci.* **36**, 571–584.

Miles, J. and Young, W. F. (1980). The effects on heathland and moorland soils in Scotland and northern England following colonization by birch (*Betula* spp.). *Bull. Ecol.* **11**, 233–242.

Montes, R. A. and Christensen, N. L. (1979). Nitrification and succession in the Piedmont of North Carolina. *For. Sci.* **25**, 287–297.

Morrison, I. K. (1983). Composition of percolate from reconstructed profiles of two jack pine forest soils as influenced by acid input. *In* "Effects of Accumulation of Air Pollutants in Forest Ecosystems" (Eds B. Ulrich and J. Pankrath) pp. 195–206. D. Reidel, Boston.

Muir, J. W., Morrison, R. I., Brown, G. J. and Logan, J. (1964). The mobilization of iron by aqueous extracts of plants. *J. Soil Sci.* **15**, 220–237.

Nilsson, S. I., Miller H. G. and Miller, J. D. (1982). Forest growth as a possible cause of soil and water acidification: an examination of the concepts. *Oikos* **39**, 40–49.

Nykvist, N. (1959). Leaching and decomposition of litter: I. Experiments on leaf litter of *Fraxinus excelcior*, II. Experiments on needle litter of *Pinus silvestris*. *Oikos* **2**, 190–224.

Nykvist, N. (1963). Leaching and decomposition of water-soluble organic substances from different types of leaf and needle litter. *Studia Forestalia Suecica* **3**, 1–29.

Nykvist, N. and Rosen, K. (1985). Effect of clear-felling and slash removal on the acidity of northern coniferous soils. *For. Ecol. Managem.* **11**, 157–169.

Pearson, R. W. and Adams, F. (1967). "Soil Acidity and Liming" Agronomy Society of America, Madison, Wisconsin.

Pelisek, J. (1983). Acidification of forest soil by acid rains in the region of the Zdarske Hills in the Bohemian-Moravian Uplands. *Lesnictvi* **29**, 673–682.

Perala, D. A. and Alban, D. H. (1982). Biomass, nutrient distribution and litterfall in *Populus, Pinus*, and *Picea* stands on two different soils in Minnesota. *Plant Soil* **64**, 117–192.

Pierre, W. H. and Banwart, W. L. (1973). The excess-base and excess-base/nitrogen ratio of various crop species and of plant parts. *Agron. J.* **64**, 91–96.

Pierre, W. H., Meisinger, J. and Birchett, J. R. (1971) Cation–anion balance in crops as a factor in determining the effect of nitrogen fertilizers on soil acidity. *Agron. J.* **62**, 106–112.

Plice, M. J. (1934). "Acidity, Antacid Buffering, and Nutrient Content of Forest Litter in Relation to Humus and Soil" Cornell Univ. agr. Exp. Sta. Memoir 165, Ithaca, New York.

Postel, S. (1984). "Air Pollution, Acid Rain, and the Future of Forests" Worldwatch Paper #58, Washington, D.C.

Read, D. J., Francis, R. and Finlay, R. D. (1985). Mycorrhizal mycelia and nutrient cycling in plant communities. *In* "Ecological Interactions in Soil: Plants, Microbes and Animals" (Ed. A. H. Fitter) pp. 193–217. Blackwell Scientific, Oxford.

Read, R. A. and Walker, L. C. (1950). Influence of eastern redcedar on soil in Connecticut pine plantations. *J. For.* **48**, 337–339.

Rehfeuss, K. E. (1981). Uber die Wirkungen der sauren Niederschlage in Waldokosystemen. *Forstwissen. Centr.* **6**, 363–381.

Reuss, J. O. (1977). Chemical and biological relationships relevant to the effect of acid rainfall on the soil-plant system. *Water, Air and Soil Poll.* **7**, 461–478.

Reuss, J. O. and Johnson, D. W. (1985). Effect of soil processes on the acidification of water by acid deposition. *J. Environ. Qual.* **14**, 26–31.

Reuss, J. O. and Johnson, D. W. (1986). "Acid Deposition and Acidification of Soils and Waters". Springer-Verlag, New York.

Richter, D. D. (1986). Sources of acidity in some forested Udults. *Soil Sci. Soc. Am. J. in press.*

Richter, D. D., Ralston, C. W. and Harms, W. R. (1982). Prescribed fire: effects on water quality and forest nutrient cycling. *Science* **215**, 661–663.

Richter, D. D., Johnson, D. W. and Todd, D. E. (1983). Atmospheric sulfur deposition, neutralization, and ion leaching in two deciduous forest ecosystems. *J. Environ. Qual.* **12**, 263–270.

Robertson, G. P. (1982). Nitrification in forested ecosystems. *Phil. Trans. R. Soc. Lond.* **B 296**, 445–457.

Roy, A. B. and Trudinger, P. A. (1970). "The Biochemistry of Inorganic Compounds of Sulfur" Cambridge University Press, London.

Schnitzer, M. (1980). Effect of low pH on the chemical structure and reactions of humic substances. *In* "Effects of Acid Precipitation on Terrestrial Ecosystems" (Eds T. C. Hutchinson and M. Havas) pp. 203–222. Plenum Press, New York.

Seip, H. M. (1980). Acid snow–snowpack chemistry and snowmelt. *In* "Effects of Acid Precipitation on Terrestrial Ecosystems" (Eds T. C. Hutchinson and M. Havas) pp. 77–94. Plenum Press, New York.

Sollins, P. (1976). Preliminary annual element budgets for the 120-year pine stand at Jadraas. Memo 761221, Swedish Coniferous Forest Project, Uppsala.

Sollins, P., Cromack, K. and Li, C. Y. (1981). Role of low-molecular-weight organic acids in the inorganic nutrition of fungi and higher plants. *In* "The Fungal Community" (Eds D. T. Wicklow and G. C. Carroll) pp. 607–619. Marcel Dekker, New York.

Sollins, P., Grier, C. C., McCorison, F. M., Cromack, K., Fogel, R. and Fredriksen, R. L. (1980). The internal element cycles of an old-growth Douglas-fir ecosystem in western Oregon. *Ecol. Mon.* **50**, 261–285.

Stone, E. L. (1975). Windthrow influences on spatial heterogeneity in a forest soil. *Mitteilungen, Eidgenössiche Anstalt für das Forstliche Versuchswesen* **51**, 77–87.

Stone, E. L. (1975). Effects of species on nutrient cycles and soil change. *Phil. Trans. Royal Soc., London* **271**, 149–162.

Stuanes, A. (1980). Effects of acid precipitation on soil and forest. 5. Release and loss of nutrients from a Norwegian forest soil due to artificial rain of varying acidity. *In* "Ecological Impact of Acid Precipitation" (Eds D. Drablos and A. Tollan) pp. 198–199. SNSF Project, Oslo-As, Norway.

Swift, M. J., Heal, O. W. and Anderson, J. M. (1979). "Decomposition in Terrestrial Ecosystems" Univ. California Press, Berkeley.

Tamm, C. O. (1976). Acid precipitation: biological effects on soil and on vegetation. *Ambio* **5**, 235–238.

Thomas, G. W. and Harward, W. L. (1984). The chemistry of soil acidity. *In* "Soil Acidity and Liming" (Ed. F. Adams) pp. 3–56, Agronomy Society of America, Madison, Wisconsin.

Thomas, W. A. (1968). Decomposition of loblolly pine needles with and without addition of dogwood leaves. *Ecology* **49**, 568–571.

Ugolini, F. C. (1968). Soil development and alder invasion in a recently deglaciated area of Glacier Bay, Alaska. *In* "Biology of Alder" (Eds J. M. Trappe, J. F. Franklin, R. F. Tarrant and G. M. Hansen) pp. 115–140. USDA Forest Service, Portland, Oregon.

Ulrich, A. (1941). Metabolism of non-volatile organic acids in excised barley roots as related to cation-anion balance during salt accumulation. *Am. J. Bot.* **28**, 523–537.

Ulrich, B. (1980). Production and consumption of hydrogen ions in the ecosphere. *In* "Effects of Acid Precipitation on Terrestrial Ecosystems" (Eds T. C. Hutchinson and M. Havas) pp. 255–282. Plenum Press, New York.

Ulrich, B. (1983a). A concept of forest ecosystem stability and of acid deposition as driving force for destabilization. *In* "Effects of Accumulation of Air Pollutants in Forest Ecosystems" (Eds B. Ulrich and J. Pankrath) pp. 1–32. D. Reidel, Boston.

Ulrich, B. (1983b). Interaction of forest canopies with atmospheric constituents. *In* "Effects of Accumulation of Air Pollutants in Forest Ecosystems" (Eds B. Ulrich and J. Pankrath) pp. 33–45. D. Reidel, Boston.

Ulrich, B. and Matzner, E. (1986). Anthropogenic and natural acidification in terrestrial ecosystems. *Experientia* **42**, 344–350.

Ulrich, B. and Pankrath, J., Eds (1983). "Effects of Accumulation of Air Pollutants in Forest Ecosystems" D. Reidel, Boston.

van Breemen, N., Mulder, J. and Driscoll, C. T. (1983). Acidification and alkalinization of soils. *Plant Soil* **75**, 283–308.

Van Miegroet, H. and Cole, D. W. (1984). The impact of nitrification on soil acidification and cation leaching in a red alder ecosystem. *J. Environ. Qual.* **13**, 586–590.

Van Miegroet, H. and Cole, D. W. (1985). Acidification sources in red alder and Douglas-fir – importance of nitrification. *Soil Sci. Soc. Am. J.* **49** 1274–1279.

Vogt, K. A., Grier, C. C. and Vogt, D. J. (1986). Production, turnover, and nutrient dynamics of above- and belowground detritus of world forests. *Adv. Ecol. Res.* **15**, 303–337.

Volk, N. J. and Tidmore, J. W. (1946). The effect of different sources of nitrogen on soil reaction, exchangeable ions and yield of crops. *Soil Sci.* **61**, 462–477.

Wiklander, L. (1973). The acidification of soil by acid precipitation. *Grundfortbaettring* **26**, 155–164.

Wiklander, L. (1975). The role of neutral salts in the ion exchange between acid precipitation and soil. *Geoderma* **14**, 93–105.

Wiklander, L. and Koutler-Andersson, E. (1963). Influence of exchangeable ions on release of mineral-bound ions. *Soil Sci.* **95**, 9–18.

Will, G. M. (1984). Monocultures and site productivity. *In* "Symposium on Site and Productivity of Fast Growing Plantations" (Eds D. C. Grey, A. Schonau, C. Schutz and A. Van Laar) pp. 473–484. South African Forest Research Institute, Pretoria.

Williams, B. L. and Cooper, J. M. (1978). Effects of afforestation with *Pinus contorta* on nutrient content, acidity and exchangeable cations in peat. *Forestry* **51**, 29–35.

Wittwer, R. F. and Immel, M. J. (1980). Chemical composition of five deciduous tree species in four-year-old, closely-spaced plantations. *Plant Soil* **54**, 461–467.

Wolt, J. D. and Lietzke, D. A. (1982). The influence of anthropogenic sulfur inputs upon soil properties in the Copper region of Tennessee. *Soil Sci. Soc. Am. J.* **46**, 651–656.

Zinke, P. (1962). The pattern of influence of individual forest trees on soil properties. *Ecology* **43**, 130–133.

Zottl, H. (1960). Dynamik der stickstoffmineralisation im organischen waldbodenmaterial. III. pH-wert und mineralstickstoff-nachlieferung. *Pl. Soil* **13**, 207–223.

Ant–Plant–Homopteran Interactions

R. BUCKLEY

I. Summary . . . 53
II. Introduction . . . 54
III. The Twofold Components – Direct or Pairwise Interactions . . . 56
A. Ant–Plant . . . 56
B. Homopteran–Plant . . . 57
C. Ant–Homopteran . . . 60
IV. The Indirect Components . . . 66
A. Plants Affecting Ants via Homoptera . . . 66
B. Plants Affecting Homoptera via Ants . . . 66
C. Ants Affecting Plants via Homoptera . . . 67
D. Ants Affecting Homoptera via Plants . . . 68
E. Homoptera Affecting Ants via Plants . . . 68
F. Homoptera Affecting Plants via Ants . . . 68
V. Discussion and Conclusions . . . 70
A. Effect on Host Plant Populations, and Long-Term Persistence of APHI . . . 70
B. Abiotic Factors, and Embedding in Higher-Order Interactions . . . 72
C. Directions for Research . . . 73
Acknowledgements . . . 73
References . . . 74

I. SUMMARY

Ants, plants and Homoptera interact in a wide variety of ways. Each participant in these interactions can affect the other two either directly, or indirectly by affecting the third participant. The outcome of a particular interaction may also be affected by other species interacting with one or more of the participants, and by abiotic factors such as fire, soil and climate. The quantitative significance, and the degree of specificity, obligacy, and specialisation for each participant vary widely between interactions.

Whilst there are many interactions between ants and plants which do not

ADVANCES IN ECOLOGICAL RESEARCH Vol. 16
ISBN 0-12-013916-2

involve Homoptera at all, the direct twofold ant–plant components of ant–plant–homopteran interactions are generally negligible unless the plants also possess extrafloral nectaries or ant-attractant food bodies. In direct interactions between Homoptera and plants, the homopterans gain food, habitat and sometimes allelochemicals from the plants, while the plants suffer tissue damage, loss of metabolites, and most importantly, increased incidence of infection by homopteran-transmitted microbial pathogens: plants therefore employ a range of physical and chemical defenses against Homoptera. The direct interactions between ants and Homoptera fall into two main classes: either the ants prey on the homopterans, or they tend and/or defend them against predators and parasites (or in a few cases provide specialised habitats). Some ants may prey on some species of homopteran and tend others. In either case the homopterans provide food for the ants, either as prey or by exuding honeydew which the ants collect.

The indirect components of the overall threefold interaction may be as or more significant to the participants than the direct twofold components. In particular, modification of homopteran populations or feeding rates by ant attendance or predation can produce a significant negative or positive effect, respectively, on the host plant. Plants which suffer significantly from external herbivores, and lack an effective defence, may gain indirectly from resident Homoptera if these attract ants which defend the homopterans against predators and the plant against external herbivores. Plants with extrafloral nectaries, however, normally defended by ants against herbivores, may suffer indirectly from resident homopterans which decrease the effectiveness of the ant defence as well as taking sap directly.

The effects of ant–plant–homopteran interactions on host plant populations are discussed in particular detail, and directions for further research outlined.

II. INTRODUCTION

Analysis of an interspecific interaction or class of interactions should ideally consider: (1) its effects on the population dynamics, behaviour and energy/nutrient budgets of each participant; (2) the costs and benefits of the interaction to each participant, and whether these are similar for each participant or dissimilar; (3) how taxon-specific, and how obligate or facultative, it is for each participant; and (4) what morphological, behavioural and physiological specialisations are involved.

These aspects have been examined with various degrees of precision for a range of twofold interactions and a more limited range of higher-order interactions, such as competing predators with common prey species, mutualisms, host – parasite – hyperparasite systems, and plant – herbivore – protector systems (see, e.g., Atsatt and O'Dowd, 1976; Price *et al.*, 1980

May, 1982; Boucher, 1982; Abrams, 1983). Such threefold and higher-order interactive systems form an important field for investigation, firstly since they represent closer approximations to real ecosystems than do twofold interactions, and secondly since twofold interactions may be better understood if seen in the context of higher-order interactions of which they are part.

Ecological analysis of ant–plant–homopteran interactions in general, and ant–plant homopteran interactions (APHI) in particular, has only been attempted during the last few years. Most descriptions of ant–plant–arthropod interactions have focused on only one of the species involved, the interaction being treated as one aspect of the life history of that species. Life history information of this type was reviewed recently by Wood (1983) for tropical membracids and by Nault and Madden (1985) for *Dalbulus* leafhoppers. Wood's review covered the taxonomic and biogeographic patterns in ant–membracid interactions in considerable detail, including evolutionary implications; Nault and Madden concentrated on the "ecological strategies" of the leafhoppers. Here, however, it is interactions rather than life histories or strategies which are the central theme; and the aim of this review is simply to summarise available information on the types, mechanisms and effects of APHI. To maintain a manageable scope, it is generally limited to an ecological rather than an evolutionary time scale. The range of reproductive systems in the Homoptera, and their relatively short life cycles, render APHI particularly suitable for examining the evolution of interactions, but relevant data are currently available only in piecemeal form. By integrating such information, it is hoped that this review will provide a basis for future considerations of evolutionary implications, as well as a framework for further analysis of individual systems.

APHI form a taxonomically rather than functionally defined class, and there are many similarities with ant–plant–lepidopteran interactions such as those described by Pierce and Mead (1981), Atsatt (1981a, b), Mueller (1983), Cottrell (1984), Horvitz and Schemske (1984), Pierce (1984, 1985), Pierce and Elgar (1985) and Smiley (1985). As with other interactive systems, APHI do not operate in isolation: other species groups such as predators, parasites and pathogens of both plants and homopterans are often also involved, together with species providing alternative food sources for the ants. External abiotic controls on each of the participants, such as soil and climatic factors, may also affect such interactions, as may internal life-history factors which regulate the population dynamics of each species involved.

The main type of APHI are summarised in Fig. 1: ants, plants and homopteran bugs each affect each other directly, or indirectly through the third participant. The different types of interaction are not mutually exclusive, and the overall effect of the interaction on each participant depends on the relative magnitude of each individual effect.

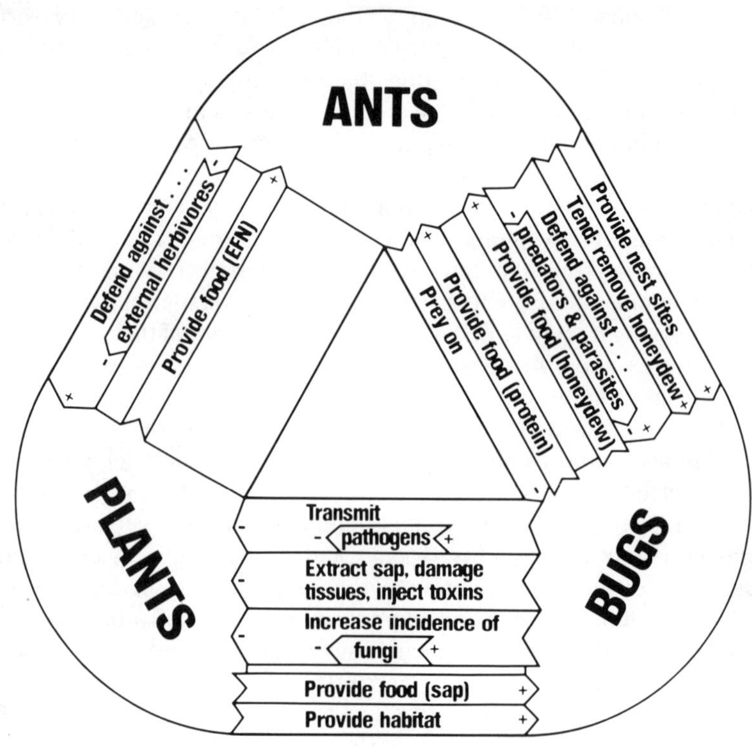

Fig. 1.

III. THE TWOFOLD COMPONENTS – DIRECT OR PAIRWISE INTERACTIONS

A. Ant–Plant

Whilst there are many interactions between ants and plants which do not involve Homoptera at all, the direct ant–plant components of APHI are generally negligible unless the plant possesses extrafloral nectaries (EFN) or ant-attractant food bodies (Buckley, 1982; Maschwitz, *et al.*, 1984). Ants involved in interactions with plants possessing domatia may also eat Homoptera or their honeydew, but the Homoptera involved are generally on different plants (Janzen, 1969, 1974; Huxley 1978, 1982; Buckley, 1982). Ant–plant interactions involving EFN but not Homoptera have been reviewed by Bentley (1977), Buckley (1982) and Bentley and Elias (1983) and will not be considered here. APHI involving plant species with EFN or food bodies generally include major indirect components and are considered later in this review.

B. Homopteran–Plant

Except for a few systems with economic significance (see, e.g., Sylvester, 1984), the direct plant–homopteran components of APHI have not generally been studied in any detail. In contrast to the direct ant–plant components, however, it seems likely that the direct plant–homopteran components of APHI are similar to twofold plant–homopteran interactions not involving ants. The principal features of such plant–homopteran interactions are therefore summarised below.

The effects of Homoptera on plants fall into two main categories: direct physiochemical effects on plant tissues, and transmission of microbial pathogens. Pathogen transmission, in particular, is of major economic importance in agriculture, horticulture and forestry, and has consequently been studied in considerable detail. Homoptera are the main vectors of plant viruses, spiroplasmas, mycoplasmas and mycoplasma-like organisms (MLO's), and rickettsias and rickettsia-like organisms (RLO's) (Maramorosch, 1980). Aphids, for example, transmit more plant viruses than any other arthropod group (Sylvester, 1984): *Myzus persicae* alone is responsible for the transmission of over 100 viruses.

Most of these pathogens do not affect their insect vectors, but there are exceptions: corn-stunting mollicutes affect *Dalbulus* and *Baldulus* leafhoppers, for example (Madden and Nault, 1983), and the MLO agent of *Prunus* western-X disease reduces the longevity and fecundity of *Colladonus montanus* (Sylvester, 1984). Many of these pathogens multiply within their insect vectors, and some are transmitted transovarially; many are also dependent on their insect vectors for survival between plant generations. Sulochana (1984) suggested that they are therefore more accurately described as insect-plant pathogens, rather than as plant pathogens transmitted by insects, even if their symptomatic expression is generally greater in their plant hosts. Species life histories, interactions, and agricultural aspects have been reviewed on a number of occasions (e.g., Harris and Maramorosch, 1977, 1979; Maramorosch and Harris, 1979; Maramorosch, 1980; D'Arcy and Nault, 1982; Plumb and Thresh, 1983; David and Alexander, 1984; Huffaker *et al.*, 1984; Mukhopadyay, 1984; Sen-Sarma, 1984; Sulochana, 1984; Sylvester, 1984).

Whilst pathogen transmission is certainly more serious from an agricultural viewpoint, direct physiochemical damage by Homoptera can also have a significant effect on plant growth and reproduction. These effects include sap extraction (e.g., Bushing and Burton, 1974; Maramorosch, 1980; Ortega *et al.*, 1980); increased water loss (e.g., Southwood, 1985), disruption of translocation patterns (e.g., Way, 1973; Nault and Phelan, 1984), and injection of toxins (e.g., Eversun and Gallun, 1980; Teetes, 1980). Harris and Maramorosch (1979) showed direct damage to alfalfa by *Therioaphis maculata* in northern U.S.A., for example; Ortega *et al.* (1980) found that heavy infestations of *Rhopalosiphum maidis* weaken maize plants and reduce pollen shedding; and Choudhury (1984) noted that aphid infestations reduce seed yield in *Pisum sativum*. Many Homoptera also cut slits in the bark of their host plants for oviposition, and Wood (1983) reported that in 5 membracid genera, females may kill leaf mid-rib tissue near the petiole when egg-laying.

Honeydew exuded by Homoptera may encourage colonisation of the plant host by fungi (Haines and Haines, 1978a, b), to its probable detriment. It has been suggested, however, that honeydew falling to the ground under the host plant may stimulate soil nitrogen fixation or mineralisation and hence benefit the plant, though this suggestion remains controversial (Owen and Wiegert, 1976, 1984; Owen, 1978; Petelle, 1980, 1984; Choudhury, 1984, 1985; Lam and Dudgeon, 1985). It is also conceivable that falling honeydew might transmit plant allelochemicals to the ground, where they could act to keep the plant's immediate surrounds free of competitors, as suggested by Silander *et al.* (1983) for beetle frass falling from *Eucalyptus*: though it is hard to see that this could be of advantage to the plant, when rainfall would be both a cheaper and safer agent of distribution.

Overall, therefore, the effects of Homoptera on plants are entirely or almost entirely detrimental, and plants hence have a range of defences against homopteran attack. Those defences may be physical, such as hairs, viscid secretions, or thick parenchyma around the vascular bundles (e.g., Gibson, 1972, 1974; Gallun *et al.*, 1975; Niles, 1980); or chemical (e.g., Eschrich, 1970; Burnett *et al.*, 1978; Maramorosch, 1980; Van Emden, 1978; Dreyer *et al.*, 1981, 1985; McClure and Hare, 1984; Dreyer and Campbell, 1984; McClure, 1985a). Physical and chemical defences are often combined, as for example in the resistance of *Pinus* spp. to *Matsucoccus matsumurae* in Japan (McClure, 1985b); and crop breeding programmes selecting for increased plant resistance (Maramorosch, 1980; Nielsen and Lehman, 1980; Kugler and Ratcliffe, 1983; Ratcliffe and Murray, 1983; Webster and Inayatullah, 1984; Webster and Stark, 1984; Messina *et al.*, 1985) have generally not discriminated between different resistance mechanisms.

Plant chemistry influences both the selection of individual host plants by Homoptera, and subsequent homopteran survival, growth and fecundity. Factors involved include plant nitrogen status (e.g., McClure, 1980; Prestidge and McNeill, 1983; Rabb *et al.*, 1984; White, 1984; Salama *et al.*, 1985) and water content, as well as a range of secondary compounds (e.g.,

Jordens-Rottger, 1979; Van Emden, 1978), and have been reviewed recently by Scriber (1984). Abiotic habitat factors, by affecting plant water and nitrogen status, may hence affect their susceptibility to homopteran attack: McClure (1985a), for example, showed that populations of *Nuculaspis tsugae* on trees at the edge of the host plant range were denser and more fecund than populations on trees in the centre of the range.

The wide range of plant secondary compounds which provide a defence against insects which feed by chewing or boring (e.g., Feeny, 1976; Rhoades and Cates, 1976; Cates, 1980; Haukioja, 1980) are generally ineffective against Homoptera: the homopteran feeding technique of extracting phloem largely by intercellular stylet penetration (Pollard, 1973; Al-Mousawi *et al.*, 1983; Kiss and Chau, 1984; Tjallingii, 1985a, b; Spiller *et al.*, 1985) enables them to avoid plant secondary compounds which are compartmentalised inside plant cells, notably most phenolics (Matile, 1984; Dreyer *et al.*, 1985). It was formerly believed that quinones, alkaloids, glycosides and terpenes did not occur in plant phloem and hence could not be involved in plant resistance to Homoptera (see Maramorosch, 1980 for review). Recently Wink and Witte (1984), however, demonstrated phloem transport of alkaloids in several *Lupinus* spp., and Dreyer *et al.*, (1985) found the indolizidine alkaloid swainsonine in the honeydew of aphids feeding on *Astragalus lentiginosus*, indicating that it must be transported in the plant phloem. Dreyer *et al.* found that another indolizidine alkaloid, castanospermine, is intensely active against *Acyrthosiphon pisum*, as are several quinolizidine alkaloids. As the concentrations of quinolizidine alkaloids reported by Wink and Witte (1984) in *Lupinus* phloem, 1–5 mg ml^{-1}, are 10 times the ED_{50} for *Acyrthosiphon pisum* as determined by Dreyer *et al.* (1985), these alkaloids could potentially provide an effective plant defence. Pyrrolizidine alkaloids may also be involved in some cases (Molyneux and Johnson, 1984; Johnson *et al.*, 1985). though not all (Dreyer *et al.*, 1985). Homopteran development and reproduction can also be affected by plant growth hormones (Bur, 1985).

A particularly interesting plant defence against Homoptera was discovered recently by Gibson and Pickett (1983): the wild *Solanum berthaultii* repels the aphid *Myzus persicae* by releasing its alarm pheromone (E)-β-farnesene (Montgomery and Nault, 1977a) from glandular hairs on the foliage.

Homoptera possess a range of strategies to overcome host-plant defences (see, e.g., Wallace and Mansell, 1976; Brattsten, 1979; Rosenthal and Janzen, 1979; Maxwell and Jennings, 1980). In particular, a number of Homoptera can sequester secondary compounds from the host plant, and use them in their own defence. Examples of this were reviewed by Duffey (1980) and Van Emden (1980), and Wink *et al.* (1982) have subsequently shown that *Aphis cytisorum* can sequester and metabolise quinolizidine alkaloids.

There are major differences between species in all aspects of plant–homopteran interactions, including specificity, host–plant selection, homopteran success, and plant resistance (e.g., Maramorosch, 1980; Wood, 1980; Collins and Scott, 1982; Service and Lenski, 1982; Dolva and Scott, 1982; Wood and Guttman, 1982, 1983; Berger *et al.*, 1983). Individual plants of *Pinus ponderosa*, for example, differ in their susceptibility to the black pine-leaf scale *Nuculaspis californica*; and different populations of *Nuculaspis*, in turn, appear to be physiologically adapted to particular individual pines, and poorly adapted to others (Edmunds and Alstad, 1978). Similarly Service (1984a) found that when the aphid *Uroleucon rudbeckiae* is grown on 4 different clones of *Rudbeckia laciniata*, aphid fitness is affected by both genotypic and phenotypic differences in the host plants, and by the interaction of aphid genotype and plant genotype; and Kidd (1985) showed that differences between individual host plants had a greater effect on the survival and growth of *Cynara pinea* than differences in aphid density. Kiss and Chau (1984) found that the membracid *Vanduzea arquata* is more host-specific and selective than *Enchenopa binotata*, and Weber (1985) showed considerable variation in host plant adaptation among 1000 individual clones of the green peach aphid *Myzus persicae*. Such variation is of major practical importance in agriculture and horticulture, and has hence sparked considerable research effort, as shown, for example, by two recent bibliographies (Webster and Inayatullah, 1985; Khan and Saxena, 1985).

Characteristics of the host plant may affect the susceptibility of Homoptera to parasitism (Gilbert and Gutierrez, 1973; Van Emden, 1978; Salto *et al.*, 1983). Van Emden (1978) for example, suggested that while plant isothiocyanates may provide a defence against aphid attack, aphid parasites may use the concentration of volatile isothiocyanates as an indicator of likely aphid host plants. Hence aphids on less resistant plants suffer heavier parasitism, so that overall, aphid survival may be greater on the more resistant plants than the less resistant ones. A lepidopteran analogue has also been described: Mueller (1983) showed that the plant on which *Heliothis* larvae feed is an important factor in determining both the likelihood of attack by the parasitoid *Microplitis croceipes* and the probability of successful parasitoid establishment.

C. Ant–Homopteran

The direct interactions between ants and Homoptera are diverse. A wide range of ant species prey on Homoptera, particularly on coccids and aphids (Fol'kina, 1978; Morrill, 1978; Thomas *et al.*, 1980; Skinner and Whittaker 1981). Some ants prey on one homopteran and tend another: *Formica rufa* for example, preys on *Drepanosiphum platanoidis* but tends *Periphyllu testudinaceus* (Skinner and Whittaker, 1981). An ant–plant–lepidopteran analogue has been described recently by Smiley (1985).

Ant predation can cause significant reductions in homopteran prey populations, even leading to local extinctions, as recorded for *Quadraspidiotus perniciosus* attacked by *Crematogaster subdentata* on pear trees (Fol'kina, 1978). Though occasionally species-specific (Skinner and Whittaker, 1981), ant predation on Homoptera appears to be generally facultative and opportunistic, and there is no indication that any ant species is dependent on homopteran prey. Some Homoptera may respond to ant attack by evasive behaviour; others employ chemical deterrence of various kinds. Deterrents effective against ants may not necessarily provide protection against other predators: coccids in the genus *Dactylopius*, for example, protect themselves against *Monomorium destructor* with poisonous carminic acid; but larvae of the pyralid moth *Laetilia coccidivora* eat the coccids and retain the carminic acid in the gut, regurgitating it as ant-repellent gut droplets if attacked by *Monomorium* (Eisner *et al.*, 1980; Huheey, 1984).

Ant attendance has been recorded for coccids and pseudococcids (e.g., Nixon, 1951; Burns, 1973; Janzen, 1974; Miller and Kosztarab, 1979; Stout, 1979; Hill and Blackmore, 1980; Yensen *et al.*, 1980); aphids (e.g., Banks and Nixon, 1958; El-Ziady, 1960; Rothcke *et al.*, 1967; Bradley and Hinks, 1968; Wood-Baker, 1977; O'Neill and Robinson, 1977; Addicott, 1978, 1979; Dash, 1978; Sterling *et al.*, 1979; Ebbers and Barrows, 1980; McLain, 1980; Skinner and Whittaker, 1981; Tilles and Wood, 1982; Reilly and Sterling, 1983a, b; Sudd, 1983); membracids (e.g., Funkhouser, 1915; Way, 1963; 1973; Kitching and Filshie, 1974; Clark and Yensen, 1976; Cookson and New, 1980; Skinner and Whittaker, 1981; Messina, 1981; Wood, 1979, 1980, 1982, 1983; Wood and Guttman, 1982; Fritz, 1982, 1983; Kopp and Tsai, 1983; Buckley, 1983; Bristow, 1983, 1984); and psyllids (e.g., Skinner and Whittaker, 1981).

Ant attendance commonly involves defence against or interference with predators and parasitoids, as suggested by Flanders (1951), Wellenstein (1952), Way (1954), El Ziady and Kennedy (1956), Bartlett (1961), Banks (1962), Leston (1978), Wood (1977, 1979, 1982, 1983), McEvoy (1979), McLain (1980), Bristow (1984), and Nechols and Seibert (1985). Such defence may operate at various stages of the predator's or parasitoid's life cycle. Ants may prevent oviposition, eat eggs, or attack larvae or adults venturing near the Homoptera. *Lasius niger*, for example, attacks both eggs and larvae of the coccinellid *Adalia bipunctata*, and repels adult coccinellids from aggregations of *Aphis fabae* (Banks, 1962; El Ziady and Kennedy, 1956); and *Camponotus modoc* attacks eggs and small larvae of the coccinellid *Neomysia oblonguttata*, but not larger ones (Tilles and Wood, 1982). *Camponotus maculatus*, attending *Icerya seychellarum* on Aldabra, apparently reduces populations of the diaspid scale predator *Chilocorus nigritus* (Hill and Blackmore, 1980). Clark and Yensen (1976) suggested that two *Formica* spp. defend *Campylenchia* sp. on *Artemisia dracunculus* against the spider *Phidippus clarus*. Tilles and Wood (1982) found more predators on unattended *Cinara occidentalis* colonies than on those tended by *Camponotus modoc*, and Fritz (1982) found a similar pattern for *Vanduzea arquata* attended by *Formica subsericea*. Ants can also provide a defence against parasites or parasitoids: Nichols and Siebert (1985) reported that *Technomyrmex albipes* decreases the proportion of *Nipaecoccus vastator* parasitised by the encyrtid *Anagyrus indicus* on *Leucaena leucocephala*. An interesting parallel occurs in the case of the membracids *Tritropidea alticollum* and *Enchenopa* sp. on a *Vismia* sp. in Colombia, which are guarded against ants by the wasp *Parachartergus richardsi* (Schremmer, 1978).

Alternatively or additionally, ant attendance of Homoptera may involve reduction of fungal attack by removal of honeydew, as suggested by Majer (1982) and Collins and Scott (1982) for *Pulvinariella mesembryanthemi* on *Carpobrotus edulis*; Dolva and Scott (1982) for *Pseudococcus macrozamiae* on *Macrozamia reidlei*; and Firempong (1981) for *Planococcoides njalensis* on *Theobroma cacao*. According to Firempong, colonies of *P. njalensis* not attended by ants are often completely exterminated by fungi.

Besides defence against predators and parasitoids and reduction of fungal attack, benefits to Homoptera attended by ants can include transport to particularly palatable or unprotected portions of the plant, to overwintering areas, or to new areas for colonisation. Richerson and Jones (1982), for example, found that *Crematogaster lineolata* moves *Aphis lugentis* from the roots to the foliage of *Senecio douglasii* during the midday sun, and back again in the late afternoon; and Strickland (1958) and Way (1963) found that ants carry *Pseudococcus* spp. to new habitats. Attendant ants may also remove dead individuals from homopteran populations (e.g., Nixon, 1951). Where these individuals contain parasites or parasitoids, their removal (and probably consumption) could benefit the remaining population by reducing the incidence of parasites and parasitoids. This has yet to be tested experimentally.

The quantitative significance of ant attendance to homopteran populations varies widely. Attendance by *Anoplolepis longipes* and *Formica rufa* respectively leads to significant population increases by *Ceroplastes rubens* in the Seychelles (Haines and Haines, 1978a, b) and *Periphyllus testudinaceus* in the U.K. (Skinner and Whittaker, 1981). Attendant ants enhance survival of *Entylia bactriana* (Wood, 1977), *Publilia concava* (McEvoy, 1979; Messina, 1981), *Vanduzea arquata* (Fritz, 1983), and *Sextius virescens* (Buckley, 1983). Attendance by *Dolichoderus taschenbergi* raises the survival rate of tuliptree scales in Pennsylvania from 8 to 47% (Burns, 1973); and attendance by *Formica lugubris* increases population densities of the aphid *Symydobius oblongus* by over 3000% (Fowler and Macgarvin, 1985). Attendant *Iridomyrmex* and *Camponotus* species do not reduce mortality of *Pulvinarius mesembryanthemi* or *Pseudococcus macrozamiae* in Western Australia, however (Majer, 1982; Collins and Scott

1982; Dolva and Scott, 1982), despite significant numbers of the coccinellid *Cryptolaemus montrouzieri*, chrysopid larvae, and spiders and pseudoscorpions in association with the mealybug populations. Similarly, an attendant *Iridomyrmex* species, though reducing late-season mortality of the univoltine *Sextius virescens* near Canberra, has no effect on mid-season mortality (Buckley, 1983).

An ant defence may be a significant factor in the success of individual homopteran populations, as suggested by Way (1963) and Wood (1977). Survival of *Cinara occidentalis* populations on white fir in the western U.S.A. depends on attendance by *Camponotus modoc* (Tilles and Wood, 1982), which protects it against a range of insect predators, notably the coccinellid *Neomysia oblonguttata*; and survival of the scale *Toumeyella numismaticum* depends on protection by *Formica obscuripes* against the coccinellid *Hyperaspis congressis* (Bradley, 1973). Survival of the membracid *Vanduzea arquata* is significantly greater when tended by *Formica subsericea* than when ants are excluded (Fritz, 1982). McEvoy (1979) found that removal of attendant ants decreased survival of *Publilia concava* nymphs 20-fold, and Nechols and Siebert (1985) reported higher survival of *Nipaecoccus vastator* on *Leucaena leucocephala* branches tended by *Technomyrmex albipes* than on untended branches. Some aphid species also compete for ant attendance (Addicott, 1978). For most Homoptera ant attendance is facultative, however, as noted by Kitching and Filshie (1974), Cookson and New (1980) and Buckley (1983) for *Sextius virescens*. Exclusion of ants does not affect establishment or survival of *Pseudococcus macrozamiae* or *Pulvinarius mesembryanthemi* colonies, for example (Collins and Scott, 1982; Dolva and Scott, 1982).

Besides affording protection for eggs and nymphs, ant attendance may increase homopteran fecundity directly. Bristow (1983) found that female *Publilia concava* without attendant ants lay only one clutch of eggs and then tend that brood, whereas females with attendant ants leave their offspring to be defended by the ants and are therefore able to lay several egg clutches per season. Ant attendance has no effect on growth or fecundity of *Aphis vernoniae* on *Vernonia noveboracensis*, however, though it does reduce the development time of the membracid *Publilia reticulata* (Bristow, 1984).

The benefits of ant attendance for a particular homopteran may depend on the ant species concerned. Bristow (1984), studying *Aphis vernoniae* and *Publilia reticulata* attended by *Tapinoma sessile, Myrmica americana* and *M. lobicornis fracticornis* on *Vernonia noveboracensis*, found that *Aphis* colonies benefited more from attendance by *Tapinoma*, but *Publilia* colonies more from *Myrmica*.

Homopteran honeydew can provide a significant item of diet for many ants (Wellenstein, 1952; Way, 1963; Carroll and Janzen, 1973; Janzen, 1974; Haines and Haines, 1978a, b) and a few ants are largely dependent on it

(Weber, 1944; Carroll and Janzen, 1973). According to Kiss (1981), *Lasius niger* "feeds exclusively on honeydew" but dietary data were not presented, and it seems unlikely that homopteran honeydew alone would contain sufficient protein for an ant diet.

Less common ant–homopteran interactions include that between the red fire ant *Solenopsis invicta* and *Oliarius vicarius*, a planthopper which inhabits unused galleries in its nest mounds (Sheppard *et al.*, 1979); and that between a *Myrmelachista* species which nests in hollow stems of *Ocotea pedalifolia* in Costa Rica, and two *Dysmicoccus* species which can only reach the plant phloem from inside the stems and are dependent on the ants to provide access holes (Stout, 1979).

Overall, therefore, ant-homopteran interactions are extremely variable, ranging from specific to general and from obligate (e.g., Das, 1956; Way, 1954) to facultative (e.g., Dolva and Scott, 1982; Strickland, 1958; Way, 1963). Some have significant effects on homopteran populations and others do not; some ants rely heavily on homopteran prey or "herds" and others do not. The effects of ant–homopteran interactions on the host plant are therefore likely to be equally variable. Studies by Samways *et al.* (1982) on South African citrus orchards provide a quantitative example: of 123 ant species in the orchards, 25 collect homopteran honeydew; and of these 25, only 2 species precipitate major coccid outbreaks, with another 3 species responsible for smaller or more localised increases in coccid populations.

A range of homopteran adaptations associated with ant attendance have been recorded. Only one morphological adaptation has apparently been described – the anal whip of *Sextius virescens* (Kitching and Filshie, 1974) – but both aphids and membracids show coupled physiological and behavioural adaptations. Aphids, for example, defend themselves with cornicle secretions in the absence of ants, though not always successfully (Goff and Nault, 1974; Nault and Phelan, 1984) and with evasive behaviour triggered by alarm pheromones (Dahl, 1971; Kislow and Edwards, 1972; Nault *et al.*, 1974, 1976; Phelan *et al.*, 1976; Montgomery and Nault, 1977a, b, 1978; Calabrese and Sorensen, 1978; Burnett *et al.*, 1978; Roitberg and Myers, 1978; Nault and Montgomery, 1979; Nault and Phelan, 1984). (E)-β-farnesene is the most common alarm pheromone, occurring in at least 19 genera (Bowers *et al.*, 1972, 1977a; Edwards *et al.*, 1973; Wohlers, 1981; Nault and Phelan, 1984), but (–)-germacrene-A and two pinenes are also active in particular genera (Bowers *et al.*, 1977b; Pickett and Griffiths, 1980), and some species possess multiple pheromones (Nault and Bowers, 1974; Pickett and Griffiths, 1980). Alarm pheromones have apparently not been identified chemically for other families of Homoptera, but have been shown to exist in the membracids *Publilia concava* and *Entylia bactriana* (Nault *et al.*, 1974; Wood, 1977).

Evasive behaviour in non-myrmecophilous aphids varies between species

(Wood, 1976, 1979; Nault and Phelan, 1984) but typically involves dispersal from the pheromone source by walking along the plant or by falling or jumping off it. Myrmecophilous aphids show at least four behavioural adaptations to ant attendance, as follows. (1) They aggregate into dense clusters, minimising the "tending territory" of the ants and facilitating defence and honeydew collection. This aggregation is related to ant attendance rather than to the localisation of optimal feeding spots; is assisted by a pheromone; and increases aphid size and fecundity (Way and Cammell, 1970; Nault and Phelan, 1984). (2) They secrete honeydew in response to ant palpation. (3) They emit alarm pheromone under lesser provocation than non-myrmecophilous species. (4) They remain on the plant if their alarm pheromone is emitted, rather than jumping off.

Many ant-attended membracids also live in dense aggregations which attract more ant protection than dispersed populations (McEvoy, 1979; Wood and Guttman, 1982). The egg froth of *Enchenopa binotata* contains an oviposition attractant (a lipid) which leads to egg clumping and thereby to aggregations of nymphs, often with associated adults (Wood and Guttman, 1982). These attract ants which ensure protection for eggs and nymphs as well as adults. Possible membracid pheromones, analogous to aphid sesquiterpenes and monoterpenes, have not been investigated, but homopteran organs paralleling the Malicky's organs in lepidopteran larvae (Malicky, 1970; Pierce and Mead, 1981) should perhaps be sought.

Homopteran honeydews may also contain specific ant attractants. *Lasius niger*, for example, prefers the trisaccharides melezitose and glucosucrose, produced by aphids, to a wide range of mono- and disaccharides (Duckett, 1974), and Kiss (1981) suggested that the aphids produce melezitose specifically as an ant attractant. Whilst a low amino-acid content would generally be anticipated for homopteran honeydews, since the honeydew is the residue remaining once the Homoptera have extracted the protein component from carbohydrate-rich plant sap, it is possible that homopteran honeydews may contain amino-acids secreted specifically as ant-attractants. This possibility merits further investigation.

Ant adaptations appear to be primarily behavioural and largely unspecialised. The various behavioural patterns associated with defence of homopteran "stock" against predators and parasitoids are generally interpreted as unspecialised "ownership behaviour" (Way, 1963) associated with defence of valuable sugar resources. Ants attending myrmecophilous aphids, however, show a very specific adaptation: on release of alarm pheromones by the aphids, they palpate the aphids to prevent them dispersing, and they search for and attack aphid predators or parasitoids (Nault *et al.*, 1976; Montgomery and Nault, 1977; Nault and Phelan, 1984). This response, comparable to that shown by ants attending lycaenid larvae (Pierce and Mead, 1981; Atsatt, 1981b), indicates a high degree of physiological and behavioural specialisation, notably the ability to recognise specific terpenes or sets of terpenes. Another behavioural specialisation was reported by David and Wood (1980), who showed that each spring, *Camponotus modoc* in giant sequoia forest re-established its previous season's trails to branches bearing aphids: the aphid larvae would therefore presumably be protected more consistently, and from an earlier age, than if they were located from scratch each season.

It has also been noted that alate production is often delayed in ant-attended homopteran populations (El Ziady and Kennedy, 1956; Johnson, 1959, Tilles and Wood, 1982). Unattended populations of *Cinara occidentalis*, for example, produce twice as many alatiform nymphs as populations tended by *Camponotus modoc* (Tilles and Wood, 1982). This delay may be induced by a secretion from the ants' mandibular glands (Kleinjan and Mittler, 1975; Tilles and Wood, 1982). If so, this could indicate a specific physiological adaptation by the ants.

IV. THE INDIRECT COMPONENTS

A. Plants Affecting Ants via Homoptera

Plants involved in APHI can affect the ants indirectly by providing a food source for the Homoptera and hence for the ants. The significance of such effects depends on the host-specificity of the Homoptera and on the specificity and obligateness of the ant-homopteran interaction. They have not been quantified and are probably very variable.

Plants can also affect the ants indirectly by modifying the production or composition of homopteran honeydew. Sudd (1983), for example, found that *Cinara pini* and *C. pinea* are tended by *Formica lugubris* on *Pinus sylvestris*. The *Cinara* spp. occasionally also occupy *Pinus contorta*, but on the latter they produce no honeydew and are hence rarely attended by ants.

B. Plants Affecting Homoptera via Ants

The plants could affect the Homoptera indirectly if they attracted ants independently, as for example by producing extrafloral nectar, and if the Homoptera were to be affected by the presence of those ants independently of any interaction with ants which the Homoptera may themselves attract by their presence on the plant. In particular, the Homoptera may preferentially colonise ant-occupied plants, as in the case of ant-tended Lepidoptera (Atsatt, 1981a, b; Cottrell, 1984; Pierce and Elgar, 1985; Pierce *et al.*, unpubl.). Female *Publilia concava* oviposit on *Solidago* stems near *Formica* nests, probably because of limited dispersal by the previous years' adult females (McEvoy, 1979; Messina, 1981). Similarly, gravid females of the *Enchenopa binotata* complex from a given host species preferentially select that plant species for oviposition (Wood, 1980; Guttman *et al.*, 1981; Wood and Guttman, 1982). Individual *Sextius virescens* moved from ant-attended to unoccupied *Acacia decurrens* plants are found and tended by ants within a few hours, however (Buckley, 1983), and ant attendance therefore seems unlikely to determine host choice by gravid females in this species, particularly since their dispersal seems to be largely passive.

C. Ants Affecting Plants via Homoptera

The ants can affect the host plants indirectly by modifying the populations or feeding rates of the Homoptera (Carroll and Janzen, 1973). Predation by *Formica rufa* on aphids and lepidopteran larvae in limestone woodland reduces defoliation from 8 to 1% (Skinner and Whittaker, 1981); whereas in South African citrus orchards, attendance by *Anoplolepis custodiens* or *Pheidole megacephala* greatly increases the populations of commercially damaging coccids (Samways *et al.*, 1982). Ant attendance may increase the frequency of honeydew excretion by aphids – by up to 7 times in the case of *Aphis craccivora* attended by *Iridomyrmex glaber* (Takeda *et al.*, 1982). Attendance by *Iridomyrmex* has no significant effect on droplet diameter or production rate by the membracid *Sextius virescens*, however (Buckley, 1983). Total honeydew production even of tended Homoptera may still be small: Clark and Yensen (1976), for example, measured total honeydew production by *Campylenchia* sp. tended by two *Formica* spp. on *Artemisia dracunculus* as approximately 65 $\pm$ 20 ml day^{-1} ha^{-1}.

Ants sometimes transport their homopteran "herds" to the most vigorously growing parts of the plant (Wheeler, 1910; Hough, 1922; Way, 1963; Rosengren, 1971; Richerson and Jones, 1982; Carroll and Janzen, 1973), which may disadvantage the plant. *Formica* apparently produces an overall positive effect on European pine plantation production, however, despite tending large aphid populations (Wellenstein, 1952; Gosswald, 1954, 1958; Carroll and Janzen, 1973).

Ants transporting Homoptera may contribute to the transmission of plant pathogens, as in the case of cacao swollen-shoot virus, carried by mealybugs (Sylvester, 1984). They can also affect the host plant indirectly by modifying the deposition of sugary honeydew and thereby the incidence of fungi. Citrus and cinnamon trees attended by *Anoplolepis longipes* in the Seychelles, for example, are heavily infected by sooty mould (Haines and Haines, 1978), and Fokkema *et al.* (1983) found that addition of aphid honeydew to fungal inocula increased fungal infection of wheat by 2·5–10 times, by stimulating spore germination and germ-tube growth.

D. Ants Affecting Homoptera via Plants

In some cases ants can affect Homoptera indirectly by protecting their host plant from external herbivores. Messina (1981), for example, points out that by protecting the *Solidago* stems inhabited by *Publilia concava, Formica fusca* prevents their defoliation by *Trirhabda* species; and since defoliation would destroy the membracid feeding and oviposition sites, the ant defence of the host plant indirectly benefits the homopteran.

E. Homoptera Affecting Ants via Plants

Indirect effects of Homoptera on ants would not be expected unless the ants are dependent on the plant for food or domicile. In that case, if heavy homopteran infestations decreased the production rate or amino-acid or sugar content of extra-floral nectar or food bodies, or decreased the frequency or growth rate of domatia, then associated ants could be affected indirectly by the Homoptera. Such effects have apparently not been recorded to date, but have probably not been sought.

F. Homoptera Affecting Plants via Ants

Homoptera can affect their plant hosts indirectly by attracting ants which may then defend the plant against external herbivores such as Orthoptera, Lepidoptera or grazing vertebrates. The overall effect of the interaction on the plant will then depend on whether the direct damage caused by the resident Homoptera is greater or less than the reduction they bring about in damage by external herbivores. Nickerson *et al.* (1977) found that nymphs of *Spissistilus festinus* on soybean increased ant predation on the eggs of soybean pests. Stout (1979) suggested that *Ocotea* plants in lowland rainforests benefit more from defence by a *Myrmelachista* species than they lose to two *Dysmicoccus* species tended by the ants, but did not test this experimentally. Inouye and Taylor (1979) found that *Helianthella uniflora* is occupied by ants which tend aphids on the back of its involucral bracts, the same location as the EFN of the ant-defended *H. quinquenervis*, but did not quantify the effects of the interaction on the *H. uniflora* host plants.

Recently Messina (1981) demonstrated that the membracid *Publilia concava* on *Solidago altissima* attracts *Formica fusca* and *F. rubicunda*, which effectively defend the plants against the chrysomelid beetles *Trirhabda virgata* and *T. borealis*: the first experimental demonstration of such an effect. He found that *Formica* species attack adult *Trirhabda* on *Solidago* stems bearing *Publilia*, and that these stems escape defoliation, and grow faster and set more seed than neighbouring stems without ants. Where the host plants possess EFN and are effectively defended by ants in the absence of Homoptera, however, the Homoptera can affect the host plant indirectly by attracting patrolling ants away from the EFN and thereby reducing the effectiveness of the ant defence; as in the interaction between *Sextius virescens, Acacia decurrens* and an *Iridomyrmex* species near Canberra (Buckley, 1983). An analogous ant–plant–lepidopteran system has recently been described by Horvitz and Schemske (1984): larvae of the riodinid butterfly *Eurybia elvina* reduce mean seed production of *Calathea ovandensis*, a neotropical perennial with EFN, by a third if tended by ants but by two thirds if untended.

A further complication can arise when ants attack predators of a host plant's other insect herbivores as well as those herbivores themselves. Fritz (1983), studying a population of *Robinia pseudoacacia* subject to attack by the beetle *Odontata dorsalis*, found that *Robinia* branches supporting populations of the membracid *Vanduzea arquata* were defended by *Formica subsericea*, which attacked *O. dorsalis*. The *Odontata* populations, however, were largely regulated by the beetle predator *Navicula subcoleoptrata*, which was also attacked by the ants. The net effect was that *Odontata* populations were higher on *Robinia* branches with *Vanduzea* and *Formica* than those without, so that the membracids had a net negative indirect effect on their host plant.

A similar net effect can occur where homopteran-tending ants also defend the host plant's other insect herbivores directly. Two such systems have been described, both involving myrmecophilous lycaenid butterfly larvae which are also tended and defended by the ants tending Homoptera. Pierce and Elgar (1985) found that ovipositing females of the lycaenid *Jalmenus evagoras* laid significantly more eggs and egg masses on *Acacia irrorata* plants with clusters of the membracid *Sextius virescens* and many foraging ants than on plants with only a few ants. The ants also tend the *Jalmenus* larvae (Pierce and Elgar, 1985; Pierce *et al.*, unpubl.; cf. also Pierce and Mead, 1981). Hence the Homoptera, by attracting ants, increase both the initial egg load and the survival of the lycaenid larvae; and since these can have a major impact on *Acacia* foliage the overall interaction can have a significant detrimental effect on the plant.

The interactions described by Maschwitz *et al.* (1984) are slightly different. The host plants, *Macaranga* species, possess ant-attractant food bodies and are defended against herbivores by *Crematogaster* species, especially *C. borneensis*, which nests and tends *Coccus penangensis* in hollowed-out *Macaranga* internodes. Whether the coccids are necessary for ant attendance and whether they improve or inhibit the ant defence was not investigated, but at least they do not render it ineffective. The ant–plant interaction is parasitised by myrmecophilous lycaenid larvae, *Arhopala* species, which are defended by the ants rather than attacked, and are hence able to feed on the foliage, to the detriment of the plant and hence indirectly the coccids.

Three *Arhopala* species were examined, each specialising on a particular *Macaranga*. *Arhopala* numbers were small, so the effect on the host plant was minor.

V. DISCUSSION AND CONCLUSIONS

A. Effect on Host Plant Populations, and Long-Term Persistence of APHI

If ant-homopteran interactions can benefit or disadvantage their host plants, as shown by Messina (1981), Buckley (1983), and Fritz (1983), then how do they affect the population dynamics, and in the longer term, the evolutionary paths of these plants? Why are host plants disadvantaged by APHI not eliminated?

Buckley (1983) suggested a number of ecological mechanisms which might reduce or nullify the impact of the ant–homopteran interaction on the host plants in the *Sextius–Iridomyrmex–Acacia* and similar systems. The first aspect is that whilst the effects of the ants and bugs and their interaction on the host plant are statistically significant, they are quantitatively small. Hence they may simply be outweighed, in natural populations, by the effects of soil and microclimate on the host plants. Wood (1982), for example, suggested that *Enchenopa binotata* populations were simply not large enough to inflict appreciable damage on their host plants; and Maschwitz *et al.* (1984) reached a similar conclusion for an analogous ant–plant–lepidopteran system.

The second main possibility is that the interaction may not affect the host plants every year, but only occasionally. In the system studied by Buckley (1983), for example, the relative timing of fruit initiation and EFN activity in *Acacia*, hatching of *Sextius*, and transfer of attention from *Acacia* EFN to *Sextius* by the *Iridomyrmex*, may vary from one year to another, so that seed set is only affected when flowering, fertilisation and fruit initiation is late relative to *Sextius* hatching. Such a specific mechanism, of course, could apply only to a very limited set of APHI. A related but more general hypothesis is that homopteran populations may be regulated by climatic parameters, for example the strength and timing of late spring frosts, so that their effects on the host plant are externally controlled, and intermittent. To test either of these hypotheses would require experiments such as those described by Buckley (1983) to be repeated several years in succession, which has not been practicable. Visits to the site in 1984, 1985 and 1986, however, have shown that the experimental trees in field populations are still alive, still apparently healthy, still growing in size, and still occupied by membracids.

The third main aspect is that in any given season, the Homoptera may occupy only a small proportion of the host plant population. If the individual host plants concerned change randomly from one season to the next, this would then simply reduce the overall effect of the APHI on the host plant population. It appears, however, that the same individual host plants are occupied year after year not only in the *Acacia–Sextius–Iridomyrmex* system, but in a wide range of APHI. The same applies to ant–plant–lepidopteran systems (Pierce and Mead, 1981; Atsatt, 1981b; Pierce, 1984; Pierce and Elgar, 1985; Pierce *et al.*, unpubl.) The implications of this depend on the factor controlling homopteran selection of individual host plants. If the factor concerned is a host plant characteristic under genetic control, such as palatability (i.e., ineffectiveness of physical and chemical defences), then if plants with this characteristic are disadvantaged by the APHI, they should be selectively eliminated if the Homoptera and ants are present at least intermittently and if the characteristic does not confer any selective advantage in the absence of the APHI; and for systems where the plants benefit from the APHI, the reverse should apply. In ecological time this does not appear to happen – the *Acacia–Sextius–Iridomyrmex* interaction persists, for example, without eliminating the host plants concerned. It is possible that in evolutionary time the most palatable or susceptible host plant genotypes are selectively eliminated, but as this occurs, the Homoptera evolve the ability to survive on less palatable genotypes – just as in twofold plant–herbivore interactions.

Perhaps, however, host plants are selected for occupation by ant-tended Homoptera not because they are more palatable, but simply because they are close to ant nests. Assuming that proximity to ant nests is not under genetic control and hence not subject to selection, then whilst the ant–homopteran interactions could reduce the fitness of individual host plants, they would have no consistent effect on the genetic composition of the host plant population overall. Hence the APHI could persist without eliminating the host plants. The ants, however, might build their nests close to the more palatable plants, so as to have ready access to homopteran honeydew. In this case, proximity of individual plants to ant nests could be related to plant genotype, so the above argument would not necessarily apply.

These various possibilities have yet to be tested for APHI. Parallel research on ant–plant–lepidopteran interactions, however, particularly on systems involving *Glaucopsyche lygdamus* and *Jalmenus evagoras*, is further advanced (e.g., Pierce and Mead, 1981; Atsatt, 1981a, b; Kitching, 1981; Ito *et al.*, 1982; Cottrell, 1984; de Vries, 1984; Henning, 1984; Pierce, 1984, 1985; Pierce and Elgar, 1985; Kitching and Luke, 1985; Pierce *et al.*, unpubl.), and many of the approaches and conclusions from these systems are likely to be applicable to APHI.

B. Abiotic Factors, and Embedding in Higher-Order Interactions

The outcome of APHI may be affected by external abiotic factors such as fire (e.g., Dolva and Scott 1982), and probably by soil and climate, as suggested by Buckley (1983): the effects of ants, homopterans and external herbivores on a healthy plant may be very different from the effects on, say, a chlorotic or drought-stricken plant (see also Mattson, 1980; Schowalter, 1981; McClure, 1985b).

The overall outcome of an APHI for each participant may also depend not only on the other participants in the threefold interaction, but on additional species which interact with one or more of the participants: just as the outcome of a twofold interaction may depend on its context within a threefold interaction, the outcome of a threefold interaction may depend on its context within four or fivefold interactions. In the case of *Publilia concava, Formica fusca* and *Solidago altissima*, for example, Messina (1981) showed that plants with ants and membracids were at a relative advantage to those without when the *Solidago* stands were subject to severe attack by the chrysomelid beetles *Trirhabda virgata* and *T. borealis*, but suggested that they would be at a relative disadvantage when *Trirhabda* populations were low. Similarly, the conclusions reached by Buckley (1983) for the overall effect of the *Sextius–Iridomyrmex* interaction on *Acacia decurrens* depend on the level of external attack by grasshoppers and gall wasps.

Messina (1981) noted that studies of the twofold interactions among *Publilia, Formica, Solidago* and *Trirhabda* would be insufficient to predict the outcome of the overall fourfold interaction on the host plant, and the same is true of the fivefold system examined by Buckley (1983), and presumably also the system studied by Fritz (1983). Hence conclusions from studies of twofold interactions may be suspect unless these are seen in the context of higher-order interactions of which they may be part. If, as hypothesised by Schowalter (1981), plant–animal interactions are major controls of nutrient cycling and ecosystem succession, the practical consequences of this would be considerable. No study of interspecies interaction can hope to include all the participants, and any interaction studied must be treated as artificially delimited, and "embedded" in the complex higher-order interactions between the set of taxa, and their abiotic environment, which make up an ecosystem. The time is now here, however, when three, four and fivefold interactions can be examined quantitatively as well as twofold ones, and APHI provide a useful system for such studies.

C. Directions for Research

This review has mentioned a number of unsolved or unstudied aspects of APHI in the preceding sections, and there are evidently many more. In general, the broad patterns of the twofold interactions between ants, plants and Homoptera are well established, but the mechanisms involved, particularly the biochemical aspects, have only been examined in a limited number of cases. The indirect components of threefold APHI, and of higher-order interactions including APHI, have as yet been quantified only in very few cases, and many more systems need to be studied to consolidate our understanding of these threefold interactions. Such studies should, where at all possible, endeavour to place the threefold interaction under examination within the context of higher-order interactions of which it may be part. Parallels with ant–plant–lepidopteran systems and other ant–plant–arthopod systems need to be examined in more detail.

Specific questions raised earlier in this review may be summarised here for convenience: (1) Does removal of dead individuals from homopteran populations by ants decrease the incidence of parasitism in the remainder? (2) Do ant-attended Homoptera possess organs analogous to the Malicky's organs of lycaenid butterfly larvae? (3) Do homopteran honeydews contain specific amino acids or other compounds which attract ants? (4) Is the outcome of particular APHI, for each participant, approximately the same every year, or does it vary significantly from one year to the next?

The most pressing question at present still seems to be to determine the factors controlling which individual host plants are occupied by Homoptera and which are not. For example: does host plant genotype, or soil fertility, determine host plant palatability; palatability determine selection by Homoptera; and presence of Homoptera determine attendance by ants, with the ants, if need be, moving their nests near the homopteran colonies? Or does proximity of individual host plants to existing ant nests determine the likelihood that Homoptera will be attended by ants, and ant attendance control the survival of homopteran populations? Or is some other causal pathway involved? To solve this problem conclusively will require several years' concurrent study of ant, plant and homopteran populations and the many factors which may influence their dynamics, of which APHI are only one. I look forward to an answer.

ACKNOWLEDGEMENTS

Previous versions of this review were presented firstly at the Ecological Society of Australia Symposium on the Ecological Basis for Interactions

Between Organisms, held in Brisbane, Australia, on 26 November 1982; secondly at the Third European Ecological Symposium on Plant–Animal Interactions, held in Lund, Sweden, on 22–26 August 1983; and thirdly at a special seminar at The University of Adelaide, Australia, on 14 August 1984. I am particularly grateful to Dr N. E. Pierce, formerly of Harvard and currently of Oxford University, for numerous discussions over the past five years: I have drawn many parallels from her pioneering work on ant–plant–lepidopteran interactions.

REFERENCES

Abrams, P. A. (1983). Arguments in favour of higher order interactions. *Amer. Nat.* **121**, 887–891.

Addicott, J. F. (1978). Competition for mutualists: aphids and ants. *Can. J. Zool.* **56**, 2093–2096.

Addicott, J. F. (1979). A multispecies aphid–ant association: density, dependence and species specific effects. *Can. J. Zool.* **57**, 558–569.

Al-Mousawi, A. H., Richardson, P. E. and Burton R. L. (1983). Ultrastructural studies of greenbug (Homoptera: Aphididae) feeding damage to susceptible and resistant wheat cultivars. *Ann. Entomol. Soc. Amer.* **76**, 964–971.

Atsatt, P. R. (1981a). Ant-dependent food plant selection by the mistletoe butterfly *Ogyris amaryllis* (Lycaenidae). *Oecologia* **48**, 60–63.

Atsatt, P. R. (1981b). Lycaenid butterflies and ants: selection for enemy-free space. *Amer. Nat.* **118**, 638–654.

Atsatt, P. R. and O'Dowd, D. (1976). Plant defense guilds. *Science* **193**, 24–29.

Banks, C. J. (1962). Effects of the ant, *Lasius niger* on insects preying on small populations of *Aphis fabae* Scop. on bean plants. *Ann. appl. Biol.* **50**, 669–679.

Banks, C. J. and Nixon, H. L. (1958). Effects of the ant *Lasius niger* (L.) on the feeding and excretion of the bean aphid, *Aphis fabae* Scop. *J. Exp. Biol.* **35**, 703–711.

Bartlett, B. R. (1961). The influence of ants upon parasites, predators and scale insects. *Ann. Entomol. Soc. Amer.* **50**, 543–551.

Bentley, B. L. (1977). Extrafloral nectaries and protection by pugnacious bodyguards. *Ann. Rev. Ecol. Syst.* **8**, 407–427.

Bentley, B. L. and Elias, T. (1983). "The Biology of Nectaries" Columbia UP.

Berger, P. H., Toler, R. W. and Harris, K. F. (1983). Maize dwarf virus transmission by greenbug, *Schizaphis graminum*, biotypes. *Plant Dis.* **67**, 496–497.

Boucher, D. H. (1982). The ecology of mutualism. *Ann. Rev. Ecol. Syst.* **13**, 315–347.

Bowers, W. S., Nault, L. R., Webb, R. E. and Dutky, S. R. (1972). Aphid alarm pheromone: isolation, identification, synthesis. *Science* **177**, 1121–1122.

Bowers, W. S., Nishino, C., Montgomery, M. E. and Nault, L. R. (1977a). Structure-activity relationships of analogs of the aphid alarm pheromone, (E)-β-farnesene. *J. Insect Physiol.* **23**, 697–701.

Bowers, W. S., Nishino, C., Montgomery, M. E., Nault, L. R. and Nielsen, M. W. (1977b). Sesquiterpene progenitor, germacrene A: an alarm pheromone in aphids. *Science* **196**, 680–681.

Bradley, G. A. (1973). Effect of *Formica obscuripes* (Hymenoptera: Formicidae) on the predator-prey relationship between *Hyperaspis congressis* (Coleoptera: Coccinellidae) and *Toumeyella numismaticum* (Homoptera: Coccidae). *Can. Entomol.* **105**, 1113–1118.

Bradley, G. A. and Hinks, J. D. (1968). Ants, aphids and jack pine in Manitoba. *Can. Entomol.* **100**, 40–50.

Brattsten, L. (1979). Biochemical defense mechanisms in herbivores against plant allelochemicals. *In* "Herbivores: their Interaction with Secondary Plant Metabolites" (Eds G. A. Rosenthal and D. H. Janzen), pp. 199–270. Academic Press, New York.

Bristow, C. M. (1983). Treehoppers transfer parental care to ants: a new benefit of mutualism. *Science* **220**, 532–533.

Bristow, C. M. (1984). Differential benefits from ant attendance to 2 species of Homoptera on New York iron weed. *J. Anim. Ecol.* **53**, 775–726.

Buckley, R. C. (1982). Ant-plant interactions – a world review. *In* "Ant–Plant Interactions in Australia" (Ed. R. C. Buckley), pp. 111–142. Junk, The Hague.

Buckley, R. C. (1983). Interaction between ants and membracid bugs decreases growth and seed set of host plant bearing extrafloral nectaries. *Oecologia* **58**, 132–136.

Bur, M. (1985). Influence of plant growth hormones on development and reproduction in aphids (Homoptera, Aphidinea, Aphididae). *Entomol. Gen.* **10**, 183–200.

Burnett, W. C., Jones, S. B. and Mabry, T. J. (1978). The role of sesquiterpene lactones in plant-animal coevolution. *In* "Biochemical Aspects of Plant and Animal Coevolution" (Ed. J. B. Harborne), pp. 233–258. Academic Press, New York.

Burns, D. P. (1973). The foraging and tending behaviour of *Dolichoderus taschenbergi* (Hymenoptera, Formicidae). *Can. Entomol.* **105**, 97–104.

Bushing, R. W. and Burton, V. E. (1974). Leafhopper damage to silage corn in California. *Econ. Entomol.* **67**, 656–658.

Calabrese, E. J. and Sorensen, A. J. (1978). Dispersal and recolonization by *Myzus persicae* following aphid alarm pheromone exposure. *Ann. Entomol. Soc. Amer.* **71**, 181–182.

Carroll, C. R. and Janzen, D. H. (1973). Ecology of foraging by ants. *Ann. Rev. Ecol. Syst.* **4**, 231–257.

Cates, R. G. (1980). Feeding patterns of monophagous, oliphagous, and polyphagous insect herbivores: the effect of resource abundance and plant chemistry. *Oecologia* **46**, 22–31.

Choudhury, D. (1984). Aphids and plant fitness: a test of Owen and Wiegert's hypothesis. *Oikos* **43**, 401–403.

Choudhury, D. Aphid honeydew: a reappraisal of Owen and Wiegert hypothesis. *Oikos* **45**, 287–289.

Clark, W. H. and Yensen, N. P. (1976). Tending behaviour of the ants *Formica neoclara* Emery and *Formica obscuriventris clivia* Creighton, and honeydew production of the treehopper *Campylenchia* sp. (Hymenoptera: Formicidae; Homoptera: Membracidae). *J. Idaho Acad. Sci.* **12**, 3–7.

Collins, L. and Scott, J. K. (1982). Interaction of ants, predators and the scale insect *Pulvinariella mesembryanthemi* on *Carpobrotus edulis*, an exotic plant naturalised in Western Australia. *Aust. Entomol. Mag.* **8**, 73–78.

Cookson, L. and New, T. R. (1980). Observations on the biology of *Sextius virescens* (Fairmaire) (Homoptera, Membracidae) on *Acacia* in Victoria. *Aust. Entomol. Mag.* **7**, 4–10.

Cottrell, C. B. (1984). Aphytophagy in butterflies: its relation to myrmecophily. *Zool. J. Linn. Soc.* **79**, 1–57.

Dahl, M. L. (1971). Uber einen Schreckstoff bei Aphiden. *Deut. Entomol. Z.* **18**, 121–128.

D'Arcy, C. J. and Nault, L. R. (1982). Insect transmission of plant viruses, mycoplasmalike and rickettsialike organisms. *Plant Dis.* **66**, 99–104.

Das, G. M. (1956). Observations on the association of ants with coccids on tea. *Bull. Entomol. Res.* **45**, 437–447.

Dash, A. R. (1978). Ants attending the maize aphid, *Rhopalosiphum maidis*. *Prakruti Utkal. Univ. J. Sci.* **12**, 139–140.

David, C. T. and Wood, D. L. (1980). Orientation to trails by a carpenter ant, *Camponotus modoc* (Hymenoptera: Formicidae) in a giant sequoia forest. *Can. Entomol.* **112**, 993–1000.

David, H. and Alexander, K. C. (1984). Insect vectors of virus diseases of sugarcane. *Proc. Ind. Acad. Sci. (Anim. Sci.)* **93**, 339–348.

De Vries P. J. (1984). Of crazy ants and Curetinae: are *Curetis* butterflies tended by ants? *Zool. J. Linn. Soc.* **79**, 59–66.

Dolva, J. M. and Scott, J. K. (1982). The association between the mealybug *Pseudococcus macrozamiae*, ants, and the cycad *Macrozamia reidlei* in a fire-prone environment. *J. Roy. Soc. West. Aust.* **65**, 33–36.

Dreyer, D. L. and Campbell, B. C. (1984). Association of the degree of methylation of intercellular pectin with plant resistance to aphids and with induction of aphid biotypes. *Experientia* **40**, 224–226.

Dreyer, D. L. Jones, K. C. and Molyneux, R. J. (1985). Feeding deterrency of some pyrrolizidine, indolizidine and quinolizidine alkaloids towards pea aphid (*Acyrthosiphum pisum*) and evidence for phloem transport of indolizidine alkaloid swainsonine. *J. Chem. Ecol.* **11**, 1045–1051.

Dreyer, D. L., Reece, J. C. and Jones, K. C. (1981). Aphid feeding deterrents in sorghum: bioassay, isolation and characterisation. *J. Chem. Ecol.* **7**, 273–284.

Duckett, D. P. (1974). Further studies of ant-aphid interactions. PhD Thesis, Imperial College, Univ. of London.

Duffey, S. S. (1980). Sequestration of natural products by insects. *Rev. Entomol.* **25**, 447–448.

Ebbers, B. C. and Barrows, E. M. (1980). Individual ants specialise on particular aphid herds (Hymenoptera, Formicidae: Homoptera, Aphididae). *Proc. Entomol. Soc. Wash.* **82**, 405–407.

Edmunds, G. F and Alstad, D. N. (1978). Coevolution in insect herbivores and conifers. *Science* **199**, 941–945.

Edwards, L. J., Siddall, J. B., Dunham, L. L., Uden, P. and Kislow, C. J. (1973). Trans-β-farnesene, alarm pheromone of the green peach aphid, *Myzus persicae* (Sulzer). *Nature* **241**, 126–127.

Eisner, T., Nowicki, S., Goetz, M. and Meinwald, J. (1980). Red cochineal dye (carminic acid): its role in nature. *Science* **208**, 1039–1042.

El-Ziady, S. and Kennedy, J. S. (1956). Beneficial effects of the common garden ant, *Lasius niger L.*, on the black bean aphid, *Aphis fabae* Scopoli. *Proc. Roy. Entomol. Soc. Lond. (A)* **31**, 61–65.

El-Ziady, S. (1960). Further effects of *Lasius niger L.* on *Aphis fabae* Scopoli. *Proc Roy. Entomol. Soc. Lond. (A)* **35**, 30–38.

Eschrich, W. (1970). Biochemistry and fine structure of phloem in relation to transport. *Ann. Rev. Plant Physio.* **21**, 193–214.

Everson, E. H. and Gallun, R. L. (1980). Breeding approaches in wheat. *In* “Breeding Plants Resistant to Insects” (Eds F. G. Maxwell and P. R. Jennings), pp. 513–533. Wiley, New York.

Feeny, P. (1976). Plant apparency and chemical defense. *Rec. Adv. Phytochem.* **10**, 1–40.

Firempong, S. (1981). A method of rearing the mealybug *Planococcoides njalensis* on cocoa seedlings. *Rev. Theobroma* **11**, 229–232.

Flanders, S. E. (1951). The role of the ant in the biological control of homopterous insects. *Can. Entomol.* **83**, 93–98.

Fokkema, N. J., Riphagen, I., Poot, R. J., and de Jong, C. (1983). Aphid honeydew, a potential stimulant of *Cochliobolus satirus* and *Septoria nodorum* and the competitive role of saprophytic mycoflora. *Trans. Br. Mycol. Soc.* **81**, 355–363.

Fol’kina, M. Y. (1978). A case of mass extermination of San-Jose scale *Quadraspidiotus perniciosus* by the ant *Crematogaster subdentata. Zool. Zh.* **57**, 301.

Fowler, S. V. and Macgarvin, M. (1985). The impact of hairy wood ants, *Formica lugubris*, on the guild structure of herbivorous insects on birch, *Betula pubescens. J. Anim. Ecol.* **54**, 847–855.

Fritz, R. S. (1982). An ant-treehopper mutualism: effects of *Formica subsericea* on the survival of *Vanduzea arquata. Ecol. Entomol.* **7**, 267–276.

Fritz, R. S. (1983). Ant protection of a host plant’s defoliator: consequences of an ant-membracid mutualism. *Ecology* **64**, 789–797.

Funkhouser, W. D. (1915). Life history of *Vanduzea arquata* Say. *Psyche* **22**, 183–198.

Gallun, R. L., Starks, K. J. and Guthrie, W. D. (1975). Plant resistance to insects attacking cereals. *Ann. Rev. Entomol.* **20**, 337–357.

Gibson, R. W. (1972). The distribution of aphids on potato leaves in relation to vein size. *Entomol. exp. appl.* **15**, 213–223.

Gibson, R. W. (1974). Aphid-trapping glandular hairs on hybrids of *Solanum tuberosum* and *S. bertahultii. Potato Res.* **17**, 152–154.

Gibson, R. W. and Pickett, J. A. (1983). Wild potato repels aphids by release of aphid alarm pheromone. *Nature* **302**, 608–609.

Gilbert, N. and Gutierrez, A. P. (1973). A plant–aphid–parasite relationship. *J. Anim. Ecol.* **42**, 323–340.

Goff, A. M. and Nault, L. R. (1974). Aphid cornicle secretions ineffective against attack by parasitoid wasps. *Environ. Entomol.* **3**, 565–566.

Gosswald, K. (1954). Uber die Wirtschaftlichkeit des Masseneinsatzes der Roten Waldameise. *Z. Angew. Zool.* pp. 145–185.

Gosswald, K. (1958). Neue Erfahrungen uber Einwirkung der Roten Waldameise auf den Massenwechsel von Schadinsekten sowie einige methodische Verbesserungen bei ihrem praktischen Einsatz. *Proc. X Int. Congr. Entomol., Montreal* **4**, 567–571.

Guttman, S. I., Wood, T. K. and Karlin, A. A. (1981). Genetic differentiation along host plant lines in the sympatric *Enchenopa binotata* Say complex (Homoptera: Membracidae). *Evolution* **35**, 205–217.

Haines, I. H. and Haines, J. B. (1978a). Colony structure, seasonality and food requirements of the crazy ant *Anoplolepis longipes* in the Seychelles. *Ecol. Entomol.* **3**, 109–118.

Haines, I. H. and Haines, J. B. (1978b). Pest status of the crazy ant *Anoplolepis longipes* (Hymenoptera, Formicidae) in the Seychelles. *Bull. Entomol. Res.* **68**, 627–638.

Harris, K. F. and Maramorosch, K. (1977). "Aphids as Virus Vectors" Academic Press, New York.

Harris, K. F. and Maramorosch, K. (1979). "Vectors of Plant Pathogens" Academic Press, New York.

Haukioja, E. (1980). On the role of plant defenses in the fluctuation of herbivore populations. *Oikos* **35**, 202–213.

Henning, S. F. (1984). The effect of ant association on lycaenid larval duration (Lepidoptera: Lycaenidae). *Entomol. Rec.* **96**, 99–102.

Hill, M. G. and Blackmore, P. J. M. (1980). Interactions between ants and the coccid *Icerya seychellarum* on Aldabra Atoll. *Oecologia* **45**, 360–365.

Horvitz, C. C. and Schemske, D. W. (1984). Effects of ants and an ant-tended herbivore on seed production of a neotropical herb. *Ecology* **65**, 1369–1378.

Hough, W. S. (1922). Observations on two mealybugs, *Trionymus trifolii* Forbes, and *Pseudococcus meritimus* Ehrh. *Entomol. News* **33**, 171–176.

Huffaker, C. B., Dahlsten, D. L., Janzen, D. H. and Kennedy, G. G. (1984). Insect influences in the regulation of plant populations and communities. *In* "Ecological Entomology" (Eds C. B. Huffaker and R. L. Rabb), pp. 659–691. Wiley, New York.

Huheey, J. E. (1984). Warning coloration and mimicry. *In* "Chemical Ecology of Insects" (Eds W. J. Ball and R. T. Carde), pp. 257–209. Chapman & Hall, London.

Huxley, C. R. (1978). The ant-plants *Myrmecodia* and *Hydnophytum* (Rubiaceae) and the relationships between their morphology, ant occupants, physiology and ecology. *New Phytol.* **80**, 231–268.

Huxley, C. R. (1982). Ant-epiphytes of Australia. *In* "Ant–Plant Interactions in Australia" (Ed. R. C. Buckley), pp. 63–74. Junk, The Hague.

Inouye, D. W. and Taylor, O. R. (1979). A temperate region plant–ant–seed predator system: consequences of extra floral nectar secretion by *Helianthella quinquenervis*. *Ecology* **60**, 1–7.

Ito, Y., Tsubaki, Y. and Osada, M. (1982). Why do *Luehdorfia* butterflies lay eggs in clusters? *Res. Popul. Ecol.* **24**, 375–387.

Janzen, D. H. (1969). Allelopathy by myrmecophytes: the ant *Azteca* as an allelopathic agent of *Cecropia*. *Ecology* **50**, 147–153.

Janzen, D. H. (1974). Epiphytic myrmecophytes in Sarawak, Indonesia: mutualism through the feeding of plants by ants. *Biotropica* **6**, 237–259.

Johnson, A. E., Molyneux, R. J. and Merrill, G. B. (1985). Chemistry of toxic range plants. Variation in pyrrolizidine alkaloid content of *Senecio, Amsinckia*, and *Crotalaria* species. *J. Agric. Food Chem.* **33**, 50–55.

Johnson, B. (1959). Ants and form reversal in aphids. *Nature* **184**, 740–741.

Jordens-Rottger, D. (1979). The role of phenolic substances for host-selection behaviour of the black bean aphid. *Aphis fabae*. *Entomol. exp. appl.* **26**, 49–54.

Khan, Z. R. and Saxena, R. C. (1985). A selected bibliography of the whitebacked planthopper *Sogatella furcifera* (Horvath) (Homoptera, Delphacidae). *Insect Sci. Applic.* **6**, 115–134.

Kidd, N. A. C. (1985). The role of the host plant in the population dynamics of the large pine aphid, *Cynara pinea*. *Oikos* **44**, 114–122.

Kislow, C. J. and Edwards, L. J. (1972). Repellent odour in aphids. *Nature* **235** 108–109.

Kiss, A. (1981). Melezitose, aphids and ants. *Oikos* **37**, 382.

Kiss, A. and Chau, R. (1984). The probing behaviour of nymphs of *Vanduzea arquata* and *Enchenopa binotata* (Homoptera, Membracidae) on host and non-host plants. *Ecol. Entomol.* **4**, 429–435.

Kitching, R. L. (1981). Egg clustering and the southern hemisphere lycaenids: comments on a paper by N. E. Stamp. *Amer. Nat.* **118**, 423–425.

Kitching, R. L. and Filshie, B. K. (1974). The morphology and mode of action of the anal apparatus of membracid nymphs with special reference to *Sextius virescens* (Fairmaire) (Homoptera). *J. Entomol.* (*A*). **49**, 81–88.

Kitching, R. L. and Luke, B. (1985). The myrmecophilous organs of the larvae of some British Lycaenidae (Lepidoptera): a comparative study, *J. Nat. Hist.* **19**, 259–276.

Kleinjan, J. E. and Mittler, T. E. (1975). A chemical influence of ants on wing development in aphids. *Entomol. exp. appl.* **18**, 384–388.

Kopp, D. D. and Tsai, J. H. (1983). Systematics of genus *Idioderma* (Homoptera: Membracidae) and biology of *Idioderma virescens*. *Ann. Entomol. Soc. Amer.* **76**, 149–157.

Kugler, J. L. and Ratcliffe, R. H. (1983). Resistance in alfalfa to a red form of the pea aphid (Homoptera: Aphididae). *J. Econ. Entomol.* **76**, 74–76.

Lam, P. K. S. and Dudgeon, D. (1985). Fitness implication of plant-herbivore "mutualism". *Oikos* **44**, 360–361.

Leston, D. (1978). A neotropical ant mosaic. *Ann. Entomol. Soc. Amer.* **71**, 649–653.

Madden, L. V. and Nault, L. R. (1983). Differential pathogenicity of corn stunting mollicutes to leafhopper vectors in *Dalbulus* and *Baldulus* species. *Phytopathol.* **73**, 1608–1614.

Majer, J. D. (1982). Ant-plant interactions in the Darling Botanical District of Western Australia. *In* "Ant–Plant Interactions in Australia" (Ed. R. C. Buckley), pp. 45–62. Junk, The Hague.

Malicky, H. (1970). New aspects on the association between lycaenid larvae (Lycaenidae) and ants (Formicidae, Hymenoptera). *J. Lep. Soc.* 24, 190–202.

Maramorosch, K. (1980). Insects and plant pathogens. *In* "Breeding Plants Resistant to Insects" (Eds F. G. Maxwell and P. R. Jennings), pp. 138–156. Wiley, New York.

Maramorosch, K. and Harris, K. F. (1979). "Leafhopper Vectors and Plant Disease Agents" Academic Press, New York.

Maschwitz, U., Schroth, M., Hanel, H. and Pong, T. Y. (1984). Lycaenids parasitising symbiotic plant-ant partnerships. *Oecologia* **64**, 78–80.

Matile, P. (1984). Das toxische Kompartiment der Pflanzenzelle. *Naturwissenschaften* **71**, 18–24.

Mattson, W. J. (1980). Herbivory in relation to plant nitrogen content. *Ann. Rev. Ecol. Syst.* **11**, 119–161.

Maxwell, F. G. and Jennings, P. R. (Eds) (1980). "Breeding Plants Resistant to Insects" Wiley, New York.

May, R. M. (1982). Theoretical models for mutualism and some applications. *Amer. Assoc. Adv. Sci. Abstr. Pap. Nat. Meet.* **148**, 35.

McClure, M. S. (1980). Foliar nitrogen: a basis for host suitability for elongate hemlock scale, *Fiorinia externa* (Homoptera: Diaspididae). *Ecology* **61**, 72–79.

McClure, M. S. (1985a). Patterns of abundance, survivorship, and fecundity of *Nuculaspis tsugae* (Homoptera: Diaspididae) on *Tsuga* species in Japan in relation to elevation. *Environ. Entomol.* **14**, 413–415.

McClure, M. S. (1985b). Susceptibility of pure and hybrid stands of *Pinus* to attack by *Matsucoccus matsumurae* in Japan (Homoptera: Coccoidea: Margarodidae). *Environ. Entomol.* **14**, 535–538.

McClure, M. S. and Hare, J. D. (1984). Foliar terpenoids in *Tsuga* species and the fecundity of scale insects. *Oecologia* **63**, 185–193.

McEvoy, P. B. (1979). Advantages and disadvantages to group living in treehoppers (Homoptera, Membracidae). *Misc. Pub. Entomol. Soc. Amer.* **11**. 1–13.

McLain, D. K. (1980). Relationships among ants, aphids and coccinellids on wild lettuce, *Lactuca canadensis*. *J. Ga. Entomol. Soc.* **15**, 417–418.

Messina, F. J. (1981). Plant protection as a consequence of an ant-membracid mutualism: interactions on goldenrod, *Solidago* sp. *Ecology* **62**, 1433–1440.

Messina, F. J., Renwick J. A. A. and Barmore, J. L. (1985). Resistance to *Aphis craccivora* (Homoptera: Aphididae) in selected varieties of cowpea. *J. Entomol. Sci.* **20**, 263–269.

Miller, D. R. and Kosztarab, M. (1979). Recent advances in the study of scale insects. *Ann. Rev. Entomol.* **24**, 1–27.

Molyneux, R. J. and Johnson, A. E. (1984). Extraordinary levels of production of pyrrolizidine alkaloids in *Senecio riddellii*. *J. Nat. Prod.* **47**, 1030–1032.

Montgomery, M. E. and Nault, L. R. (1977a). Comparative responses of aphids to the alarm phenomone (E) β-farnesene. *Entomol. Exp. Appl.* **22**, 236–242.

Montgomery, M. E. and Nault, L. R. (1977b). Aphid alarm pheromones: dispersion of *Hyadaphis erysimi* and *Myzus persicae*. *Ann. Entomol. Soc. Amer.* **70**, 669–672.

Montgomery, M. E. and Nault, L. R. (1978). Effects of age and wing polymorphism on the sensitivity of *Myzus persicae* to alarm pheromone. *Ann. Entomol. Soc. Amer.* **71**, 788–790.

Morrill, W. L. (1978). Red imported fire ant predation on the alfalfa weevil and pea aphid. *J. Econ. Entomol.* **71**, 867–868.

Meuller, T. F. (1983). The effects of plants on the host relations of a specialist parasitoid of *Heliothis* larvae. *Entomol. exp. appl.* **34**, 78–84.

Mukhopadyay, S. (1984). Interactions of insect vectors with plants in relation to transmission of plant viruses. *Proc. Ind. Acad. Sci.* (*Anim. Sci.*) **93**, 349–358.

Nault, L. R. and Bowers, W. S. (1974). Multiple alarm pheromones in aphids. *Entomol. exp. appl.* **17**, 455–457.

Nault, L. R. and Madden, L. V. (1985). Ecological strategies of *Dalbulus* leafhoppers. *Ecol. Entomol.* **10**, 57–63.

Nault, L. R. and Montgomery, M. E. (1979). Aphid alarm pheromones. *Misc. Pub. Entomol. Soc. Amer.* **11**, 23–31.

Nault, L. R., Montgomery, M. E. and Bowers, W. S. (1976). Ant–aphid association: the role of aphid alarm pheromone. *Science* **192**, 1349–1351.

Nault, L. R. and Phelan, P. L. (1984). Alarm pheromones and sociality in pre-social insects. *In* "Chemical Ecology of Insects" (Eds W. J. Bell and R. T. Carde), pp. 237–256. Chapman & Hall, London.

Nault, L. R., Wood, T. K. and Goff, A. M. (1974). Treehopper (Membracidae) alarm pheromones. *Nature* **149**, 387–388.

Nechols, J. R. and Seibert, T. F. (1985). Biological control of the spherical mealybug, *Nipaecoccus vastator* (Homoptera: Pseudococcidae): assessment by ant exclusion. *Environ. Entomol.* **14**, 45–47.

Nickerson, J. C., Kay, C. A. R., Buschman, L. L. and Whitcomb, W. H. (1977). The presence of *Spissistilus festinus* as a factor affecting egg predation by ants in soybeans. *Fla. Entomol.* **60**, 193–199.

Nielson, M. W. and Lehman, W. F. (1980). Breeding approaches in alfalfa. *In*

"Breeding Plants Resistant to Insects" (Eds F. G. Maxwell and P. R. Jennings), pp. 277–311. Wiley, New York.

Niles, G. A. (1980). Breeding cotton for resistance to insect pests. *In* "Breeding Plants Resistant to Insects" (Eds F. G. Maxwell and P. R. Jennings), pp. 337–369. Wiley, New York.

Nixon, G. E. J. (1951). "The Association of Ants with Aphids and Coccids" Comm. Inst. Entomol., London.

O'Neill, M. C. and Robinson, A. G. (1977). Ant-aphid associations in the province of Manitoba. *Man. Entomol.* **11**, 74–88.

Ortega, A., Vasal, S. K., Mihm, S. K. and Hershey, C. (1980). Breeding for insect resistance in maize. *In* "Breeding Plants Resistant to Insects" (Eds F. G. Maxwell and P. R. Jennings), pp. 370–419. Wiley, New York.

Owen, D. F. (1978). Why do aphids synthesise melezitose? *Oikos* **31**, 264–267.

Owen, D. F. and Wiegert, R. G. (1976). Do consumers maximise plant fitness? *Oikos* **27**, 488–492.

Owen, D. F. and Wiegert, R. G. (1984). Aphids and plant fitness: 1984. *Oikos* **43**, 403.

Petelle, M. (1980). Aphids and melezitose: a test of Owen's 1978 hypothesis. *Oikos* **35**, 127–128.

Petelle, M. (1984). Aphid honeydew sugars and soil nitrogen fixation. *Soil Biol. Biochem.* **16**, 203–206.

Phelan, P. L., Montgomery, M. E. and Nault, L. R. (1976). Orientation and locomotion of apterous aphids dislodged from their hosts by alarm pheromone. *Ann. Entomol. Soc. Amer.* **69**, 1153–1156.

Pickett, J. A. and Griffiths, D. C. (1980). Composition of aphid alarm pheromones. *J. Chem. Ecol.* **6**, 349–360.

Pierce, N. E. (1984). Amplified species diversity: a case study of an Australian lycaenid butterfly and its attendant ants. *Symp. Roy. Entomol. Soc. Lond. II.*

Pierce, N. E. (1985). Lycaenid butterflies and ants – selection for nitrogen-fixing and other protein-rich food plants. *Amer. Nat.* **125**, 888–895.

Pierce, N. E. and Elgar, M. A. (1985). The influence of ants on host plant selection by *Jalmenus evagoras*, a myrmecophilous lycaenid butterfly. *Behav. Ecol. Sociobiol.* **16**, 209–222.

Pierce, N. E. and Mead, P. S. (1981). Parasitoids as selective agents in the symbiosis between lycaenid butterfly larvae and ants. *Science* **211**, 1185–1187.

Pierce, N. E., Kitching, R. L., Buckley, R. C., Taylor, M. J. F. and Benbow, K. F. (Ms). The costs and benefits of cooperation between the Australian lycaenid butterfly, *Jalmenus evagoras*, and its attendant ants.

Plumb, R. T. and Thresh, J. M. (Eds) (1983). "Plant Virus Epidemiology" Blackwell, Oxford.

Pollard, D. G. (1973). Plant penetration by feeding aphids (Hemiptera, Aphidoidea): a review. *Bull. Entomol. Res.* **62**, 631–714.

Prestidge, R. A. and McNeill, S. (1983). Role of nitrogen in the ecology of grassland Auchenorrhyncha. *In* "Nitrogen as an Ecological Factor" (Eds J. A. Lee, L. McNeill and J. H. Rorison), pp. 257–281. Blackwell, Oxford.

Price, P. W., Bouton, C. E., Gross, P., McPheron, B. A., Thompson, J. N. and Weis, A. E. (1980). Interactions among three tropic levels: influence of plants on interactions between insect herbivores and natural enemies. *Ann. Rev. Ecol. Syst.* **11**, 41–65.

Rabb, R. L., Defoliart, G. K. and Kennedy, G. G. (1984). An ecological approach to managing insect populations. *In* "Ecological Entomology" (Eds C. B. Huffaker and R. L. Rabb), pp. 697–728. Wiley, New York.

Ratcliffe, R. H. and Murray, J. J. (1983). Selection for greenbug (Homoptera: Aphididae) resistance in Kentucky bluegrass cultivars. *J. Econ. Entomol.* **76**, 1221–1224.

Reilly, J. J. and Sterling, W. L. (1983a). Dispersion patterns of the red imported fire ant (Hymenoptera: Formicidae), aphids, and some predaceous insects in East Texas cotton fields. *Environ. Entomol.* **12**, 380–385.

Reilly, J. J. and Sterling, W. L. (1983b). Interspecific association between the red imported fire ant (Hymenoptera: Formicidae), aphids, and some predaceous insects in a cotton agroecosystem. *Environ. Entomol.* **12**, 541–545.

Rhoades, D. F. and Cates, R. G. (1976). Toward a general theory of plant anti-herbivore chemistry. *Rec. Adv. Phytochem.* **10**, 168–313.

Richerson, J. V. and Jones, R. D. (1982). An ant-aphid association on threadleaf groundsel in the Davis Mountains area of west Texas. *Southwest Nat.* **27**, 466.

Roitberg, B. D. and Myers, J. H. (1978). Adaptations of alarm pheromone responses to the pea aphid, *Acyrthosiphon pisum. Can. J. Zool.* **56**, 103–108.

Rosengren, R. (1971). Route fidelity, visual memory and recruitment behaviour in foraging wood ants of the genus *Formica. Acta Zool. Fenn.* **133**, 1–106.

Rosenthal, G. A. and Janzen, D. H. (1979). "Herbivores: their Interaction with Secondary Plant Metabolites" Academic Press, New York.

Rothcke, B., Hamrum, C. L. and Gloss, H. W. (1967). Observations on the interrelationships among ants, aphids and aphid predators. *Mich. Entomol.* **1**, 169–173.

Salama, H. S., Elsherif, A. F. and Megahed, M. (1985). Soil nutrients affecting the population density of *Parlatoria zizyphus* (Lucas) and *Icerya purchasi* Mask (Homopt., Coccoidea) on citrus seedlings. *Z. Angew. Entomol.* **99**, 471–475.

Salto, C. E., Eikenbary, R. D. and Starks, K. J. (1983). Compatibility of *Lysiphlebus testaceipes* (Hymenoptera: Braconidae) with greenbug (Homoptera: Aphididae) biotypes "C" and "F" reared on susceptible and resistant oat varieties. *Environ. Entomol.* **12**, 603–604.

Samways, M. J., Nel, M., Prins, A. J. (1982). Ants (Hymenoptera; Formicidae) foraging in citrus trees and attending honeydew-producing Homoptera. *Phytophylactica* **14**, 155–157.

Schowalter, T. D. (1981). Insect herbivore relationship to the state of the host plant: biotic regulation of ecosystem nutrient cycling through ecological succession. *Oikos* **37**, 126–130.

Schremmer, F. (1978) A neotropical wasp species (Hymenoptera, Vespidae) which guards treehopper larvae (Homoptera, Membracidae) and collects their honeydew. *Entomol. Ger.* **4**, 183–186.

Scriber, J. M. (1984). Host-plant suitability. *In* "Chemical Ecology of Insects" (Eds W. J. Bell and R. T. Carde), pp. 159–202. Chapman & Hall, London.

Sen-Sarma, P. K. (1984). Mycoplasma and allied diseases of forest trees in India and vector-host-pathogen interactions. *Proc. Indian Acad. Sci.* (*Anim. Sci.*) **93**, 323–334.

Service, P. M. and Lenski, R. E. (1982). Aphid genotypes, plant phenotypes, and genetic diversity: a demographic analysis of experimental data. *Evolution* **36**, 1276–1282.

Service, P. (1984a). Genotypic interactions in an aphid-host plant relationship: *Uroleucon rudbeckiae* and *Rudbeckia laciniata. Oecologia* **61**, 271–276.

Service, P. (1984b). The distribution of aphids in response to variation among individual host plants: *Uroleucon rudbeckiae* (Homoptera: Aphididae) and *Rudbeckia laciniata* (Asteraceae). *Ecol. Entomol.* **9**, 321–328.

Sheppard, C., Martin, P. B. and Mead, F. W. (1979). A planthopper (Homoptera, Cixiidae) associated with red imported fire ant (Hymenoptera, Formicidae) mounds. *J. Ga. Entomol. Soc.* **14**, 140–144.

Silander, J. A., Trembath, B. R. and Fox, L. R. (1983). Chemical interference among plants mediated by grazing insects. *Oecologia* **58**, 415–417.

Skinner, G. J. and Whittaker, J. B. (1981). An experimental investigation of interrelationships between the wood ant (*Formica rufa*) and some tree-canopy herbivores. *J. Anim. Ecol.* **50**, 313–326.

Smiley, J. T. (1985). *Heliconius* caterpillar mortality during establishment on plants with and without attending ants. *Ecology* **66**, 845–849.

Southwood, T. R. E. (1985). Interactions of plants and animals: patterns and processes. *Oikos* **44**, 5–11.

Spiller, N. J., Kimmins, F. M. and Llewellyn, M. (1985). Fine structure of aphid stylet pathways and its use in host plant resistance studies. *Entomol. exp. appl.* **38**, 293.

Sterling, W. L., Jones, D. and Dean, D. A. (1979). Failure of the red imported fire ant to reduce entomophagous insect and spider abundance in a cotton agroecosystem. *Environ. Entomol.* **8**, 976–981.

Stout, J. (1979). An association of an ant, a mealybug, and an understorey tree from a Costa Rican rain forest. *Biotropica* **11**, 309–311.

Strickland, A. H. (1958). The entomology of swollen shoot of cacao. I. The insect species involved, with notes on their biology. *Bull. Entomol. Res.* **41**, 725–748.

Sudd, J. H. (1983). The distribution of foraging wood-ants (*Formica lugubris* Zett.) in relation to the distribution of aphids. *Insectes soc.* **30**, 298–307.

Sulochana, C. B. (1984). Leafhopper and planthopper transmitted viruses of cereal crops. *Proc. Ind. Acad. Sci.* (*Anim. Sci.*) **93**, 335–338.

Sylvester, E. S. (1984). Insects as disseminators of other organisms, especially as vectors. *In* "Ecological Entomology" (Eds C. B. Huffaker and R. L. Rabb), pp. 633–658. Wiley, New York.

Takeda, S., Kinomura, K. and Sakurai, H. (1982). Effects of ant attendance on the honeydew excretion and larviposition of the cowpea aphid. *Appl. Entomol. Zool.* **17**, 133–135.

Teetes, G. L. (1980). Breeding sorghums resistant to insects. *In* "Breeding Plants Resistant to Insects" (Eds F. G. Maxwell and P. R. Jennings), pp. 457–486. Wiley, New York.

Thomas, M. J., Pillai, K. B. and Nair, N. R. (1980). *Solenopsis geminata* (Formicidae, Hymenoptera) as a predator of the brown plant hopper *Nilaparvata lugens*. *Agric. Res. J. Kerala* **18**, 145.

Tilles, D. A. and Wood, D. L. (1982). The influence of carpenter ant (*Camponotus modoc*) (Hymenoptera: Formicidae) attendance on the development and survival of aphids (*Cinara* spp.) (Homoptera: Aphididae) in a Giant Sequoia forest. *Can. Entomol.* **114**, 1133–1142.

Tjallingii, W. F. (1985a). Electrical nature of recorded signals during stylet penetration by aphids. *Entomol. exp. appl.* **38**, 177–186.

Tjallingii, W. F. (1985b). Membrane potentials as an indication for plant cell penetration by aphid stylets. *Entomol. exp. appl.* **38**, 187–193.

Van Emden, H. F. (1978). Insects and secondary plant substances – an alternative viewpoint with special reference to aphids. *In* "Biochemical Aspects of Plant and Animal Coevolution" (Ed. J. B. Harborne), pp. 309–323. Academic Press, New York.

Wallace, J. and Mansell, R. (Eds) (1976). "Biochemical Interactions Between Plants and Insects. Rec. Adv. Phytochem. 10" Plenum, New York.

Way, M. J. (1954). Studies on the association of the ant *Oecophylla longinoda* (Latr.) (Formicidae) with the scale insect *Saissetia zanzibarensis* Williams (Coccidae). *Bull. Entomol. Res.* **45**, 113–135.

Way, M. J. (1963). Mutualism between ants and honeydew-producing Homoptera. *Ann. Rev. Entomol.* **8**, 307–344.

Way, M. J. (1973). Population structure in aphid colonies. *In* "Perspectives in Aphid Biology" (Ed. A. D. Lowe), *Bull.* **2**, Entomol. Soc. N.Z., Auckland.

Way, M. J. and Cammell, M. (1970). Aggregation behaviour in relation to food utilisation by aphids. *In* "Animal Populations in Relation to their Food Resources" (Ed. A. Watson), pp. 229–247. Blackwell, Oxford.

Weber, N. A. (1944). The neotropical coccid-tending ants of the genus *Acropyga* Roger. *Ann. Entomol. Soc. Amer.* **37**, 89–122.

Weber, G. (1985). Genetic variability in host plant adaptation of the green peach aphid. *Myzus persicae. Entomol. exp. appl.* **38**, 49–56.

Webster, J. A. and Inayatullah, C. (1984). Greenbug (Homoptera: Aphididae) resistance in triticale. *Environ. Entomol.* **13**, 444–447.

Webster, J. A. and Stark, K. J. (1984). Sources of resistance in barley to two biotypes of the greenbug *Schizaphis graminum* (Rondoni), Homoptera: Aphididae. *Protec. Ecol.* **6**, 51–55.

Webster, J. A. and Inayatullah, C. (1985). Aphid biotypes in relation to plant resistance: a selected bibliography. *Southwest. Entomol.* **10**, 116–125.

Wellenstein, G. (1952). Sur Ernahrungsbiologie der Roten Waldameise (*Formica rufa L.*) *Z. Pflanzenkr.* **59**, 430–451.

Wheeler, W. M. (1910) "Ants, their Structure, Development and Behaviour" Columbia Univ. Press, New York.

White, T. C. R. (1984). The abundance of invertebrate herbivores in relation to the availability of nitrogen in stressed food plants. *Oecologia* **63**, 90–105.

Wink, M. and Witte, L. (1984). Turnover and transport of quinolizidine alkaloids. Diurnal fluctuations of lupanine in the phloem sap, leaves and fruits of *Lupinus albus* L. *Planta* **161**, 519–524.

Wink, M., Hartmann, T., Witte, L. and Rheinheimer, J. (1982). Interrelationships between quinolizidine alkaloid producing legumes and infecting insects: exploitation of the alkaloid containing phloem sap of *Cytisus scoparius* by the broom aphid *Aphis cytisorum. Z. Naturforsch.* **37C**, 1091–1086.

Wohlers, P. (1981). Effects of the alarm pheromone (E)-β-farnesene on dispersal behaviour of the pea aphid *Acyrthosiphon pisum. Entomol. exp. appl.* **29**, 117–124.

Wood, T. K. (1976). Alarm behaviour of brooding female *Umbonia crassicornis* (Homoptera: Membracidae). *Ann. Entomol. Soc. Amer.* **69**, 340–344.

Wood, T. K. (1977). Role of parental females and attendant ants in maturation of the treehopper, *Entylia bactriana* (Homoptera: Membracidae). *Sociobiol.* **2**, 257–272.

Wood, T. K. (1979). Sociality in the Membracidae (Homoptera). *Misc. Publ. Entomol. Soc. Amer.* **11**, 15–22.

Wood, T. K. (1980). Divergence in the *Enchenopa binotata* Say complex (Homoptera: Membracidae) effected by host plant adaptation. *Evolution* **34**, 147–160.

Wood, T. K. (1982). Ant-attended nymphal aggregations in the *Enchenopa binotata* complex (Homoptera: Membracidae) *Ann. Entomol. Soc. Amer.* **45**, 649–653.

Wood, T. K. (1983). Life history patterns of tropical membracids (Homoptera: Membracidae). *Sociobiol.* **8**, 299–343.

Wood, T. K. and Guttman, S. T. (1982). Ecological and behavioural basis for reproductive isolation in the sympatric *Enchenopa binotata* complex (Homoptera: Membracidae). *Evolution* **36**, 233–242.

Wood, T. K. and Guttman, S. T. (1983). *Enchenopa binotata* complex: sympatric speciation? *Science* **220**, 310–312.

Wood-Baker, C. S. (1977). An interesting plant, aphid and ant association. *Entomol. Mon. Mag.* **113**, 205.

Yensen, N., Yensen, E. and Yensen, D. (1980). Intertidal ants from the Gulf of California, Mexico. *Ann. Entomol. Soc. Amer.* **13**, 266–269.

Vegetation, Fire and Herbivore Interactions in Heathland

R. J. HOBBS and C. H. GIMINGHAM

I. Introduction . . . 87
II. Vegetation Dynamics and Productivity . . . 90
A. The Dynamics of *Calluna vulgaris* . . . 90
B. Production and Organic Matter Accumulation . . . 95
C. Community Dynamics . . . 103
III. Effects of Fire on Habitat and Vegetation . . . 112
A. Fire Characteristics . . . 112
B. Ecosystem Effects . . . 117
C. Post-Fire Vegetation Development . . . 120
IV. Herbivore Dynamics and Interactions with Vegetation . . . 129
A. Red Grouse . . . 129
B. Red Deer . . . 137
C. Mountain Hares . . . 140
D. Sheep . . . 142
E. Invertebrates . . . 144
V. Management and Conservation of Heathlands . . . 147
A. Grazing and Burning . . . 149
B. Recreation . . . 152
C. Conservation . . . 152
VI. Conclusions . . . 155
VII. Summary . . . 157
Acknowledgements . . . 158
References . . . 158

I. INTRODUCTION

The unifying features of world heathlands are their evergreen sclerophyllous nature, the presence of characteristic heath families and their restriction to soils low in plant nutrients (Specht, 1979). The definition of heathland

ADVANCES IN ECOLOGICAL RESEARCH Vol. 16
ISBN 0-12-013916-2

followed here is that given by Gimingham (1972) to describe the European heathlands:

> territories in which tall trees are sparse or absent, and in which the dominant lifeform is that of the evergreen dwarf shrub, particularly as represented in Ericaceae.

It is taken to include all heath types over a broad gradient of soil moisture.

This paper considers only N.W. Europe, where heathlands form a significant proportion of the natural, or semi-natural landscape. We devote particular attention to Scotland and northern England, where heathlands, in one form or another, cover much of the land area. Most of the heathlands considered in this paper have the ericaceous shrub *Calluna vulgaris* as a major component. This plant is the dominant species on approximately 25% of the unimproved land in Scotland, or about 1·2 million hectares (Tivy, 1973), and has a wide distribution throughout N.W. Europe (Fig. 1). These *Calluna*-dominated heathlands are probably the most intensively studied

Fig. 1. Map of the main area of distribution of *Calluna vulgaris* in Europe (out-lyin stations omitted). ----- approximate limits of distribution; -·-·-·- area withi which *Calluna* is commonly a community dominant; —— approximate extent o ecologically optimal habitats for *Calluna*. From Gimingham (1972).

heath systems in the world, and in this review we draw together recent research on them.

Apart from some heaths found above the tree-line, at the coast and on some bogs these heathlands are a mainly artificial vegetation type, produced initially by forest clearance, and perpetuated by management of one sort or another (Gimingham, 1972; Gimingham and de Smidt, 1983). The history of heathland development and subsequent exploitation in Britain has been reviewed on several occasions elsewhere (e.g. Dimbleby, 1962; McVean and Lockie, 1969; Tivy, 1973; Gimingham, 1972), and will not be discussed in detail here. It is sufficient to state that heathlands have been used mainly as grazings for at least the past 250 years, with sheep as the main grazing animal. In Britain, however, increasing use has been made of heathlands for sporting purposes, mainly the shooting of the red deer (*Cervus elaphus* L.) or game birds. The major bird species hunted is the red grouse, *Lagopus lagopus scoticus* L. (Lath.), and the management of heathland as grouse moor is sometimes considered the most lucrative form of upland management (Airlie, 1971; Miller, 1980).

In other parts of Europe, heathlands were often integrated into the agricultural system with the heath acting as a source of nutrients for crop growth, either through sod cutting, vegetation cutting or a rotation system which involved creating fields in new parts of the heath (Gimingham and de Smidt, 1983). These traditional management practices, discussed more fully in Section V, are now less common but have played an important role in the formation of heath vegetation in different parts of Europe.

The heathland systems of N.W. Europe have been studied extensively, and, because they are relatively simple structurally and floristically and are subjected to a variety of management practices, have proved ideal systems to observe and manipulate experimentally. From these observations and experiments a remarkably detailed picture has emerged of the dynamics of the individual ecosystem components, and, perhaps more importantly, of the interaction between these components. Much of the information on heathland systems has been reviewed previously by Gimingham (1972, 1975) and Gimingham *et al.* (1979), and more recently Miles (1981a) has discussed current problems in heathland ecology. In this paper we do not intend to reiterate the material discussed in these works: instead we discuss recent advances in the understanding of the dynamics of plant and animal populations and communities in heathlands and advance some new ideas on the functioning of heathland systems. We divide the heath system into three broad components, vegetation, fire and herbivores, and discuss each component in turn and consider the interactions between them. Since European heathlands are almost all managed systems, the effects of management are discussed throughout the paper. However, a final section discusses the possible land-use, management and conservation options available.

Nomenclature throughout this review follows Tutin *et al.* (1964–80) for vascular plants, Smith (1978) for mosses and Hawksworth *et al.* (1980) for lichens. Authorities for animal nomenclature are given in the text.

II. VEGETATION DYNAMICS AND PRODUCTIVITY

A. The Dynamics of Calluna vulgaris

Any consideration of the vegetation dynamics of N.W. European heathlands has to devote particular attention to the behaviour of *Calluna*. It has a wide geographical distribution and is a dominant plant in a large proportion of heath types (Gimingham, 1972; Gimingham *et al.*, 1979). It occurs in a broad spectrum of communities ranging from montane heath, tundra and wet bogs to dry and coastal sandy heaths, illustrating its broad environmental tolerance. It also demonstrates an immense variety of growth and morphological responses under differing environmental conditions (e.g. Grant and Hunter, 1962; Gimingham, 1975; Marrs and Bannister, 1978). In fact, Caughley and Lawton (1981) have commented that

> as a woody perennial, with mixed sexual and vegetative reproduction, ill-defined individuals and variable allocation of resources to roots, shoots and flowers under different grazing regimes, the plant is a paradigm of the problems facing plant population dynamicists.

Despite this, *Calluna* is a much studied plant!

The concept of the "*Calluna* cycle" has been central to the development of ecological studies on *Calluna*. This developed from the observations of Watt (1947, 1955, 1964) that many plant communities appeared to follow a phasic development, and in particular that *Calluna* passed through four distinct phases during its life history (i.e., pioneer, building, mature and degenerate, followed by death). A phasic or, if repeated, cyclical pattern of *Calluna* development has been accepted in most studies on *Calluna* heaths since then, and the idea has been developed further by Gimingham (1960, 1972, 1975). While the four phases are subjectively delimited and are in reality part of the continuous growth process of *Calluna*, they are defined sufficiently well to be readily recognised in the field. Definition is based primarily on morphology, but phases can be characterised in terms of their biomass, production and associated microenvironmental features (Fig. 2; Barclay-Estrup and Gimingham, 1969; Barclay-Estrup, 1970, 1971; Gimingham 1975). These phases have been observed to occur simultaneously in uneven-aged heath stands, but more usually are seen covering large areas in even-aged stands created by management.

These even-aged stands are often virtually monospecific and might there-

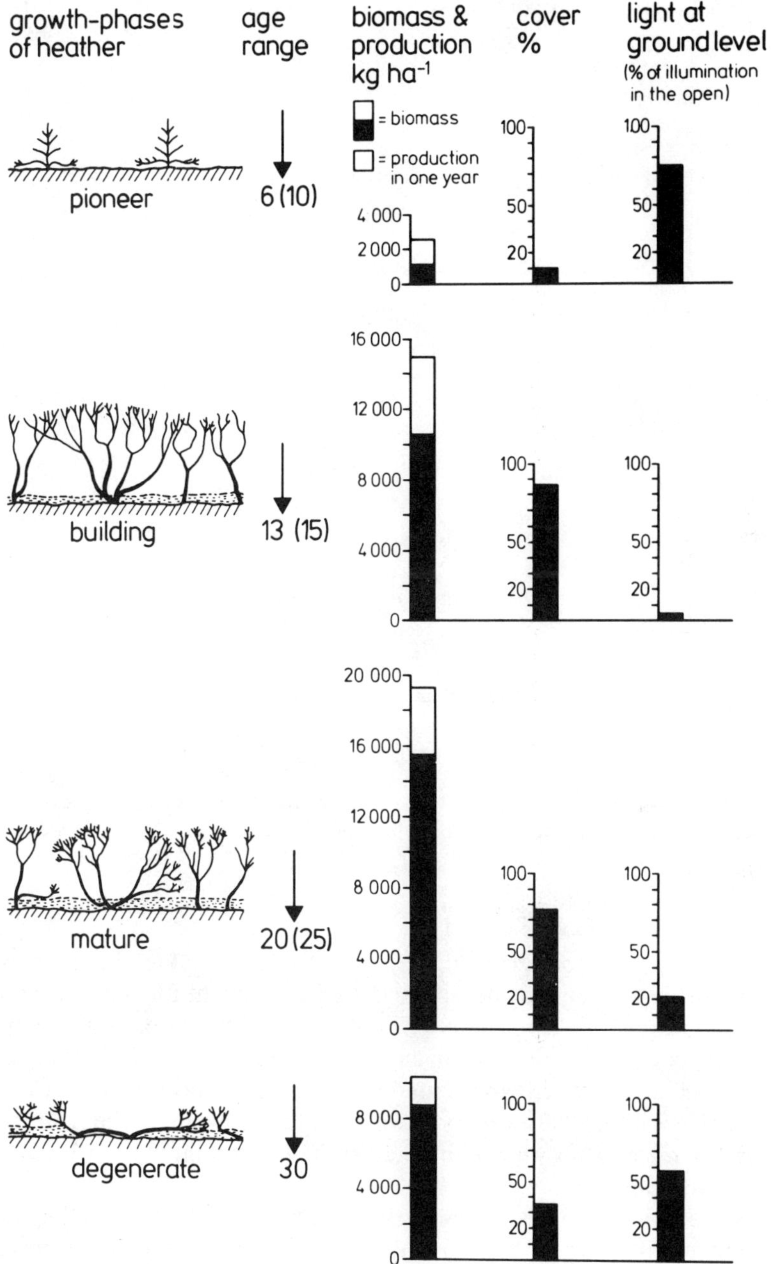

Fig. 2. Diagram illustrating the four growth phases of *Calluna* and associated changes in biomass and production, cover and illumination at ground level. From Gimingham (1975).

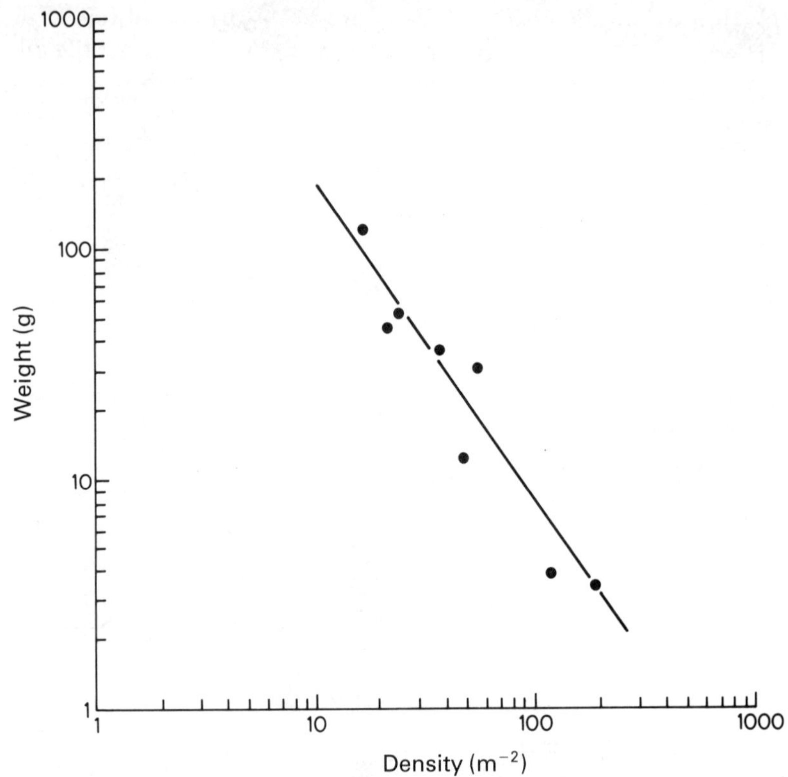

Fig. 3. Relationship between weight of individual *Calluna* plants (above-ground stems) and density from dry heath stands of different ages in N.E. Scotland. (log weight = 3·703 − 1·410 log density, r = 0·945, $p < ·001$, d.f. = 6). From Hobbs (1981).

fore be expected to behave like other monocultures. Fig. 3 illustrates the relationship between log density and log plant weight for stands consisting almost entirely of *Calluna* in N.E. Scotland. It is clear from this graph that *Calluna* undergoes self thinning during its development, the thinning curve matching almost exactly that found for other monospecific stands (Yoda *et al.*, 1963; White, 1980), with both log K and a (constants in the relationship between mean plant weight and density) falling within the range found by other studies.

While distinct growth phases can be seen in many heath types, the cyclical nature of the process has generally been inferred rather than observed directly. Barclay-Estrup and Gimingham (1969) gave evidence for a cyclical development of *Calluna* in a study of an uneven-aged stand in N.E. Scotland where there was apparent replacement of old shrubs by young, pioneer plants. However, no long-term studies have been carried out to test the validity of the *Calluna* cycle and it is difficult to assess how widespread cyclic development would be in the absence of other factors. Miles (1981a) has even suggested that the situation observed by Watt in the Breckland may, in fact, be somewhat anomalous, and extrapolation to other heath types may not therefore be valid. At present there are very few uneven-aged *Calluna* stands in Scotland. This is mainly due to current management practices which aim to produce young even-aged stands of *Calluna* to increase herbivore production (see subsequent sections). It is likely that the formation of a true uneven-aged stand would require at least 50 years without disturbance, a rare phenomenon indeed in today's landscape. It is doubtful, however, whether uneven-aged *Calluna* stands would persist if seed of tree species such as *Betula* (*pendula* or *pubescens*) and *Pinus sylvestris* were able to invade. In these cases there would be a relatively quick change from heath to woodland, as has happened in areas where management ceased after World War II. This would not, of course, occur above the natural tree-line (ca. 600m), where heathland is probably the climax vegetation type. At lower elevations, however, *Calluna* stand development should possibly be included in a larger cycle embracing all the components of the natural upland landscape, i.e., heath, birchwood and pinewood. In the absence of interference by man, heathlands and their component phases occurring below 600 m elevation were almost certainly seral stages with ultimate development towards woodland dominated by *Betula* or *Pinus sylvestris*. Heathland would then appear again only after the death of the woodland or following disturbance: for instance, fire is thought to have been a natural component of the pine forest, and pine regeneration may even be enhanced by fire (McVean and Lockie, 1969; Carlisle, 1977; Miles and Kinnaird, 1979). A fire frequency considerably lower than that found today would then produce a "moving mosaic" of heath and woodland in a variety of developmental phases, as suggested by Gimingham (1977). Today, however, the fragmentation of the natural landscape renders such large-scale developmental patterning virtually impossible, and the lack of trees to act as seed sources, coupled with grazing of seedlings by sheep, mean that large areas of heathland are not invaded.

Phasic development of *Calluna* is not observed in some types of community. In wet heath or bog, in the absence of disturbance an uneven-aged stand results, not from death and recolonisation, but by constant rejuvenation of the above-ground stem population. Forrest (1971) showed that *Calluna-Eriophorum* bog reached an apparent "steady-state" approximately 20 years after disturbance. In this case *Calluna* does not degenerate, but is constantly rejuvenated by the burial of older stems by young growth of *Sphagnum* moss. *Calluna* shoots root adventitiously and the older stems eventually become non-functional and decay. Keatinge (1975) has also suggested that *Calluna* and *Eriophorum vaginatum* interact in a way that prevents obvious phasic development of either species. Wallén (1980) also reported that the vegetation on a sand dune heath in Sweden reached a steady-state with a stable uneven-age structure maintained through similar processes of burial and adventitious rooting. Wind-pruning at high elevations in Scotland may also prevent senescence by producing a wave-type growth form, with constant rejuvenation at the leeward edge (Bayfield, 1984).

There is some observational evidence that adventitious rooting of *Calluna* also occurs in drier heaths in Scotland, and that this may be a major mechanism leading to the recolonisation of gaps in the centre of old *Calluna* bushes (Miles, 1981a; Hobbs, unpublished). In this case, decumbent shoots from surrounding shrubs or from the lowermost branches of the original shrub grow into the gaps, root adventitiously and recolonise the gap vegetatively. This mode of reinvasion may be more effective than seedling establishment since the degenerate-phase gap usually contains mats of pleurocarpous mosses and litter which are poor substrates for seed germination (Miles, 1974a,b; Mallik, Hobbs and Legg, 1984). de Hullu and Gimingham (1984) have shown that seedlings may sometimes be quite numerous in degenerate stands, but that few survive to become established. Hence, if a cyclical process such as that described by Watt does occur, it is more likely to result from vegetative propagation by "layering" than by the establishment of seedlings.

A further factor influencing the development of *Calluna* stands is periodic infestation by the heather beetle (*Lochmaea suturalis* Thomson, Coleoptera: Chrysomelidae). Although this beetle is present in Scotland, it seldom has the same devastating effects as in the European and S. English heaths. The effects of *Lochmaea* have prompted de Smidt (1977) to comment that:

> after many years of observations on permanent quadrats in the Netherlands, serious doubts must be expressed as to the actual dying of old age of *Calluna*.

In these Dutch heaths, *Calluna* may never complete its growth cycle (i.e. through to senescence and death) because of death through beetle infestation at an earlier age. Long-term observations by Diemont and Heil (1984) suggest that cyclical changes may nevertheless occur in heathlands affected by heather beetle. They discussed, for instance a case where *Calluna* and *Deschampsia flexuosa* appeared to occupy patches alternately, but this too was based mostly on spatial rather than temporal observations. The effects of heather beetles are discussed further in Section IV.

It is clear at this stage that the classical *Calluna* cycle may, in fact, be a hypothetical framework into which heathland studies have been forced, and is perhaps best considered as one of a greater number of possible developmental pathways for heath communities dominated by *Calluna*.

B. Production and Organic Matter Accumulation

Gimingham *et al.* (1979) have reviewed the large amounts of detailed information on production in dry heaths in N.E. Scotland and S. England. Fewer studies have been carried out on other European heaths, although Tyler *et al.* (1973) and, more recently, Wallén (1980) presented data from Swedish heaths. In addition, the IBP Tundra Biome studies produced extensive data on wetter heath and blanket bog types (e.g., Heal and Perkins, 1978). Production studies on dry heath have, of necessity, followed the dynamics of organic matter accumulation over the growth cycle of *Calluna*, and all estimates of production must be related to the age of the vegetation. Studies on wet heath, on the other hand, have been able to assume a "steady state" for the purposes of investigating production (Gore and Olson, 1967; Forrest, 1971). Here, we consider briefly production in dry heaths in relation to stand dynamics and then compare that with heaths in which *Calluna* stand ageing does not take place.

1. Heaths with Phasic Calluna *Development*

Studies of the primary production of *Calluna*-dominated heathlands have had to consider the pattern of shoot development shown by *Calluna*. This has been described by Mohamed and Gimingham (1970) and is summarised in Fig. 4. The current year's growth of shoot material is in the form of new lateral growth, long shoots which may bear flowers and increments to previous year's short shoots. Miller (1979) has illustrated the seasonal course of organic matter accumulation in *Calluna* in N.E. Scotland (Fig. 5), and the physiological responses of *Calluna* to the environment have been studied in detail by Grace and Woolhouse (1970, 1973 a, b) working on blanket bog. They found that net photosynthesis depended on such factors as irradiance and direction of light, present and previous temperatures, leaf age, and flowering behaviour (Fig. 6). Grace and Woolhouse (1974) were able to use this information to construct an empirical model of *Calluna* growth, and Fig. 7 illustrates the seasonal pattern of production predicted by the model.

Most studies of production in dry heath have been conducted in areas where management has created a patchwork of relatively homogeneous stands of different ages. This has allowed consideration of the relationship between vegetation age and production. In particular the studies of Chapman *et al.* (1975 a, b) and Miller (1979) have produced excellent data on stands in S. England and N.E. Scotland respectively which span the age range 0–40 years old in increments of 2–3 years or less (Fig. 8). Similar data are available for annual production of shoots, wood and litter, and estimates of net primary production have been possible (Fig. 9).

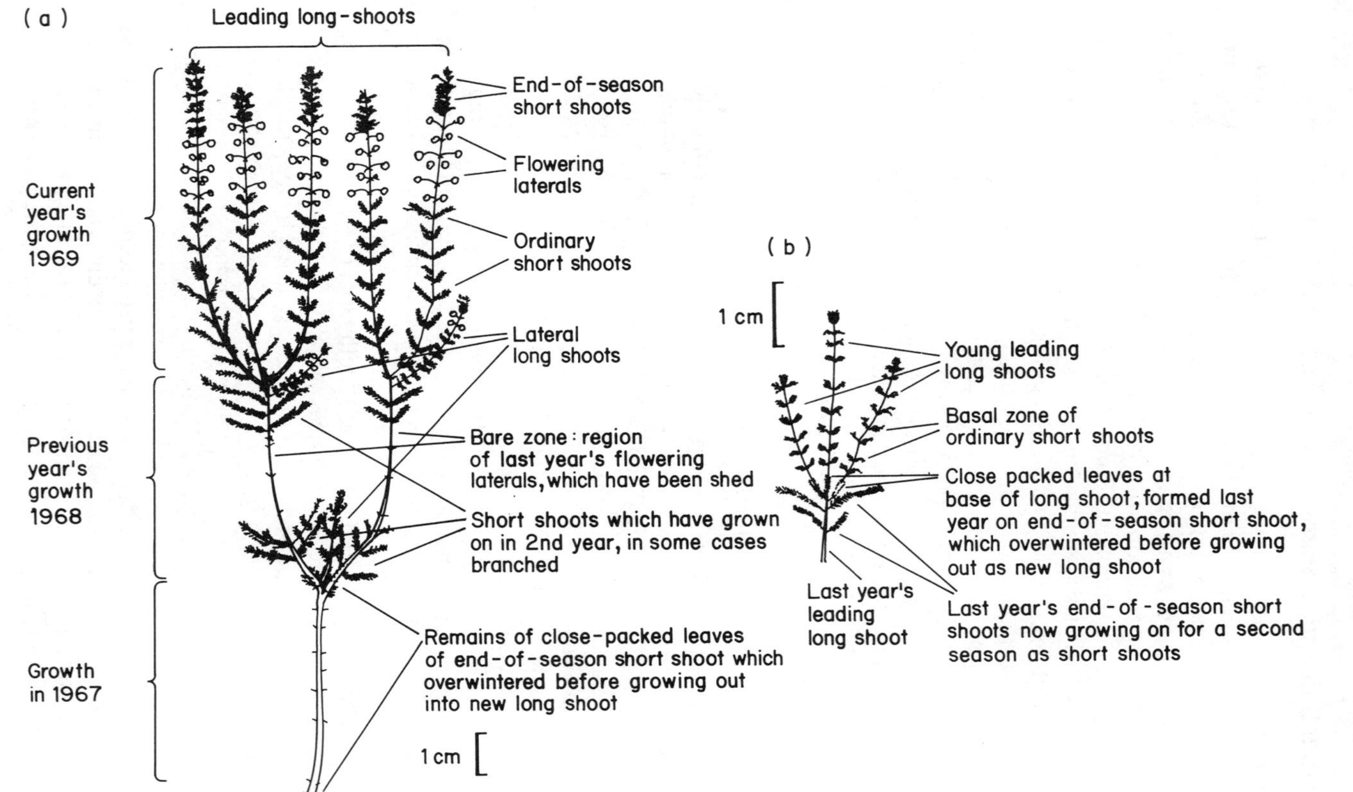

Fig. 4. Annual growth and branching in *Calluna vulgaris*. (a) Condition at the end of the growing season (early October); (b) shoot

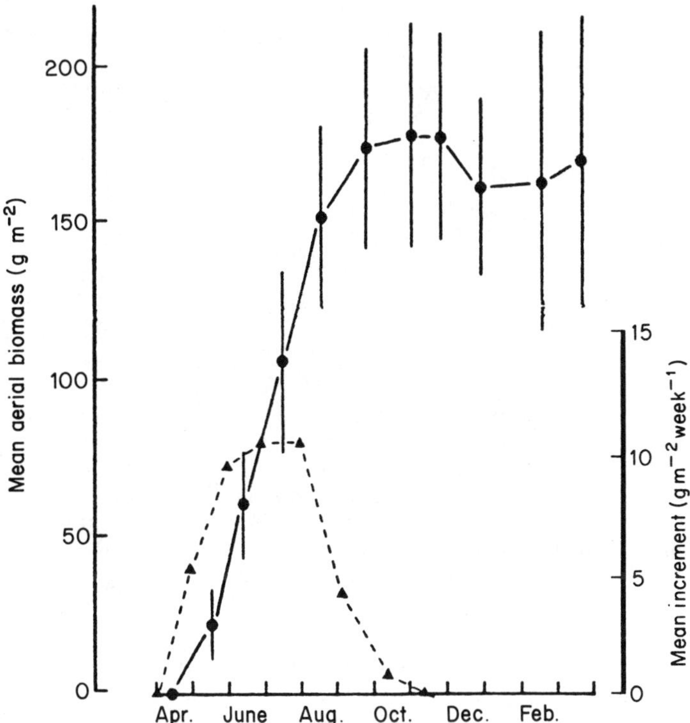

Fig. 5. Seasonal changes in the aerial biomass of current year's growth (-●—●-) and rate of increase (-▲---▲-) of the current year's growth of *Calluna* shoots in N.E. Scotland, 1964–65. Vertical lines are 95% confidence limits. From Miller (1979).

Figure 10 provides a summary of the relationship between production and organic matter accumulation and the phases of *Calluna* growth. There is an initial "post-burn" or pioneer phase where standing crop increases rapidly and net above-ground production continues to increase until the *Calluna* plants start to degenerate (i.e. after 25–30 years). At this stage organic matter accumulation slows down and the biomass of shoots begins to decline. The proportion of woody material in the stand increases rapidly as the stand ages (Miller, 1979; Hobbs and Gimingham, 1984a).

Although the relationships discussed above are generally valid, it must be noted that considerable variation may be found in the production of stands of similar ages. Miller (1979) has shown that annual shoot production can vary greatly depending on the rapidity and nature of the regeneration following fire.

Information is also available on the nutrient contents of *Calluna* of different ages (Robertson and Davies, 1965; Chapman, 1967; Miller, 1979).

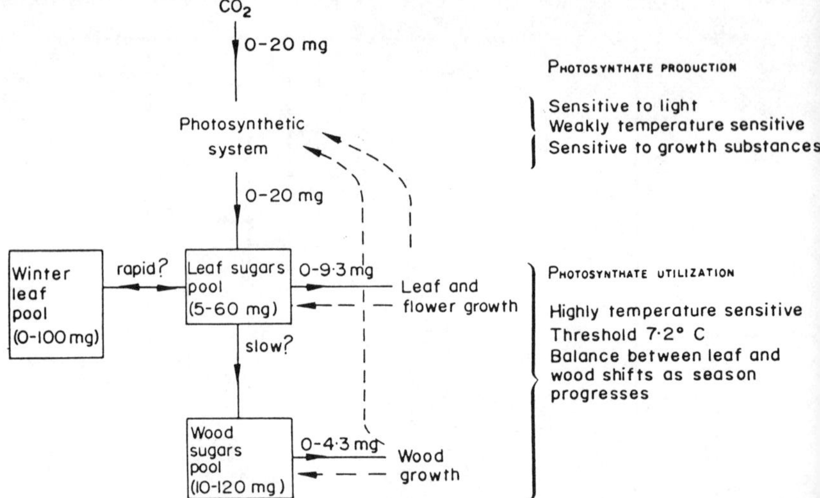

Fig. 6. Tentative scheme to show factors controlling photosynthate production and utilisation in *Calluna vulgaris*. Solid lines show fluxes of carbon; broken lines denote movements of regulating growth substances. Units: pool sizes, per g leaf; fluxes, per g leaf per day. From Grace and Woolhouse (1973b).

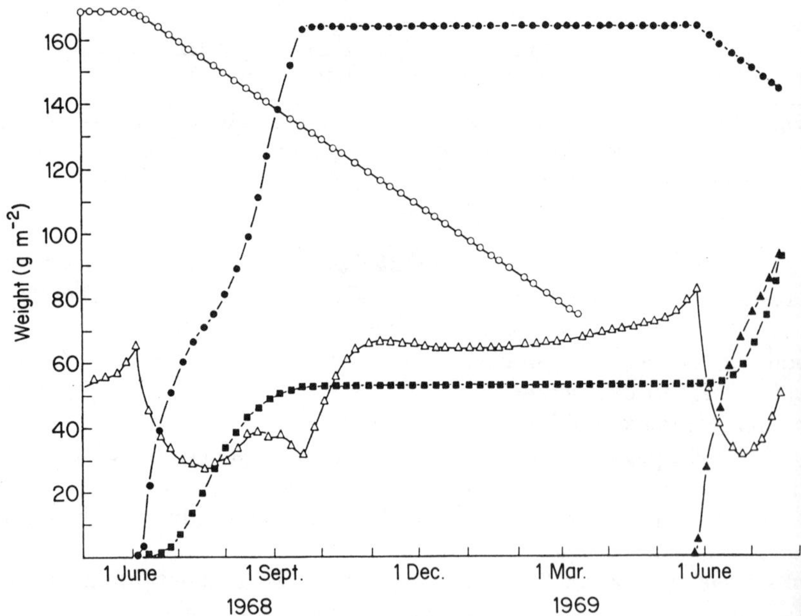

Fig. 7. The seasonal pattern of shoot and wood production and content of soluble sugars in *Calluna vulgaris*, generated from a computer model for the climatic responses and partitioning of assimilates in *Calluna*. ○, 1967 leaves; ●, 1968 leaves; ▲, 1969 leaves; ■, wood; △, sugar. From Grace and Woolhouse (1974).

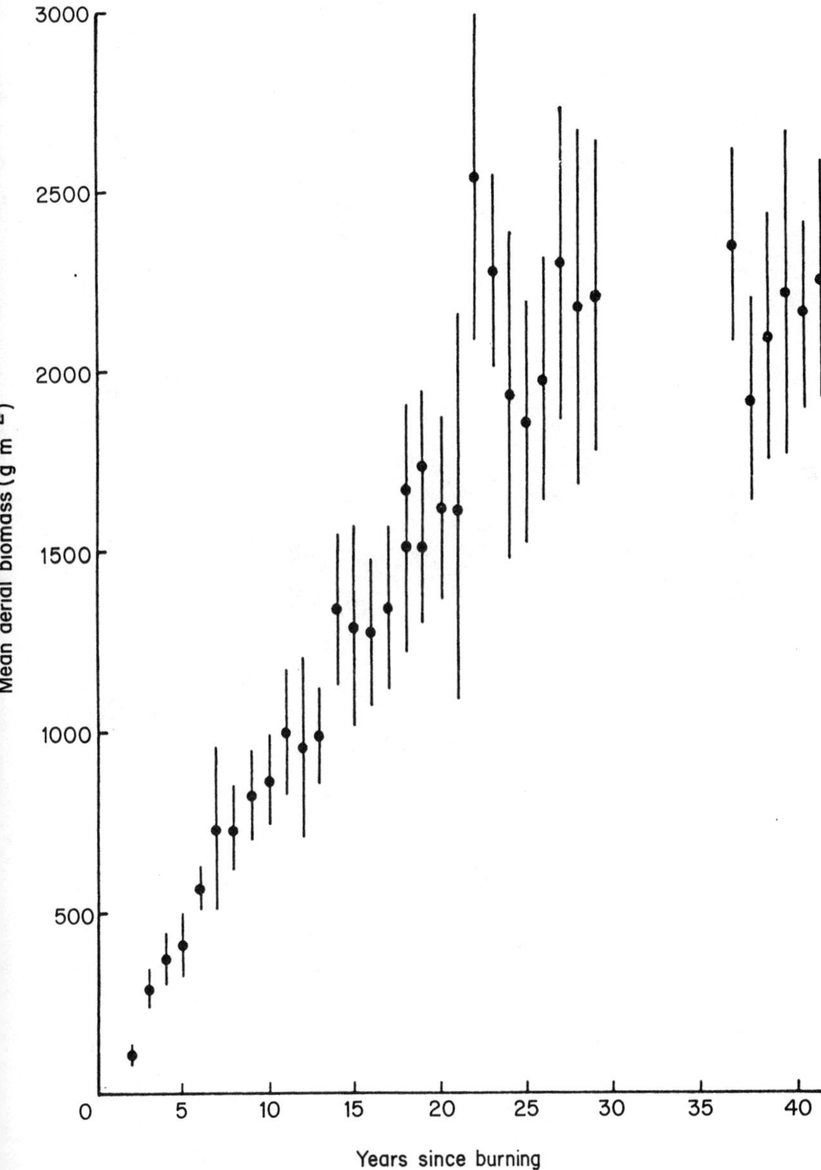

ig. 8. Increase in the aerial biomass of *Calluna* in relation to time since burning. Vertical lines are 95% confidence limits. From Miller (1979).

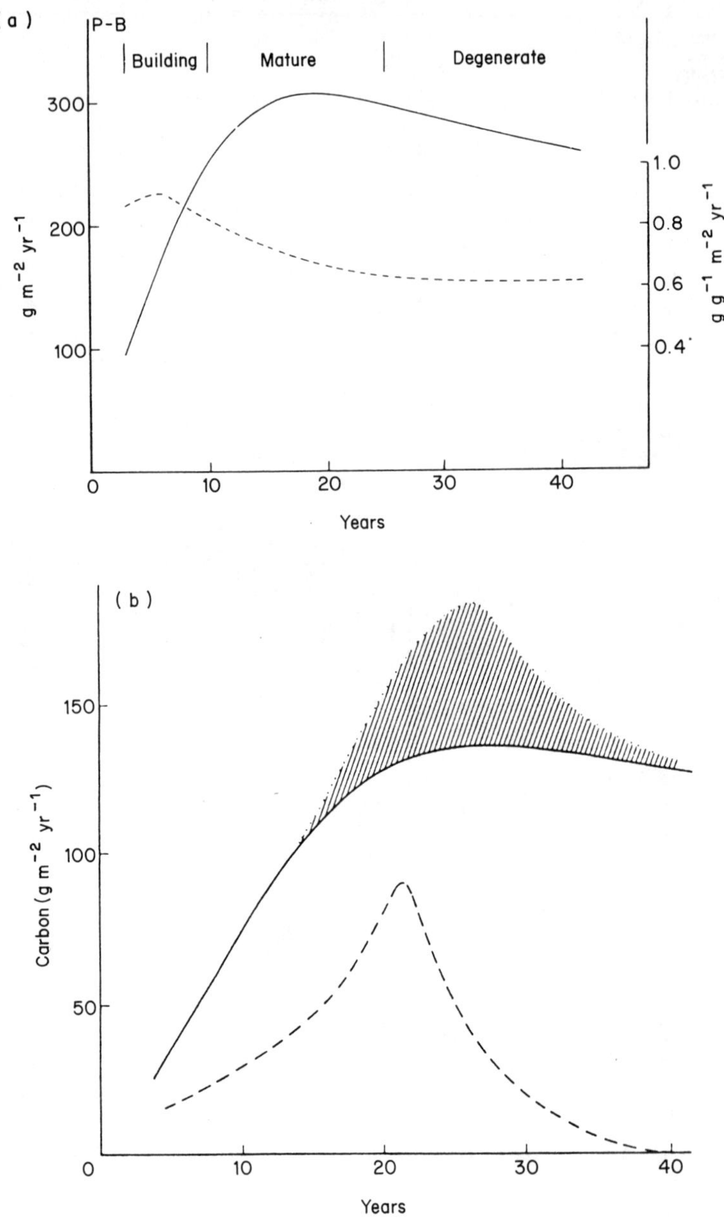

Fig. 9. Aboveground production and litterfall of dry heath in Dorset in relation to age of heathland. (a) ——, Net aboveground production; ----, relative production. (b) -----, Rate of accumulation and ——, rate of input of carbon to the litter. The dotted line (. . .) and hatched area represents the input of carbon by way of roots in the litter layer. From Chapman *et al*. (1975b) and Gimingham *et al*. (1979).

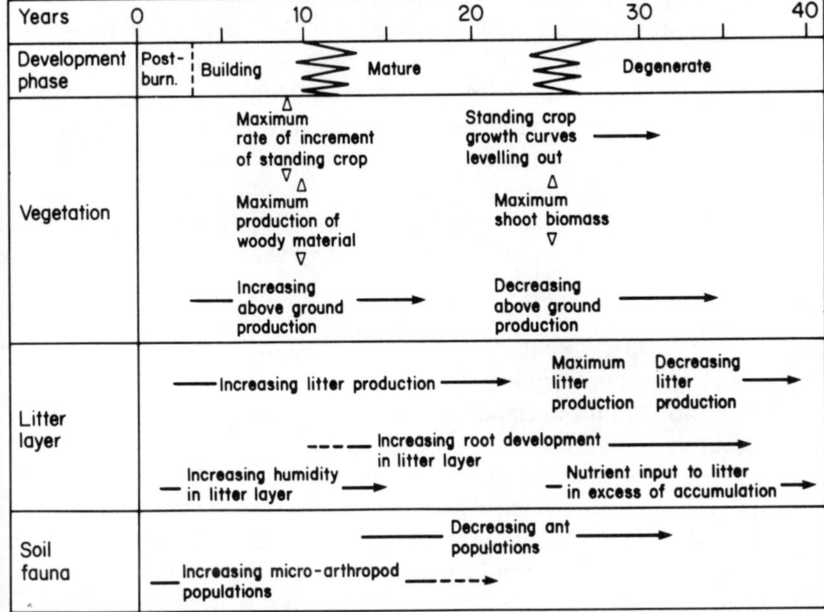

Fig. 10. A summary of the relationship between production and accumulation and the phases of development of Dorset heathlands. From Chapman and Webb (1978).

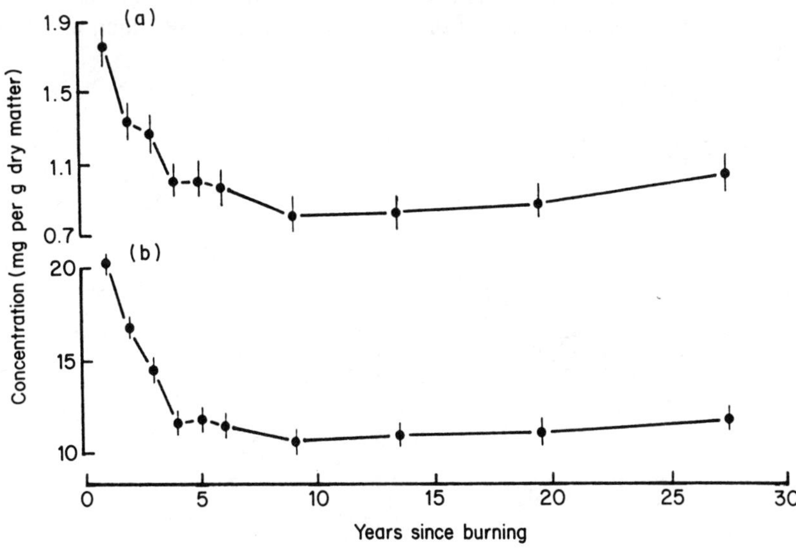

Fig. 11. Mean nutrient contents of 1cm tips of current year's shoots of *Calluna* in relation to time since burning; (a) phosphorous, (b) nitrogen. Vertical lines are 95% confidence limits. From Miller (1979).

Generally the total amounts of N and P increase up to about 15 years but level off thereafter. Miller (1979) gives values of approximately 8·0 g m^{-2} for N and 0·6 g m^{-2} for P at Kerloch Moor in N.E. Scotland. Other nutrients increased up to 25 years old (levels reached: c. 4·0 g m^{-2} Ca, 5·0 g m^{-2} K and 1·4 g m^{-2} Mg), although the increases were not proportional to dry matter accumulation due to the increased proportion of wood in older stands. Nutrient contents of current year's shoots (the part of the plant most usually consumed by herbivores) are generally highest immediately after burning, and decline rapidly reaching a steady level after 4 years (Fig. 11).

2. *Heaths in Steady-State*

The assumption that a steady-state exists in some heath systems alters the emphasis of production studies from following trends with stand development to trends within the stand. An uneven-age structure means that the above-ground population is composed of stems at a variety of developmental stages. These stems may be clonal, and it is not clear what constitutes an "individual" from a physiological point of view. Estimating production on an area basis will average the responses of individual stems. This has been done by Forrest (1971), Forrest and Smith (1975) and Wallén (1980), and an annual dry matter budget for *Calluna* on blanket bog is given in Fig. 12. This stand had an estimated total *Calluna* biomass of approximately 1100 g m^{-2} and an above-ground production of 183 g m^{-2} yr^{-1} (Forrest, 1971). However, if the stem population is divided into its different ages, the relative contribution of each age class to the overall biomass and production can be

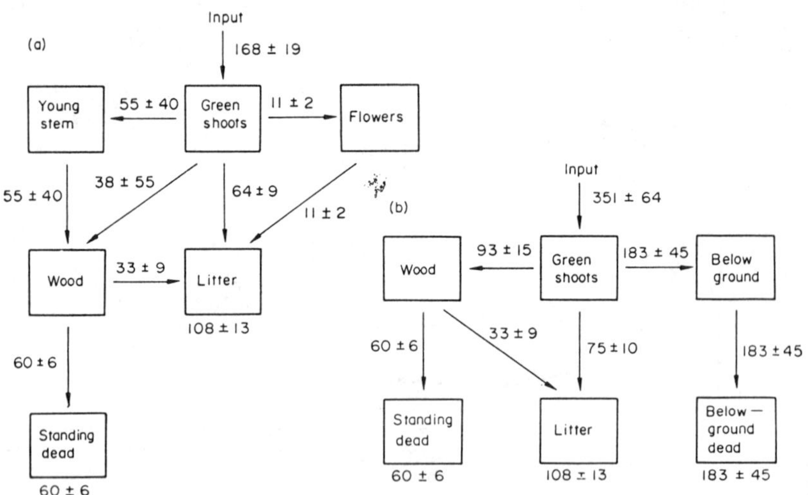

Fig. 12. Flow diagrams for dry matter in *Calluna* on blanket bog. Transfer rates are given in g m^{-2} yr^{-1}, with 95% confidence limits. An important assumption is that the ratio of biomass/production below ground is equal to that above ground. (a) Above ground transfers; (b) Total transfers. From Forrest (1971).

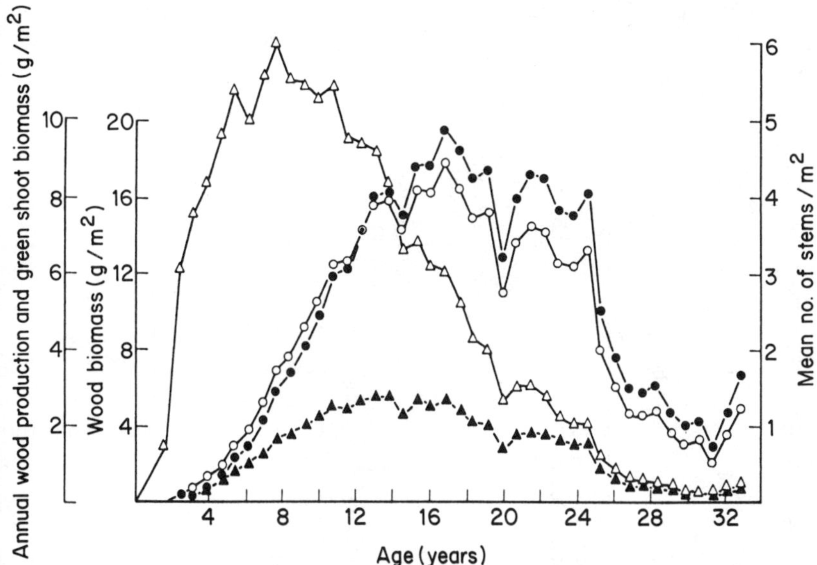

Fig. 13. Distribution of *Calluna* parameters between age classes of stems at ground level on blanket bog; △, mean number of stems m^{-2}; ○, green shoot biomass; ●, woody biomass; ▲, annual wood production. From Forrest (1971).

assessed (Fig. 13). The most productive section of the population was in the range 10–18 years old. Interestingly, this compares well with the estimated period of maximum production in dry heaths (Fig. 9b), indicating that, although the overall dynamics of the two systems may be quite different, the same basic processes are operating in both.

In wet heaths, peat accumulation and the growth of bogs are important aspects of the system, especially since peat provides a readily available fuel in the wetter parts of N.W. Britain. Accumulation of peat depends on the relative rates of production, decomposition and compression. Models which simulate peat bog growth have been formulated by Clymo (1978) and Jones and Gore (1978).

C. Community Dynamics

1. *Seasonal Dynamics*

Factors affecting the phenology of *Calluna* and other ericaceous heath species have been discussed by Woolhouse and Kwolek (1981), and Table 1 gives a summary of the phenology of these species. The physiological responses of *Calluna* have been intensively studied by Grace and Woolhouse (1970, 1973a, b, 1974), and it is known that dormancy occurs below about [illegible]°C, with growth recommencing above a similar temperature (7·2°C in

Table 1

A summary of some of the main phenophases in a selection of dominant species of European heathlands. From Woolhouse and Kwolek (1981).

Species	New shoot production	Commencement of main flowering periods	Main periods of root production	Dormancy
Calluna vulgaris	March–May	late August	unknown	definite period of dormancy approximately from October to mid-February
Erica tetralix	April–May	June	unknown	dormancy phenomena uncertain (*Erica tetralix*, *Erica cinerea*, *Vaccinium vitis-idaea*)
Erica cinerea	April–May	July	unknown	
Vaccinium vitis-idaea	March–April	two flowering periods per season; (a) spring (b) early summer (*Vaccinium vitis-idaea*, *Vaccinium myrtillus*)	rhizome growth and branching in both;	
Vaccinium myrtillus	March–April		(a) spring (b) autumn	dormant in winter (deciduous)
Empetrum nigrum	April		unknown	probably has a period of winter dormancy

Grace and Woolhouse's studies) and increasing with increasing temperatures. Figure 7 summarises the results of their studies in the form of the output from their computer model. Further information on the seasonal activity of *Calluna* in terms of carbon translocation has been provided by Wallén (1983). Information is also available on the phenology and physiological responses of other species found in the blanket bog community: e.g. for *Rubus chamaemorus* (Grace and Marks, 1978), *Eriophorum vaginatum* (Robertson and Woolhouse, 1984a, b), and *Sphagnum capillifolium* (Daggitt, 1981). Unfortunately, there is little phenological information for dry heath communities, although Barclay-Estrup and Gimingham (1975) investigated the timing of seed release in *Calluna*, and recent studies by Reader (1984) and Reader *et al.* (1983) provided data on flowering periods and timing of shoot growth for a spectrum of species in heaths in N.E. Scotland. Figure 14 shows the flowering periods for a range of these species, and it is interesting to note marked separation of flowering times over the relatively short growing season. For instance, the ericaceous species present at both

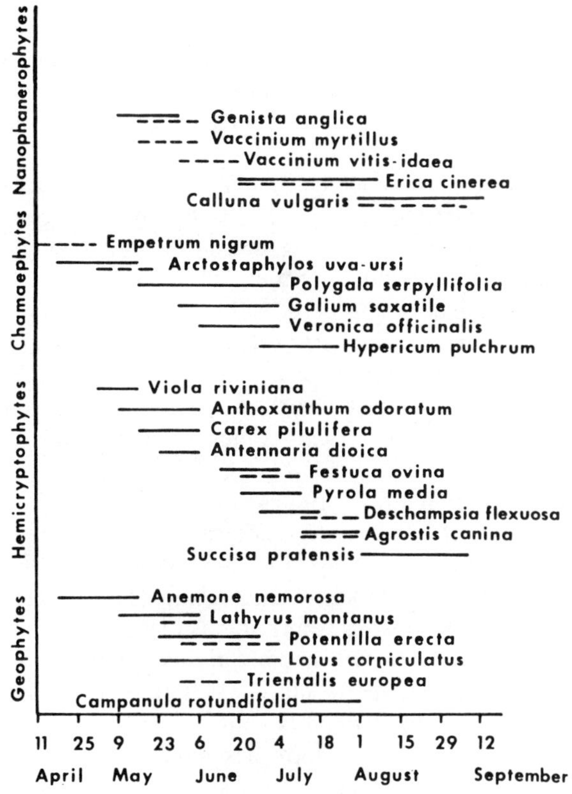

Fig. 14. Flowering periods for species present at two dry heathland sites in N.E. Scotland. Muir of Dinnet, ——; Balmenach, – – – – –. From Reader (1984).

sites had almost mutually exclusive flowering times. This suggests the possibility of temporal separation of flowering times to prevent competition for pollinators, although there is no evidence at present for this hypothesis. The pollination ecology of heathland species is a subject which has received little attention and is an important area for future research.

Reader *et al.* (1983) found that the timing of shoot extension for the same suite of species differed markedly between life forms. Species whose renewal buds were near the ground surface started growth earlier in spring than those with renewal buds above the surface. Geophytes exhibited the most rapid early growth in spring and ceased growth earliest (mid-June), with chamaephytes and hemicryptophytes continuing growth for another few weeks. The phanerophytes continued growth until mid-August. Such patterns of growth within the community are probably related to the positioning of growing points with respect to temperature gradients at different times of year, although again there is no direct evidence for this.

2. *Dynamics in Relation to* Calluna *Development*

All aspects of the dynamics of heathland vegetation are bound very closely to the management regime imposed on the vegetation. Hence a full discussion of community dynamics must consider the effects of grazing, fire and other disturbances. These factors are considered in more detail in subsequent sections. In this section we consider mainly the responses of heathland species to changes in environmental conditions occurring during the development of the *Calluna* stand, as a background for later discussions. Such changes occur either in single-aged stands brought about by management or in uneven-aged stands, but the scale of patterning is different in each. In the uneven-aged stands described by Barclay-Estrup (1970, 1971) and Barclay-Estrup and Gimingham (1969) changes in species composition take place as individual *Calluna* bushes pass through the various phases, with a relatively small-scale patchwork of species abundance patterns. In even-aged stands, changes in species composition are likely to be in synchrony over the whole stand.

Since *Calluna* is the dominant species in many heathlands, the ability of other species to co-exist with *Calluna* is determined to a large extent by their response to the *Calluna* growth cycle (Gimingham, 1975, 1978). Barclay-Estrup and Gimingham (1969) and Gimingham (1978) have shown that species diversities are highest in young stands of *Calluna*, decrease in the building and mature phases (when the *Calluna* canopy is densest), and increase again in the degenerate phase. Most species reach their peak abundances at times when *Calluna* is young and the canopy has not yet closed, or when it is old and the canopy of senescing bushes opens up again, Gimingham (1972) recognised a number of species groups which responded in different ways to the development of *Calluna*:

(a) Competitors. Species having a general similarity in size, form, structure

and development to *Calluna* itself, existing alongside it throughout the sequence of growth-phases: e.g., *Vaccinium myrtillus*.

(b) Species with "complementary strategies". (i) Subordinates – shade-tolerant species forming lower strata beneath a *Calluna* canopy, present throughout the sequence of growth-phases: e.g., *Erica cinerea, Hypnum jutlandicum*. (ii) Species requiring open phases (e.g. pioneer or degenerate *Calluna*) for establishment, capable of surviving throughout the sequence but with reduced vigour in the building and mature phases: e.g., *Festuca ovina*. (iii) Species surviving throughout the whole sequence in one spot by dying back each winter to an underground perennating organ, from which climbing or straggling shoots are produced every year, bringing foliage to the periphery of the *Calluna* canopy: e.g., *Potentilla erecta*. (iv) Species requiring open phases (e.g. pioneer or degenerate) for establishment, but spreading as the *Calluna* canopy develops by means of rhizomes or creeping stems which extend laterally into new gaps: e.g., *Vaccinium vitis-idaea, Arctostaphylos uva-ursi, Pteridium aquilinum*.. (v) Small, low-growing species which appear only in the open phases (pioneer or degenerate, or both): e.g., *Polygala serpyllifolia*, certain lichens and bryophytes.

(iv) and (v) are the "strategies" particularly associated with the gap phases; the remainder to a greater or lesser degree confer ability for co-existence with *Calluna* throughout the whole sequence of phases.

Coppins and Shimwell (1971) produced similar groupings for cryptogams found in Yorkshire heath.

Figure 15 illustrates the general course of species composition during *Calluna* stand development for a species-rich heath in N.E. Scotland, derived from a series of even-aged stands in managed heath. The short period of abundance of non-ericaceous species early in the sequence is highlighted. These are mostly species in groups b(ii) and b(v) above and are quickly replaced by species from groups b(i) and b(iv). Development of the *Calluna* canopy is slower in this species-rich heath than in pure *Calluna* stands where vegetative regrowth is rapid, but the effect on subordinate species of canopy closure in the building phase is clear. The collapse of the *Calluna* canopy in the degenerate phase is accompanied by an increase in gap-phase species.

3. *Species Interactions*

(a) Competition. Miles (1981a) remarked that

> probably no other process is so widely invoked yet so poorly understood under field conditions as competition.

Nevertheless, it is clear that the interactions between *Calluna* and its associated species discussed above must in part be explained by *Calluna*'s competitive ability. Thus most grass and forb species are able to survive only where *Calluna* growth is least vigorous and where its canopy is not complete.

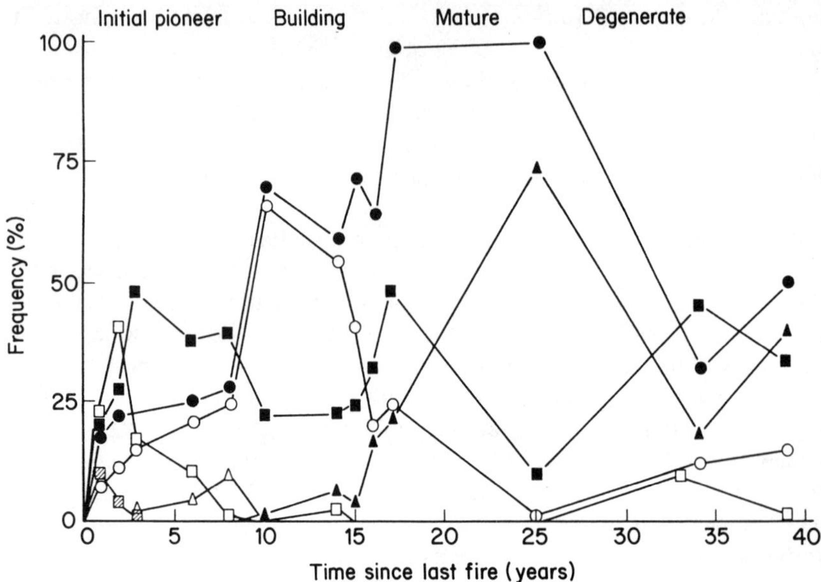

Fig. 15. Relationship between stand age and composition of major species categories in species-rich heath in N.E. Scotland. Percentage frequency of dominance or co-dominance by each category in 128 subplots (each 10 × 10 cm) in each stand. ○, *Arctostaphylos uva-ursi*; ●, *Calluna vulgaris*; ■, *Erica cinerea*; ▨, forbs; □, grasses; △, lichens; ▲, pleurocarpous mosses. From Hobbs *et al.* (1984).

The production of a dense canopy during the building phase reduces the amount of light reaching lower canopy levels (Fig. 2), and presumably *Calluna*'s extensive root system effectively reduces nutrient supplies. However, little is known about the actual mechanisms of competition between *Calluna* and its subordinates.

Nevertheless, the work of Bannister (1964 a–d, 1976) and Kashimura (1985) has elucidated some of the mechanisms affecting the interactions between *Calluna* and other ericaceous species. Extensive physiological studies have revealed the means by which *Erica cinerea* and *E. tetralix* can tolerate drier and wetter conditions respectively than *Calluna*. This allows these species to compete successfully with *Calluna* on the margins of its range (Fig. 16). While all three species have physiological optima in the same range of soil moistures, *Calluna* is the more aggressive competitor and occupies the optimum range to the exclusion of the others. Gimingham (1972) has suggested that another species, *Empetrum nigrum*, may have two distinct ecotypes adapted to wet and dry conditions respectively.

Studies on competition between other heath species have indicated complex relationships between soil water and nutrient availability. Berendse and Aerts (1984) have suggested that increasing nutrient availability or improved soil aeration leads to *Molinia caerulea* being able to crowd out

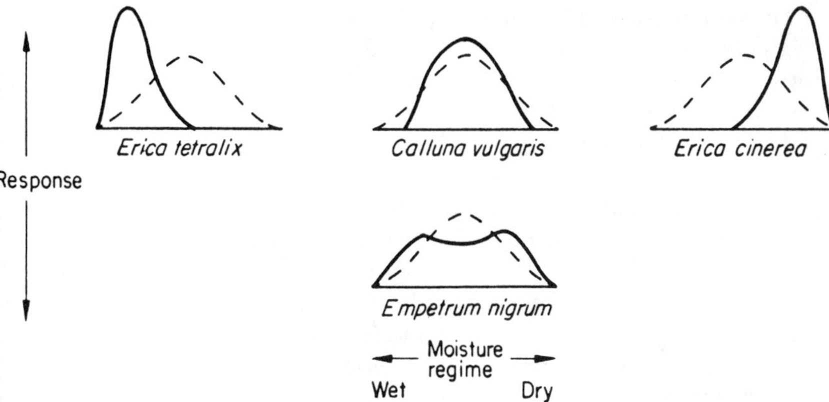

Fig. 16. A diagrammatic comparison of ecological, ——; and physiological, -----, amplitude for four heath shrubs with respect to soil moisture. From Bannister (1976).

Erica tetralix. Fertilisation experiments in Holland indicated that increased nutrient supplies may similarly allow *Festuca ovina* to compete successfully with *Calluna* (Heil and Diemont, 1983).

Competition may often be most important in the establishment phase, and Hobbs and Gimingham (1984b) have shown that the speed with which *Calluna* regenerates after fire may determine whether *Vaccinium myrtillus* is able to compete successfully with *Calluna* and become the canopy dominant (see also Section III, Fig. 20).

(b) Allelopathy. The possibility that chemical interactions could be important in heath vegetation was first suggested by observations that extracts from bushes and humus of *Calluna* could inhibit the growth of forest trees such as *Pinus sylvestris* (Braathe, 1950; Pearman, 1959). In Britain, during reafforestation of upland heaths dominated by *Calluna*, it was noted that growth of several species including the much-planted *Picea sitchensis* was considerably reduced or "checked" where *Calluna* growth was vigorous. This was thought to be due to the release of a fungitoxin from living mycorrhizal roots of *Calluna* which inhibited the growth of the tree's mycorrhizae and thus severely affected its mineral nutrition (Handley, 1963; Robinson, 1971, 1972). Other studies have provided evidence suggesting an inhibitory effect of *Calluna* on the germination or growth of grass and herb species (Roff, 1964; Mantilla *et al*., 1975; Read and Jalal, 1980), although Gong (1976) found no inhibitory effects of *Calluna* on *Betula* spp.

Evidence for the production of substances toxic to other plants has been presented for several other species common in heathland; e.g. *Erica* spp. (Deleuil, 1950; Ballester *et al*., 1977, and papers quoted therein), *Vaccinium myrtillus* (Trepp, 1961), *Arctostaphylos uva-ursi* (Winter, 1961), *Deschampsia flexuosa* (Jarvis, 1964) and *Pteridium aquilinum* (Torkildsen, 1950; Gliessman, 1976; Glass, 1976).

Hobbs (1984a) conducted a series of laboratory and greenhouse experiments on co-occurring heath species. Generalised bioassays indicated that most species tested had some inhibitory effect on growth or germination of *Avena fatua*, but this effect was reduced when leachates from intact plants rather than from plant fragments were used. Few inhibitory effects on the germination of *Calluna* or *Deschampsia flexuosa* were found, although growth of *D. flexuosa* plants was strongly inhibited by leachates from ericaceous species. *Arctostaphylos uva-ursi* produced the greatest number of inhibitory effects but also severely inhibited its own growth. Similar studies on lichens occurring in heathland have indicated that *Cladonia* spp. may inhibit germination of heath species and thus maintain lichen mats within otherwise closed heath stands (Hobbs, 1985). Several other studies have shown that *Cladonia* spp. may inhibit higher plant growth, either directly or indirectly through inhibition of associated mycorrhizae (Brown and Mikola, 1974; Ramaut and Corvisier, 1975; Veinstein and Tolpysheva, 1975).

Despite the findings of these studies, the importance of chemical interactions in the field is unclear. Almost all the investigations were carried out under simplified laboratory conditions which did not take into account the complexity of the soil environment. Nevertheless, studies by Ballester *et al.* (1972) and Jalal and Read (1983a, b) have found potential phytotoxins in heath soils in sufficient amounts to cause growth inhibition in higher plants, although activity of phytotoxins may vary seasonally (Carballeira and Cuervo, 1980; Jalal and Read, 1983b). Further rigorous field experiments are needed to separate possible allelopathic effects from competition for light and/or nutrients.

4. *Development from Heath to Woodland*

The possibility that development to birch or pine woodland was originally part of a natural long-term cycle has been discussed in Section II-1. It is certainly possible to observe areas in Scotland where heathland is rapidly turning into birch or pine woodland, especially on relatively rich soil. Tree species will invade heathland only in cleared areas (e.g. after burning) or where gaps in the *Calluna* canopy exist – i.e. in the pioneer or degenerate phases. Legg (1978) has shown experimentally that germination and seedling survival of *Pinus sylvestris* is much greater in recently burned heath than in mature *Calluna* stands. Similarly, Gong (1976) found that large numbers of *Betula* seedlings can establish in recently burned areas, with little further recruitment after *Calluna* canopy closure. Where *Calluna* regrowth is rapid after fire, most *Betula* seedlings will not survive, but slower *Calluna* regrowth may permit rapid development of birch stands.

Several studies have indicated that *Calluna* has an acidifying effect on the soil and that the typically acid podzolised soils formed under today's heathlands are the result of the maintenance of a *Calluna* "monoculture"

(Dimbleby, 1962; Grubb *et al.*, 1969; McVean and Lockie, 1969; Grubb and Suter, 1971). Miles (1974a) suggests that many of the plant species present under former woodland conditions may have disappeared, partly because of these *Calluna*-induced changes in the soil. Miles (1974b, 1975) has, in fact, shown experimentally that such species cannot re-establish in heath stands.

Recent studies have, however, indicated that the establishment of *Betula* in heathland may have equally profound effects on soil characteristics as those of *Calluna*. After birch establishes, pH, exchangeable Ca, rates of nitrogen mineralisation and earthworm numbers increase. Associated with these soil changes are changes in the ground flora, with *Calluna* gradually being replaced by *Vaccinium myrtillus*, grasses and *Pteridium aquilinum* (Fig. 17) (Miles and Young, 1980; Miles 1981b). Miles (1981a, b) also

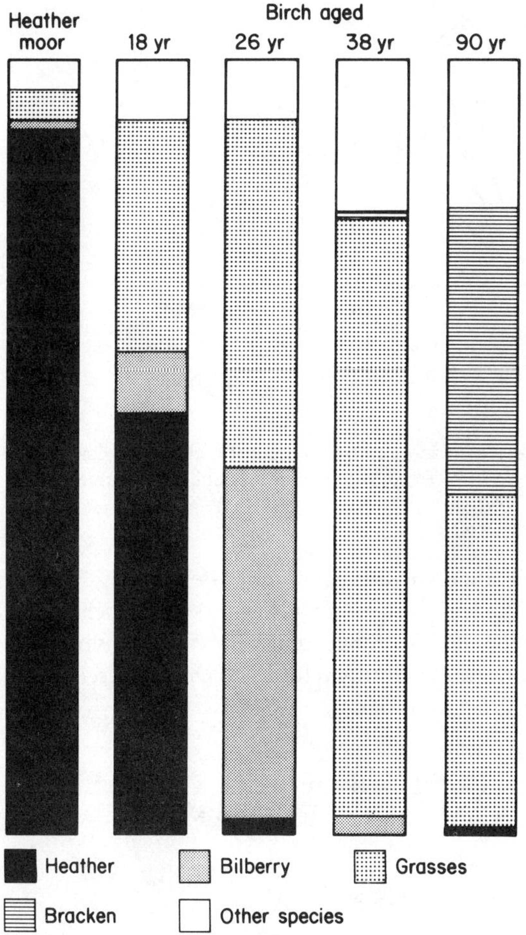

Fig. 17. Mean percentage cover of different species in heather moorland in N.E. Scotland and under adjacent birch stands of different ages. From Miles (1981b).

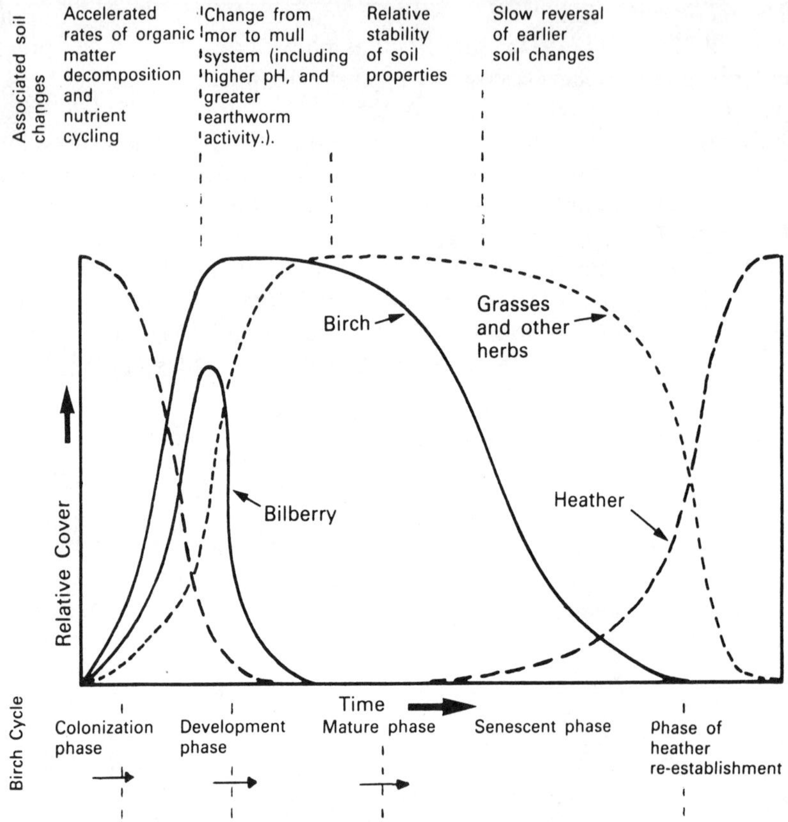

Fig. 18. Generalized sequence of vegetation changes occurring during the life cycl of a birch wood in N.E. Scotland, with associated changes in labile soil properties From Miles (1981b).

suggests that soil changes found under developing *Betula* stands tend to b reversed as the stands open up with tree senescence and death, indicating cyclical trend (Fig. 18). Such trends fit with the hypothesis that heathland were originally not an entity in themselves, but part of a constantly changin patchwork of communities.

III. EFFECTS OF FIRE ON HABITAT AND VEGETATION

A. Fire Characteristics

Fire is an important tool in the management of heathlands, especially i Britain. One of the major applications of fire is to interrupt the ageing process of *Calluna*, thus keeping it in a young, productive and nutritious state and increasing potential herbivore production (Thomas, 1934, 1937; Thomas and Dougall, 1947; Miller and Miles, 1969; Miller 1979). Fire is also used to prevent the transition from heathland to scrub or woodland and probably had a significant role in shaping the heathland landscape. For instance, Whittow (1977) concludes that:

> Today's systematic moor-burning to improve grazing value merely perpetuates the wanton forest and scrub destruction of earlier centuries which, more than anything, is responsible for the emergence of the Scottish Highlands as a landscape of moorlands rather than forests.

In the rest of N.W. Europe, fire has been used less frequently as a management tool, and instead cutting of vegetation or sods was formerly carried out (Gimingham and de Smidt, 1984). In some areas fire did, however, form part of an integrated agricultural system and was used to release nutrients locked up in dead organic matter.

The amount of care with which fire management systems are implemented varies greatly from place to place. In N.W. Scotland, for instance, often little attempt is made to burn moors systematically, some areas being burned very frequently and others not at all, depending on proximity to habitation. In comparison, in the drier north-east of Scotland fire management, especially for the red grouse [*Lagopus lagopus scoticus* L. (Lath.)], may reach considerable levels of sophistication. In addition to planned management fires, however, accidental fires also occur and management fires may go out of control. Such accidental fires may have effects quite different from the normal management fires because of their potentially increased severity and, usually, larger size.

Here we consider first the characteristics of management fires and then compare these with accidental fires. Most of the information on heath fire behaviour comes from N.E. Scotland where an intensive fire management system for grouse moors has been developed. Much of the existing information on heath fires and their effects has been made available to land managers by the Muirburn Working Party (1977). Fires are restricted to patches of limited size, which are burned on a rotation such that, ideally, all parts of an area of heathland are burned once in the course of a given period of years. The actual length of rotation may be linked to the rate of recovery of the vegetation: i.e. if *Calluna* is ready to burn at 10 years old (a rule of thumb is that *Calluna* can be burned when it reaches 20–30 cm tall), 1/10th of the total area would be burned each year. Burning is best carried out in small patches or narrow strips rather than in large blocks, especially for grouse management (See Section IV), and fires of half to one hectare in 25–30 m wide strips are common. In practice it may be impossible to adhere to this ideal system since on large estates it may require the setting of over 1000 fires per season, a task which may be beyond the capabilities of the manpower available.

Burning is allowed in Scotland only between 1 October and 15 April, with a possible extension period to 30 April (and to 15 May at high elevations). Although this is over 6 months, many factors can limit the actual number of days available for burning. Firstly, snow may lie for several months during winter and at other times the vegetation may often be too wet to ignite. If conditions are dry enough for burning, wind conditions must be right to allow for adequate control and to permit burning on to firebreaks. Often the correct combination of conditions occurs on only a few days in each season (Miller, 1980). Even then, some managers are reluctant to burn certain areas because of their juxtaposition to forests or inhabited areas. Burning very seldom takes place in winter and is usually restricted to autumn and spring. Burning must also be carried out during daylight, and days are relatively short during the burning season.

Fire behaviour in heath fires in N.E. Scotland has been studied by Whittaker (1961) using heat-sensitive paints, and Kenworthy (1963), Kayll (1966) and Hobbs and Gimingham (1984a) using thermocouples. Average maximum temperatures recorded during fires varied greatly between these studies, mainly because of differences in recording procedure. Kayll's (1966) data suggested that heath fires could be very cool, reaching only 250°C in the canopy, although these fires may have been atypical because of the wet conditions under which they were carried out. Whittaker (1961) recorded temperatures ranging from 220°C to 840°C in the canopy. The maximum temperature recorded by Kenworthy (1963) was 940°C, while Hobbs and Gimingham (1984a) recorded maximum temperatures ranging from 340°C to 790°C in the canopy and 140–840°C at the ground surface. These results indicate a great amount of variability both between and within fires. Few recordings have been made of soil temperatures during heath fires, but data from Hobbs and Gimingham (1984a) indicate that little heat is transferred beneath the soil surface. The highest temperature recorded 1 cm beneath the surface was 70°C.

Fire intensities have also been calculated by Kayll (1966) and Hobbs and Gimingham (1984a). As for fire temperatures, considerable variation was found between fires: e.g., Hobbs and Gimingham (1984a) estimated intensities of 43–1112 kW m^{-1} for fires at one site, the variation being mostly based on quantities of fuel available and rates of spread.

A number of factors may influence the behaviour of a heath fire.

1. *Windspeed*

This may affect rates of spread, intensity and effectiveness of the fire. There is some debate over the effect of windspeed on fire temperatures. Whittaker (1961) suggested that increased windspeed increases fire temperatures by fanning the fire, while Hobbs and Gimingham (1984a) found that higher windspeeds may reduce temperatures by causing the fire to travel faster and consume less fuel. Certainly in strong wind conditions often only the crowns of shrubs are ignited leaving much of the woody material unconsumed.

2. *Vegetation*

The amount of plant material present, its composition, distribution, energy content and moisture content may all be important in determining fire behaviour. In general, fires in older stands of *Calluna* may be expected to be hotter because of the great abundance of fuel present (Kenworthy, 1963; Hobbs and Gimingham, 1984a). However, Hobbs and Gimingham (1984a) suggest that fuel distribution is also important and fires in dense uniform stands of mature *Calluna* may be hotter than fires in more patchy degenerate stands. Thus, unlike Kenworthy (1963), Hobbs and Gimingham (1984a) did not find a direct relationship between stand age and fire temperature: instead the highest temperatures and longest durations of high temperatures were found in building or mature stands (Fig. 19). Temperatures were related to vegetation structure, and the variable most strongly correlated with maximum temperatures reached was vegetation height. Temperatures were also much more variable from place to place in older stands, reflecting the patchiness of fuel loading in degenerate stands (Hobbs, Currall and Gimingham, 1984). Stand structure also affected rate of spread of fire, the fire front moving more rapidly through uniform stands than through patchy degenerate stands. Fire intensities, being a function of fuel load and rate of spread, were correspondingly lower in the oldest stands.

The moisture content of the vegetation may also affect fire temperatures, although Hobbs and Gimingham (1984a) found that the effect was secondary to that of stand structure. The main reason for this seems to be that since the fires studied were management fires, they took place within a fairly limited range of moisture conditions. Wetter conditions would not allow the vegetation to burn, while burning would be avoided under drier conditions to prevent the possibility of the fire escaping. Nevertheless, variations in moisture content between different vegetation components may be important in causing variations in fire severity from place to place. In particular, variations in ground cover may cause marked differences in the amount of heat transferred beneath the soil surface: e.g., mats of mosses such as *Hypnum jutlandicum* may contain about 5 times as much moisture as *Calluna* litter (Hobbs and Gimingham, 1984a). Thus soil heating may be more rapid under litter than moss since temperatures in the upper soil layers cannot exceed 100°C until all the water is evaporated off. The insulating effects of any ground cover will depend on its moisture content at time of burning, which will in turn depend on the weather conditions prior to burning.

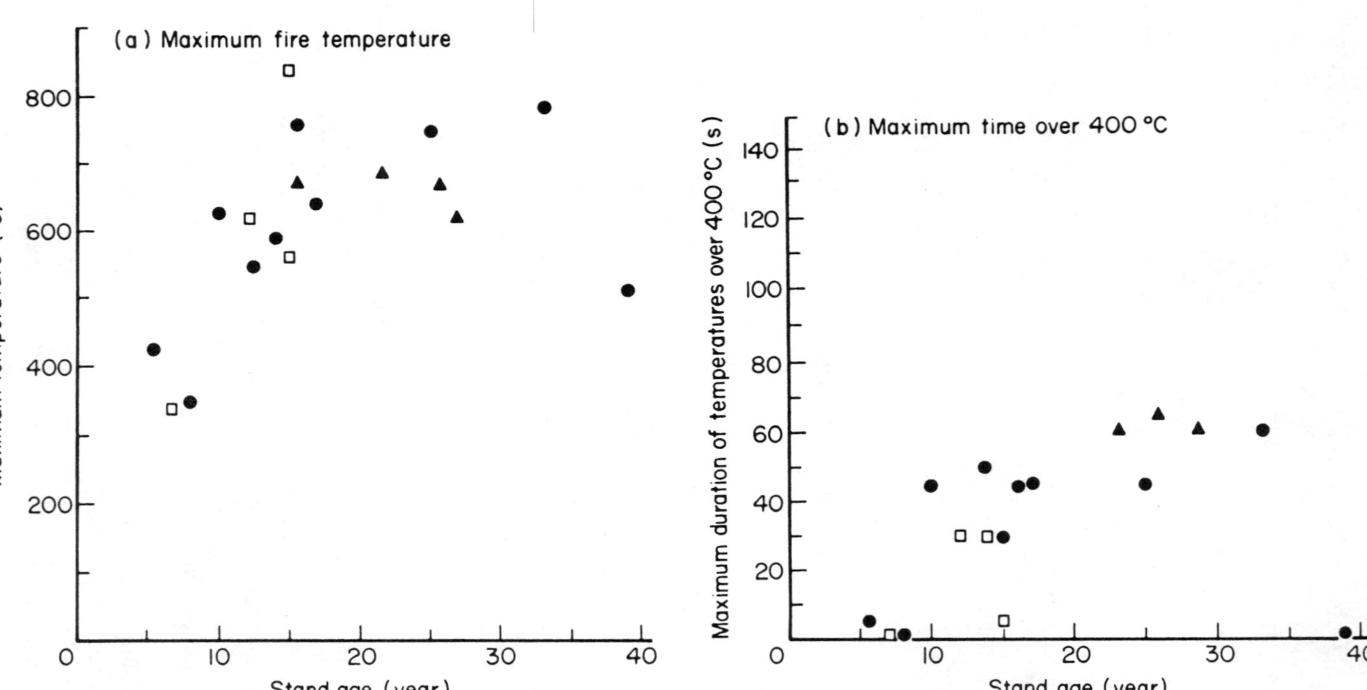

Fig. 19. Stand age in relation to (a) maximum fire temperature in each fire in Scottish heathland and (b) maximum time over 400°C in each fire. Symbols represent different sites; ●, Muir of Dinnet; □, Glensaugh; ▲, Balmenach. From Hobbs and Gimingham (1984a).

3. *Burning Technique*

Most fires are burned “with the wind”, i.e., letting the fire be carried by the wind. However, in some cases burning against the wind (back-burning) is used, and this produces a very slow-moving intense fire. This may be used for burning old *Calluna*, on wet ground or for producing clean fire breaks. Differences in temperatures and intensities reached by fires burned with and against the wind have not been documented. A further aspect of technique is burning either up- or down-slope. Burning up-slope may be expected to produce a more rapid and intense fire than burning down-slope because the fire is constantly burning up through the vegetation (Hobbs, 1981). Fire size may also influence fire severity, wider fire fronts producing more intense fires in the centre (Hobbs and Gimingham, 1984a). However, only correlation data are available for this and a causal relationship between size and intensity has yet to be demonstrated.

4. *Season of Burn*

The season may affect fire intensity primarily through its effect on fuel moisture contents: e.g., vegetation may be wetter in autumn than in spring burns. Accidental fires in summer may also be much more severe for the same reason. Again, however, there are no data available on seasonal variation in fire severity, although Kayll’s (1966) data do suggest that autumn fires might be hotter.

Little information is available on temperatures reached in accidental fires, due to the difficulties involved in preparing for such fires. However, Kayll (1966) calculated an intensity of 2430 kW m^{-1} for an experimental fire that went out of control. This figure is considerably higher than any of the intensities recorded for controlled fires by Hobbs and Gimingham (1984a). It should be noted that all studies of heath fires have considered only management fires. Such fires take place within relatively narrow limits of weather conditions, vegetation moisture etc. It is likely that factors determining fire severity in uncontrolled or accidental fires may be quite different from those in controlled management fires. A further important feature of severe uncontrolled fires is that they may ignite the humus or peat layer as well as the vegetation. Such fires may burn for long periods and cause considerable damage by removing the humus layer entirely and leading to erosion and poor vegetation regeneration (Radley, 1965; Maltby, 1980).

B. Ecosystem Effects

1. *Nutrients*

When a heath is burned a large proportion of the nutrients contained in the standing vegetation is mobilised as ash and smoke. The fate of these mineralised nutrients is important in terms of the long-term nutrient budget of the heathland system. Burning has two major effects: firstly, nutrients may be lost in smoke, both in particulate matter and, especially for N, by volatilisation, and secondly, nutrients are returned to the system in ash deposition. Nutrients are generally in a more available form in ash than in the unburned litter, and there is thus possibly a fertilising effect. However, greater mobility also leads to loss by leaching.

Marked increases in the losses of nutrients in smoke are found with increasing fire temperatures (Kenworthy, 1964; Evans and Allen, 1971). Losses of N and S especially are heavy at higher temperatures, reaching 57% and 36% respectively at 750°C in Evans and Allen's (1971) study. Substantial amounts of Na, K, Fe, Zn and Cu were also volatilised in both field and lab experiments. Allen (1964) showed that intense fires produced significantly greater losses of several elements: for instance, losses of N and P were 4–5 times greater in a fire of about 800°C than in one of around 600°C. On the other hand losses of volatile compounds (N, C and S) were only slightly higher in the hotter fire, and Kenworthy (1964) indicated significant losses of N at temperatures above 300–400°C. Roze and Forgeard (1982) found significant losses of N and C in burnt heathland in Brittany, and also estimated that net mineralisation in the soil was reduced in the burnt areas. Not all nutrients contained in smoke are lost: Allen (1964) and Evans and Allen (1971) were able to show that some fractions of the smoke condensed out not far from the fire, although they were not able to measure accurately the amounts deposited.

Nutrients not exported in smoke are deposited on the soil surface as ash. Allen (1964) and Allen *et al.* (1969) have shown that these nutrients, especially K, are readily dissolved from the ash, but that most of the dissolved nutrients are retained in the surface organic layer. Only in sandy soils did noticeable downward movement of nutrients occur. Hansen (1969) has also shown that burning may double the amounts of nutrients such as K and Mg in the surface soil, but that the fertilising effect of burning is relatively short-lived and nutrient levels return to pre-fire levels by 2 years after the fire.

A further source of nutrient loss is soil erosion by both wind and wate from the burned area, which Imeson (1971) and Kinako and Giminghar (1980) have shown may take place until vegetation cover is replaced and ma be more severe on steeply sloping ground. Kinako and Gimingham (1980 estimated that the amounts of nutrients lost through erosion in the first 1– years after fire may take a considerable time to be replaced: e.g., Na, Ca an Mg might be restored after 7 years, N and K within 14–27 years, but P onl after 67–83 years. They point out, however, that such erosion may represen only a local loss, with material being deposited in adjacent areas and th burnt area possibly regaining material from other sources in subsequen years. The same may be true for nutrients in smoke: provided they do not enter higher airstreams, most may be deposited elsewhere on a moor. However, such redistribution may not occur in very large fires where large areas are burned simultaneously: e.g. Maltby (1980) discusses a large fire which caused huge losses of surface soil material and nutrients. Material lost through erosion of upper areas of steep slopes may also not be replaced, with fires at the top of slopes leading to potentially great losses of soil in runoff. Losses from smoke and erosion may also be a problem where heathlands have been fragmented, as in S. England. Burning of one patch may then not lead to redistribution of nutrients to another patch but rather to an overall reduction in nutrient status.

Attempts have been made to produce nutrient budgets for heathlands subjected to burning so that any potentially harmful effects of fire could be recognised and De Jong and Klinkhamer (1983) have recently tried to model the effects of fire on nutrient cycling. An early study by Elliot (1953) suggested that fires progressively depleted the nutrient fund, but further studies have indicated that losses of most elements during fires (with the exception of P) may be made good by inputs in rainfall during the inter-fire years (Allen, 1964; Robertson and Davies, 1965; Chapman, 1967). However, all studies concerned with nutrient budgets have considered only low intensity fires (the temperature in the canopy not exceeding 400°C, a recommended limit for management fires), and it is possible that hotter fires could produce considerably greater losses of nutrients which may not be replaced entirely. It is likely that many (or most) heath fires reach temperatures above 400°C and it is perhaps necessary for a reconsideration of nutrient budgets using known mean fire temperatures. In addition, all studies indicated that losses of nitrogen and phosphorus might not be replaced in rainfall, even after low-temperature fires. Thus although there are often large stores of unavailable N in heath systems, some depletion of available N may occur. The activity of the soil microbial population in respect of N availability after fire has been little studied, although Maltby (1980) found an increase in numbers of bacteria on peat after fire, which might lead to a corresponding increase in humification.

Most studies also considered a fire rotation of 10 years or more in estimating inputs from rainfall. This is clearly inappropriate for areas where more frequent burning occurs: for instance, in N.W. Scotland where burning may be almost annual and losses through leaching are potentially high due to the high rainfall. We suggest, therefore, that further study of the nutrient balance of heath systems is required, in particular to assess the differences between extensive well-managed moors where there may be nutrient transfer between burned and adjacent areas, and more isolated areas where progressive nutrient loss may occur. The importance of N depletion also deserves attention.

2. *Microclimate and Soil Properties*

Although studies by Delany (1953) and Barclay-Estrup (1971) have provided information on microclimatic characteristics of heath stands, few studies have investigated the post–fire microenvironment. Kinako (1975) and Mallik (1982) have, however, provided information on certain aspects of this. The main effect of fire is to remove some or all of the above-ground vegetation, leaving an area of exposed soil or humus, usually blackened or ash-covered. The removal of the vegetation leads to greater fluctuations in temperature and moisture regimes, with increased temperatures and vapour pressure deficits at the soil surface during summer (Mallik, 1982). These increased temperatures may have the dual effect of stimulating vegetative growth (and in some cases flowering) of regenerating plants and of killing young seedlings which may also suffer from drought. Despite this, Mallik (1982) concluded that moor burning tends to conserve water in the habitat by reducing evapotranspiration and rainfall interception by the canopy and by increasing the water-holding capacity of the surface soil. Mallik, Gimingham and Rahman (1984) have shown that moisture retention in the topsoil of burned areas is increased considerably, thus reducing infiltration by up to 74%, probably as a result of the incorporation of ash particles in the surface soil. How long this phenomenon persists is not known. Mallik and Rahman (1985) have also shown that the water repellancy of surface soil increases slightly after burning but has declined to lower levels by 1 year after the fire.

One feature of the post-fire microhabitat is its heterogeneity, especially if the fire leaves some of the ground layer either unburnt or partially burned. This then leaves patches of pleurocarpous mosses and litter of varying depths, all of which dry out at different rates and represent seedbeds of varying suitability for germinating seeds (Mallik, Hobbs and Legg, 1984). The effectiveness of the fire in removing above-ground vegetation also affects the post-fire microhabitat: for instance, if large amounts of unburned woody material remain after a cool fire, the temperature regime may not be altered appreciably from that before the fire. Severe fires, on the other hand may completely alter the microhabitat by destroying or irreversibly changing the surface layers of the soil (e.g., Maltby, 1980). In such cases seed store may also be destroyed and recolonisation may be very slow and initially restricted to bryophytes: examples where this has occurred may be found in Brittany (Clément *et al.*, 1980; Clément and Touffet, 1981) and the North York Moors in England (Maltby and Legg, 1983).

C. Post-Fire Vegetation Development

Recent European studies of vegetation development in heathlands after fire have included those by Hansen (1964, 1976), Bøcher and Jørgensen (1972),

Forgeard and Touffet (1980) and Hobbs and Gimingham (1984b). In dry heath dominated by *Calluna* a generalised picture of succession following fire can be developed from these and earlier studies. The burnt ground is initially colonised by algae, acrocarpous mosses and lichens. A well defined set of mosses is known to colonise after fire, including *Ceratodon purpureus, Polytrichum juniperinum* and *P. piliferum*. The earliest lichens to colonise are usually *Lecidea granulosa* and *L. uliginosa*, followed by a variety of *Cladonia* species. Often, several species of herbs and grasses are present shortly after fire, including such species as *Deschampsia flexuosa, Agrostis canina* and *Veronica officinalis*. These then decline in abundance as the major ericaceous species become dominant. At low elevations and on relatively dry soils, *Erica cinerea* may become dominant for several years before *Calluna* regains dominance. Associated with the increase in ericaceous cover may be an increase in the cover of mat mosses such as *Hypnum jutlandicum* and *Pleurozium schreberi*.

Mallik and Gimingham (1983) and Reader *et al.* (1983) have attempted to explain the phases observed following fire in terms of the response of the species present, grouping species into different "regenerative strategies" based on life form and speed of shoot development.

The general sequence of development may be simpler or more complicated depending on the overall richness of the vegetation. However, within this general trend great variety of post-fire response is found, both in the rate of regeneration and the composition of the regrowing vegetation. Thus, in well managed dry heathland, the succession after fire is very short, and *Calluna* recovers dominance after only 1–2 growing seasons. This leaves very little time for any other species to establish, although there may be a temporary "grassy" phase (e.g., Fenton, 1949). If, however, regeneration of *Calluna* is delayed, the full sequence of "phases" described above might be observed before the *Calluna* canopy closes. Alternatively, the general direction of the vegetation development may be altered so that, for instance, what was a *Calluna* stand before fire becomes dominated by *Pteridium aquilinum* or *Vaccinium* spp. after fire (e.g., Tansley, 1939; Fenton, 1949).

Here we examine in more detail some of the variations found in post-fire response in a variety of community types and discuss the factors which may cause such variation. These factors include various attributes of the community burnt, fire severity, weather patterns in the post-fire period and the activities of herbivores.

1. *Community Composition and Species Life Histories*

A major determinant of community responses to fire is the pre-fire composition of the stand. The aggregate community response is made up of the responses of individual species present. The simplest case is, perhaps, that of managed dry heath where *Calluna* is the major species present. Here,

Calluna will usually regenerate quickly, soon covering the burned ground to the exclusion of other species. However, the rate of recovery by *Calluna* is variable, depending mainly on whether regeneration is mostly vegetative or from seed. Vegetative regeneration consists of the growth of shoots derived from axillary buds, usually at the stem base. However, the buds remain viable for a limited time only since eventually they become covered by secondary thickening of the stem, and after about 15 years old *Calluna* shrubs begin to lose the ability to regenerate vegetatively (Mohamed and Gimingham, 1970). Field studies by Kayll and Gimingham (1965) and Miller and Miles (1970) have confirmed that the capacity for vegetative regrowth declines with shrub age. A further factor in this decline is the reduction in density of stems as the stand ages (Section II) which reduces the number of potential regeneration sites present.

Where vegetative regeneration does occur, regrowth by *Calluna* is usually rapid, but where there is little vegetative regrowth regeneration must be from seed. *Calluna* produces abundant seed throughout its life (Mallik, Hobbs and Legg, 1984) and large amounts of seed are usually present in the soil. Seed germination may occur in the season following fire or may take longer (Hobbs and Gimingham, 1984b), and tends to be patchy because of different seed-bed characteristics. Germination and/or seedling survival are reduced on mats of pleurocarpous mosses and areas of deep litter (Mallik *et al.*, 1984). A recent study by Helsper and Klerken (1984) has indicated seed polymorphism in *Calluna*, with different types exhibiting differing germination responses. They also found a strong pH dependence for germination. Growth of seedlings is very slow compared with vegetative regrowth, and where both occur together the vegetative sprouts will usually overtop the seedlings. Where seedlings are the only form of regeneration, they will eventually grow out to form a *Calluna* canopy as with vegetative regrowth. The longer time to canopy closure may, however, allow other species (e.g. grasses and lichens) to become temporarily more abundant. The slower regrowth of *Calluna* may also result in other species attaining dominance and leading to a more long-term community change. For instance, Hobbs and Gimingham (1984b) have shown that where *Calluna* regrowth was slow and mostly by seed, *Vaccinium myrtillus* was able to increase in abundance and prevent *Calluna* seedling establishment (Fig. 20). The same may apply to other species if they are present in an area before fire and have the ability to spread vegetatively rapidly after the fire. In particular, *Pteridium aquilinum*, an important weedy species in heathland, has been known to invade recently burned sites where *Calluna* regrowth is slow, especially in stands which are allowed to become too old before being burned.

Stand age is thus important in determining the ability of *Calluna* to regenerate. It is also important in determining the response to fire of other species. In heaths rich in herb and grass species, the number of species

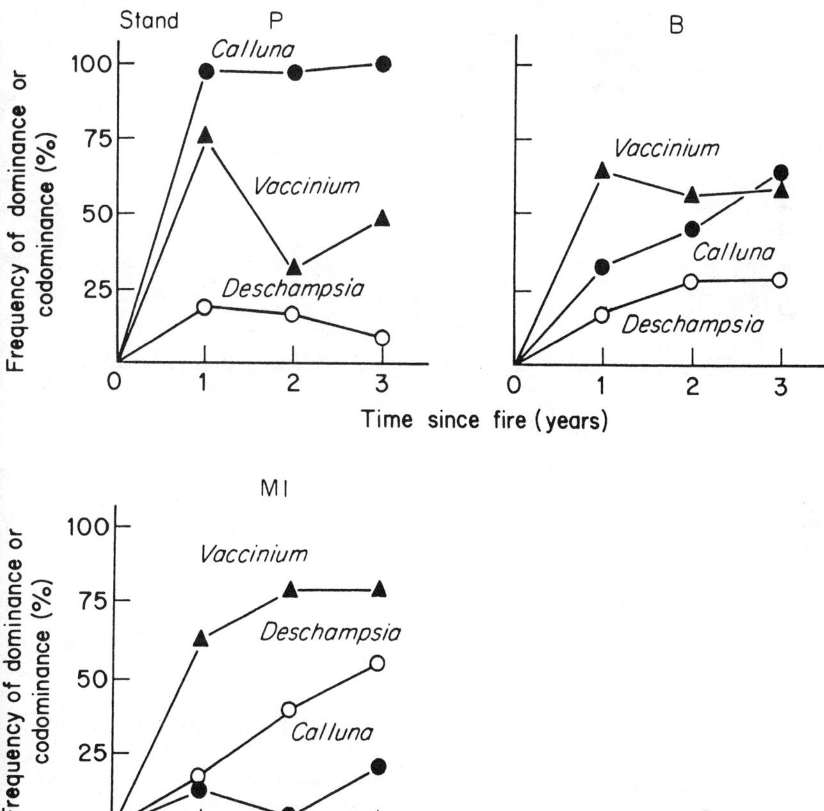

Fig. 20. Vegetation recovery in stands of species-poor heath of different ages before fire in Scotland burned in 1978. (Stand ages before fire are; P = 6 yr, B = 10 yr; M1 = 15 yr). Percentage frequency of dominance or codominance by: ●, *Calluna vulgaris*; ○, *Deschampsia flexuosa*; ▲, *Vaccinium myrtillus* in 128 subplots (each 10 × 10 cm) during 1978–80. From Hobbs and Gimingham (1984b).

present in post-fire stands declines with stage age before the fire (Fig. 21). Mallik and Gimingham (1983) and Hobbs, Mallik and Gimingham (1984) have suggested that community response to fire depends greatly on the spectrum of life history characteristics present in the community. The ability of a species to regenerate after a fire depends on the timing of the fire in relation to the timing of certain critical life history events (Fig. 22). Regeneration must be either vegetative or from seed. Several herbaceous species

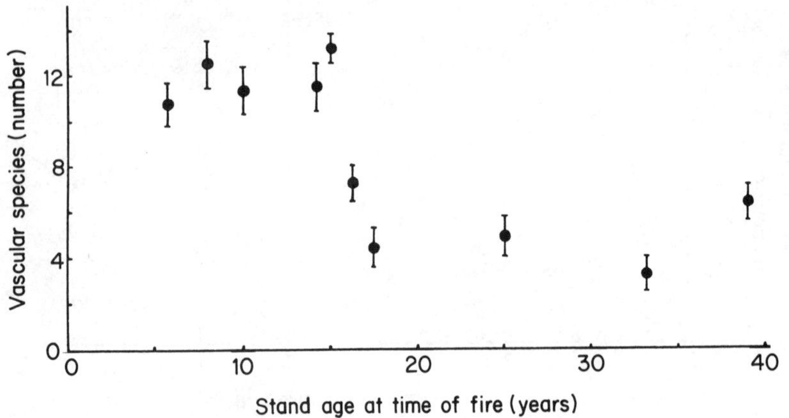

Fig. 21. Relationship between stand age at time of burning and number of vascular species present in stands of species-rich heath in Scotland one growing season after fire. Mean number in four 1 m^2 quadrats in each stand ($\pm$1.S.E.). From Hobbs *et al.* (1984).

Fig. 22. The timing of critical life history events for major species present in species-rich heath in Scotland. Symbols: o, time of disturbance; p, time at which propagules are available on site; m, time at which reproductive maturity is reached; l, local loss from the community; e, local extinction from the community. From Hobbs *et al.* (1984).

may lose the ability to regenerate vegetatively as the numbers of individuals decline when the *Calluna* canopy closes. Most species store seed in the soil, but seed of several species remains viable for relatively short periods (Mallik *et al.*, 1984). Stored seeds may also germinate under closed *Calluna* stands but subsequently fail to establish (de Hullu and Gimingham, 1984). Lack of further input to the seed store after numbers of plants decline means that there may be no viable seed present in the soil after a certain time. The species would then be locally "extinct" and would not regenerate following a fire at that time. Fig. 22 illustrates the timing of the loss of actual plants and of local extinction for species and species groups in relation to the *Calluna* cycle. A vertical line drawn through the diagram at any point would indicate which species were able to regenerate following a fire at that stage.

In wet heath, stand age affects the outcome of fire through its effect on community composition at the time of the fire. In *Calluna-Eriophiorum* bog, Hobbs (1984b) has shown that frequent burning shifts community dominance to *Eriophorum* spp. *Calluna* recovery following fire is relatively slow in this community type, and the community is dominated by *Eriophorum* spp. for several years after the fire (Fig. 23). A second fire relatively soon

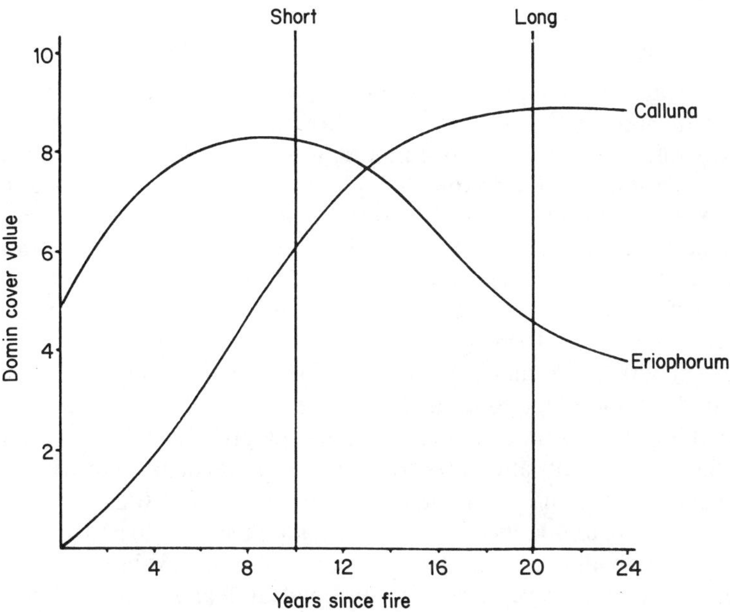

Fig. 23. Idealised representation of the post-fire development of *Calluna* and *Eriophorum vaginatum* showing the relative abundance of these species at time of burning short (10 yr) and long (20 yr) rotation experimental plots. From Hobbs (1984b).

after the first reinforces this *Eriophorum* dominance and prevents *Calluna* from re-establishing abundantly. A longer period between fires gives *Calluna* time to regain co-dominance and means that it is present in greater abundance with more potential regeneration centres. Similar effects have been found by Currall (1981) in *Molinia*-dominated communities, where very frequent burning all but eliminates *Calluna*.

Recent attempts to model the post-fire developments of heath vegetation have shown that reasonable predictions can be obtained from models using only vegetation parameters as input. Hobbs (1983) produced Markov models using data from a variety of heath types, but found that satisfactory predictions could be obtained only for the simpler heath systems. Lippe *et al.* (1985) also found that a Markov chain model was inappropriate for the dynamics of a Dutch *Empetrum nigrum* heath. However, Hobbs and Legg (1983), using a model originally developed by Legg (1978, 1980), were able to model differences in post-fire response in species-rich heath simply using variations in the composition of the regrowing vegetation. These modelling studies supported the idea that much of the variation observed in post-fire vegetation development could be explained by intrinsic community characteristics alone.

2. *Fire Severity*

It is to be expected that the response of vegetation to a fire may depend at least in part on the severity of the fire. Important in this respect is the temperature (and its duration) experienced by regeneration buds and seeds stored in the soil. Exceptionally high temperatures may be lethal to both seeds and buds, but temperatures lower than lethal may either impair or enhance regeneration. The temperatures experienced by buds and seeds depend on their positioning and degree of insulation. Temperatures at the soil surface may be very high during a fire but may remain much lower just below the surface. Below-ground buds on rhizomes or other storage structures may therefore not experience greatly elevated temperatures. Above ground, on the other hand, buds must be protected in some way, e.g., by sheathing leaf bases or by woody or bark material. Mallik and Gimingham (1985) have recently attempted to assess the effects of fire severity on a range of heath species, using treatments of 400, 600 and 800°C. Of the species studied, only *Juniperus communis* was completely sensitive to fire, failing to regenerate in all treatments because of its unprotected above-ground buds. Regeneration of several species declined at higher temperatures (e.g., *Vaccinium myrtillus, Empetrum nigrum*), while others regrew more vigorously in the high temperature treatment (*Succisa pratensis, Deschampsia flexuosa, Festuca ovina*).

Whittaker and Gimingham (1962) found that seeds of *Calluna* were killed by exposure to 200°C, but that below this level the effects depended on the period of exposure. Short exposure to temperatures of 40–160°C stimulated germination while longer exposures reduced germination. Mallik and Gimingham (1985) also showed complex interactions between temperature, length of exposure and pre- or post-treatment in the germination response of a number of heath species. For instance, seeds of *Vaccinium vitis-idaea* showed very low germination at all time/temperature combinations (50, 75 and 100°C for 0·5, 1 and 2 minutes) except at 100°C for 0·5 min where 95% germination occurred. Several species showed some stimulation of germination with heat treatment (*Erica tetralix, Genista anglica, Hypericum pulchrum*).

Clearly fire severity must therefore affect community response through its differential effects on individual species. However, Hobbs and Gimingham (1984a) have suggested that fire severity, at least in management fires, is linked to stand characteristics that themselves have an influence on the "regeneration potential" of the stand. In effect, in N.E. Scotland the most severe fires are to be expected in stands with the lowest potential for regeneration: i.e., stands in which few species can regenerate from seed or buds anyway. Despite this, fire severity must be considered in cases of anomalous fires, especially accidental summer fires; for instance Clément and Touffet (1981) have illustrated the effect of a disastrous summer fire in Brittany after which the normal regeneration by ericaceous species did not take place and mosses such as *Polytrichum juniperinum* became dominant instead.

3. *Weather Following Fire*

Little information is available on the influence of weather patterns following burning on the regeneration pattern. There must, however, be some effect due to the great variation in timing of, for instance, freezing temperatures (Reader *et al.*, 1983) and rainfall patterns. This is especially true for seedling regeneration, and seedlings may be particularly susceptible to drought in their first or second years. Mallik and Gimingham (1985) have shown that considerable summer mortality due to drought occurs in ericaceous seedlings. The extent of this mortality on different substrates will depend on the rate of drying of the substrate and length of time between rainfall periods. Grant (1968) and Grant and Hunter (1968) have reported differences in regeneration of *Calluna, Erica cinerea* and *Deschampsia flexuosa* following burns in different years which may be attributable to differences in weather. Hobbs and Gimingham (1984b) and G. R. Miller (pers. comm.) also suggested that germination of ericaceous species may vary greatly from year to year.

4. *Herbivory*

Effects of herbivores on heath vegetation are discussed more fully in Section V, but here we examine possible interactions between burning and grazing. Effects of grazing depend firstly on the herbivore involved and secondly on the grazing intensity. Grant (1968, 1971) and Grant and Hunter (1968) considered the effect of sheep on burned heather and concluded that grazing intensity may affect post-fire regeneration by influencing the rate at which *Calluna* closes canopy. Heavy grazing keeps *Calluna* short and may allow grasses such as *Deschampsia flexuosa* to persist for longer. Gimingham (1949) has also shown a strong effect of grazing pressure by sheep on the interaction between *Calluna* and *Erica cinerea* regenerating after fire (Fig. 24). In the absence of grazing, seedlings of *Erica cinerea* were much more

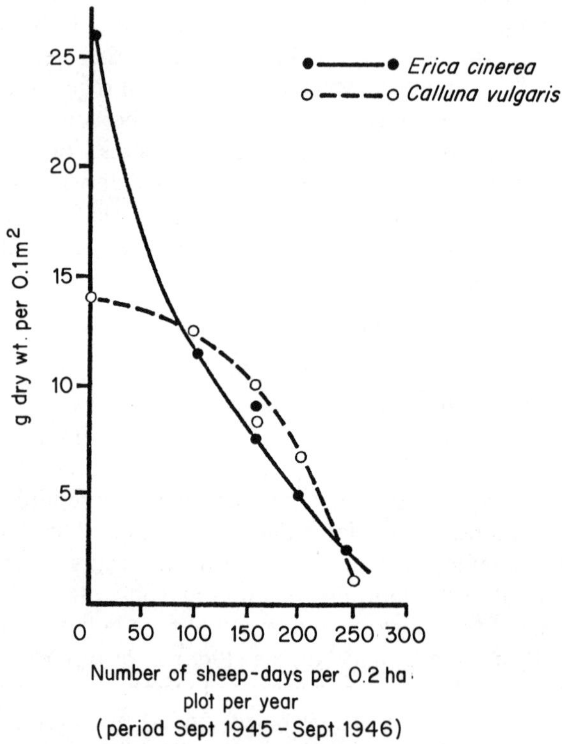

Fig. 24. Effect of the intensity of sheep-grazing on the balance between *Callun[a] vulgaris* and *Erica cinerea*, regenerating after fire. Plots were burned in sprin[g] ungrazed for one and a half years, then grazed by sheep at the intensities shown fo[r] one year before sampling. From Gimingham (1949).

abundant, but with increasing grazing pressure, *Calluna* increased in abundance. This experiment indicated a greater tolerance to grazing by *Calluna* primarily because of its ability to alter morphology from erect to plagiotropic shoots when grazed (see also Grant and Hunter, 1966; Mohamed and Gimingham, 1970). Genetically plagiotropic forms of *Calluna* are also found at high elevations (Grant and Hunter, 1962).

There is little information on the effects of grazing on regeneration of other heathland species. Selective grazing of individual species, removal of regenerating shoots and prevention of flowering may all have an effect on the post-fire regeneration. Herbivores may also concentrate their grazing on recently burned areas.

Recent studies have shown that rabbits (*Oryctolagus cuniculus* L.) can prevent heath species such as *Erica cinerea* from invading burned grasslands in Brittany by reducing its growth, flowering and seed set (Forgeard and Chapuis, 1984). Similar effects may be important in the recolonisation of burned heathland, but have yet to be examined.

The effects of seed predation by small mammals, ants and other invertebrates may also be important, but have been little studied in N.W. European heathlands (e.g., Miles 1974a, b). Studies by Brian *et al.* (1965) and Brian *et al.* (1967) have, however, shown that the ant *Tetramorium caespitum* Latreille collects and stores large quantities of seeds of *Calluna* and *Erica cinerea*. Whether this predation is significant in view of the huge seed output of these species remains to be investigated.

IV. HERBIVORE DYNAMICS AND INTERACTIONS WITH VEGETATION

In this section we examine the factors thought to control the abundance of native herbivores on heathland and investigate the interactions between herbivores and vegetation. Fire also necessarily enters the discussion since it has a major influence on vegetation structure and composition. The fauna of N.W. European heathlands has been discussed by Gimingham *et al.* (1979). Here we discuss only the major herbivores, which include 3 native vertebrates (the red grouse *Lagopus lagopus scoticus*, the red deer *Cervus elaphus* and the mountain hare *Lepus timidus* L.), the hill sheep and many invertebrate species. For the most part our discussion is restricted to northern Britain where intensive studies of these animals have been carried out.

A. Red Grouse

The red grouse is thought to have been an inhabitant of open areas within forests and above the treeline before forest clearance took place (McVean and Lockie, 1969). With deforestation the bird spread its range to the open moorland, and its survival there has perhaps been assured by its value as a game-bird. Its importance to land managers increased from about 1850 when income from hill sheep began to decline and sport provided an alternative source of income.

1. *Food Supply*

Grouse are territorial and territories are taken by males in early winter and are defended until late spring. Birds are restricted entirely to their own territories in late winter and spring when food supplies are least abundant. Territory sizes vary greatly (from 0·2 to 13·2 ha) but are mostly 2–5 ha (Watson and Miller, 1971). Detailed studies of grouse populations in N.E. Scotland indicated that population size was not determined by the observed levels of shooting or predation because of the presence of surplus birds to take over vacated territories (Jenkins *et al.*, 1967).

There is a broad correlation between the abundance of *Calluna* in an area and the abundance of grouse. Young shoots of *Calluna* form the major component of the grouse diet (Jenkins *et al.*, 1963) although they will also eat fruits of species such as *Vaccinium myrtillus* when these are available (Wilcock *et al.*, 1984), possibly forming search images for abundant fruit types.

Grouse, even at maximum densities, eat only about 5% of the annual production of dry matter by *Calluna*, and so food should never be in short supply (Savory, 1974, 1978; Miller and Watson, 1978a). However, in late winter and early spring grouse show a strong preference for young *Calluna* shoots (Moss, 1969) and it was hypothesised that the quality of diet in early spring might influence breeding success. The quantity of material from which the birds select their diet may be affected by several factors, i.e., the growth of *Calluna* in the previous summer, which affects the amount of green material present in winter; the amount of "browning" or dessication damage of *Calluna* over winter, which reduces the amount of green shoots in spring, and the timing of plant growth in the spring which also affects the quality of available material (Miller *et al.*, 1966; Watson *et al.*, 1966; Jenkins *et al.*, 1967). Miller *et al.* (1966) found that breeding success was correlated with *Calluna* growth and die-back during the preceding year.

Studies on food selection by grouse indicated that the birds preferred shoot-tips from young (but not 1-year-old) *Calluna* which has a higher N and P content than older material (Moss *et al.*, 1972). The correlation between preference and N and P contents was present only when captive birds were growing or laying eggs. Miller (1968) also found that grouse select areas fertilised with N, and Moss (1977) suggested that grouse have to consume more than they require for energy intake to obtain a sufficient quantity of N and possibly P. Field studies showed that differences in mean breeding success and mean density of grouse among different moors were usually related to differences in N and P content of the *Calluna* in spring (Moss *et al.* 1975). Such differences correspond to differences between "rich" moors on base-rich rocks and "poor" moors over acidic rocks, as discussed by Jenkins *et al.* (1967), Picozzi (1968) and Moss (1969). However, the relationship between "richness" of moor and grouse breeding success was not thought to be causal.

Moss *et al.* (1975) found that variations in breeding success from year to year were correlated with the number of days that the *Calluna* had been growing before the hens finished laying and with the standing crop of the previous summer's *Calluna* growth: so maternal nutrition may affect breeding success. Miller and Watson (1978a) thus suggested that spring is a critical time since parental food supply may determine egg quality and subsequent chick survival. Grouse may have relatively large territories, but even so it is common for hens to fail to rear any young or to raise only small broods. Because of their strongly selective feeding habits, large territories may be necessary for them to obtain sufficient plant material of adequate nutritive value. While grouse prefer feeding on pioneer heather (Miller and Watson, 1974), Miller (1968), Moss (1972) and Lance (1978a) have shown that even within this type they may be able to select for individual plants or plant parts. Such selection makes the definition of "adequate food resources" difficult.

Territory size has been shown to be related both to vegetation structure (Miller and Watson, 1978b) and to an index of the amount of N in heather shoot tips per unit area (higher N reduces territory size) (Lance, 1978b). Miller *et al.* (1970), Watson and O'Hare (1979a) and Watson *et al.* (1977) showed that fertilising areas of heathland or bog could increase grouse numbers. Further fertilisation experiments by Watson, Moss and Parr (1984) showed that adding N increased grouse numbers and breeding success while the grouse on the rest of the moor were declining and breeding poorly, but not while they were increasing and breeding well. Increased numbers were related to territory size, fertilisation reducing the size presumably through increasing the amount of N per unit area. Where densities were declining rapidly, numbers on the fertilised areas declined equally rapidly. Natural fluctuations in the grouse population on the whole moor during the course of these experiments had a greater amplitude than those due to fertilisation. These natural fluctuations could not be explained by changes in food quantity or quality.

2. *Effects of Management*

Management practices may significantly alter the heathland habitat and thus affect grouse populations. In particular, fire has an important influence. Moor burning for sheep management was widespread in the 18th and 19th centuries, but the practice declined after about 1850. At about the same time grouse stocks declined dramatically, and Lovat (1911) concluded that good stocks of grouse could be maintained only if the grouse moors were burned regularly. Numerous authors have since repeated this conclusion (e.g. Wallace, 1917; Robertson, 1957; Miller, 1964, 1980; Watson and Miller, 1976; Muirburn Working Party, 1977).

Regular burning of the moors benefits the red grouse since the practice aims to maximise the amount of young shoots of *Calluna* present. The N and P content of *Calluna* shoots is maximal in 1–3 year old *Calluna*, and there is evidence that grouse stocks are correlated with the amount of 2-year-old *Calluna* present (Miller, 1980). In addition to nutrition requirements, however, the birds also need cover and protected nest sites, and so a proportion of the moor must contain older, taller heather. Such older heather is also beneficial to sheep and deer, which graze on the taller plants in winter when snow covers other forms of food and renders it unavailable. Since grouse are territorial, however, there must be a suitable mixture of young and old heather present for suitable territories to be available. Thus, large open areas of young heather are unsuitable, and small patches of young heather interspersed with older patches provide the best conditions for grouse to live and breed. Picozzi (1968) was able to demonstrate that the best grouse stocks were found on moors with many small fire areas. Since grouse seldom venture further than 15 m from the cover of tall *Calluna* (G. R. Miller, pers. comm.), burned strips should be no more than 30 m wide to ensure maximum use of regenerating areas. A system of burning in small patches or strips has therefore become established. This system is best developed in the N.E. of Scotland, and a well managed moor has a characteristic "patchwork" appearance. In this system, a suitable proportion of the moor is burned in small strips every year to allow a complete burning rotation in a period ranging from 7 to 20 years, depending on local conditions. In this way a suitable balance is maintained between the various ages of *Calluna*, allowing potentially high densities of grouse. Such a system may not, of course, be beneficial where *Calluna* growth is not phasic such as in wetter heath: here frequent burning may shift dominance from *Calluna* to graminaceous species and hence reduce the moor's value for grouse (Rawes and Hobbs, 1979; Hobbs and Gimingham, 1980).

3. *Models of Population Changes*

Superimposed on variations in population numbers from place to place over short periods of time and because of management operations, are longer-term changes in populations. Long-term data are available on grouse populations through direct observations and in the form of shooting bags – i.e., records maintained by estates on the numbers of grouse shot each year and taken to be an index of grouse population size (Picozzi, 1968). Such data have been presented by Middleton (1934) and MacKenzie (1952) and analysed by Moran (1952) who concluded that grouse populations showed cyclical or oscillatory tendencies. More recently, Potts *et al.* (1984) have analysed grouse bags from N. England and concluded that any oscillations were so heavily damped that the data did not show clear cycles but what they termed "phase forgetting quasi-cycles" of average length 4·8 years. On the other hand, Watson *et al.* (1984) have presented a demographic model based on long-term population studies in N.E. Scotland which shows clear oscillations. Williams (1985) has also re-analysed early bag data using serial correlation and concluded that cyclic trends could be found in the majority of records (10 out of 14 recorded series), with six of these showing clear evidence of a 6-year cycle in numbers.

Based on research carried out in N.E. Scotland, as discussed previously, Watson and Moss (1972) proposed a model to explain population changes in grouse. This is illustrated in Fig. 25 and has been discussed in the light of more recent research by Moss and Watson (1980) and Watson and Moss (1980). The best general explanation for cyclic changes in red grouse densities is changes in spacing behaviour at different phases of the population cycle. This may be explained by the model shown in Fig. 25 in which changes in the quality of the hen's diet affect the eggs she lays. The proportion of chicks surviving from eggs laid by poorly-fed hens is low, but the cocks which survive from such broods are thought to be particularly aggressive when they mature and take large territories. This then causes an increase in mean territory size and hence a decline in density (Moss and Watson, 1980). However, further work, summarised by Moss and Watson (1980) and Watson and Moss (1980) suggested that this model did not operate. Instead Moss and Watson (1980) proposed that genetic differences in aggressiveness were important. Unaggressive genotypes are thought to proliferate as populations increase in density until peak density is reached when selection for aggressive genotypes commences and causes a decline in numbers.

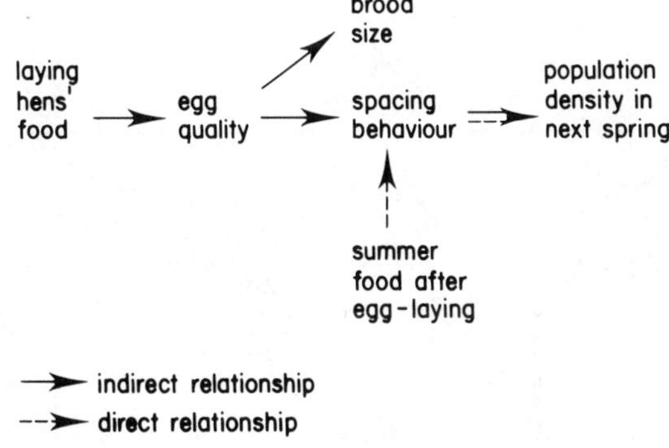

Fig. 25. Simplified version of the 1972 model of population limitation in red grouse. From Watson and Moss (1980).

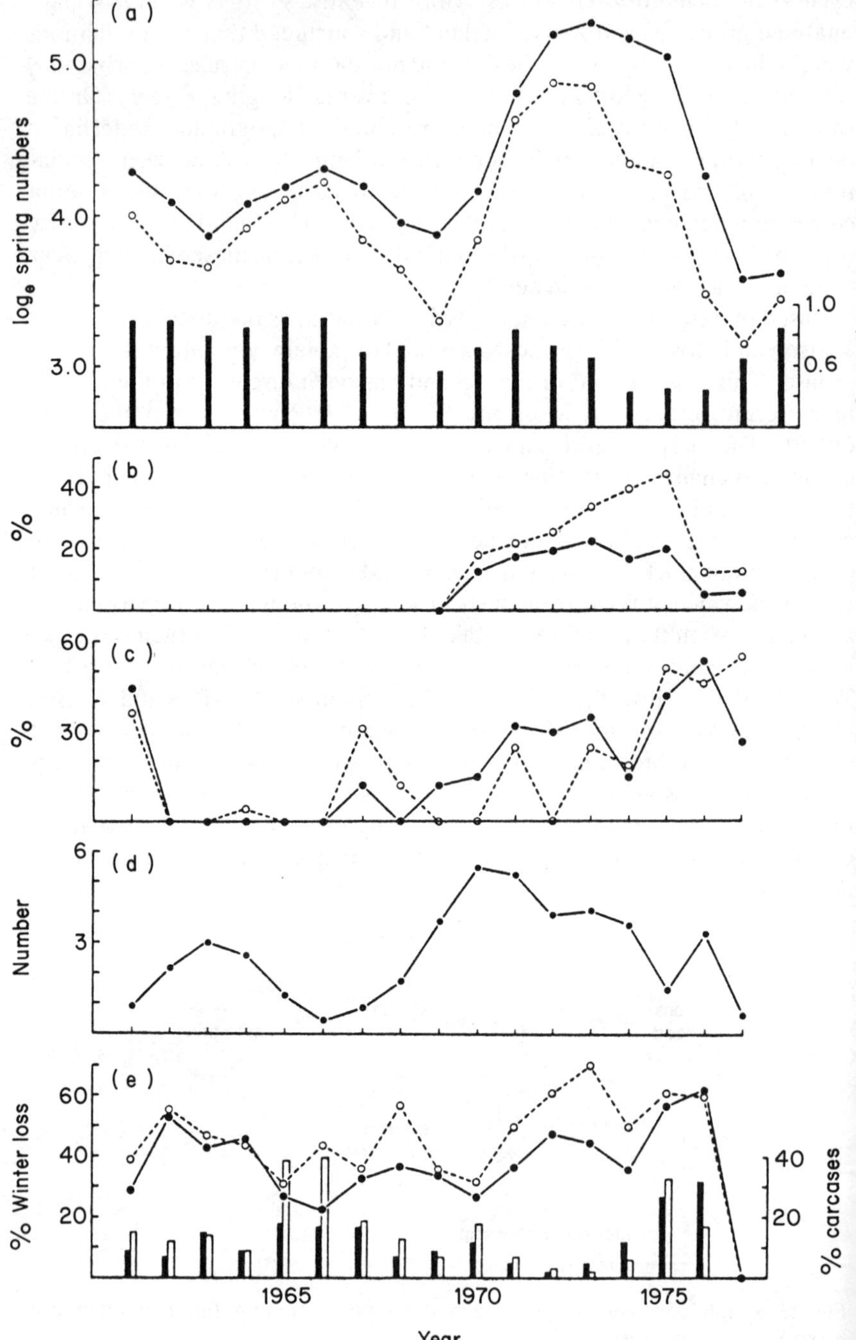
(a)
log$_e$ spring numbers
5.0
4.0
3.0
1.0
0.6
Sex ratio
(b)
%
40
20
(c)
%
60
30
(d)
Number
6
3
(e)
% Winter loss
60
40
20
40
20
% carcases
1965
1970
1975
Year

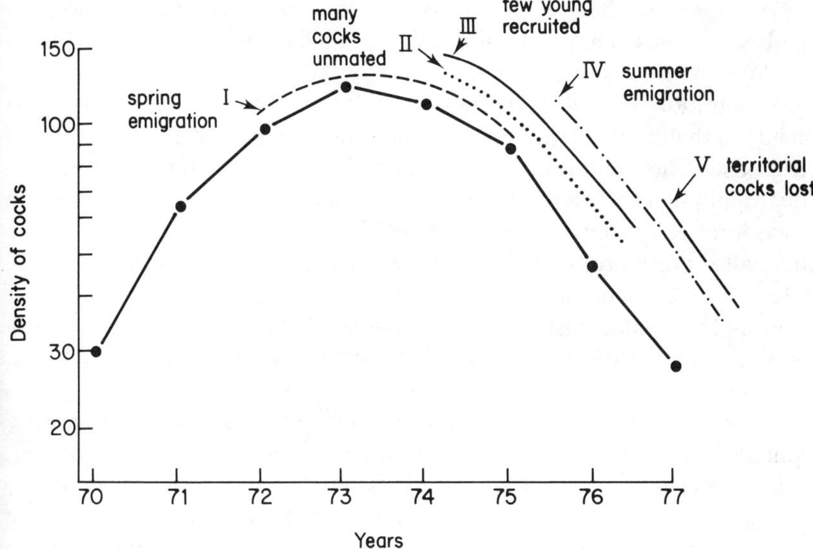

Fig. 27. Diagram of the timing of different aspects of spacing behaviour during a recent population fluctuation of red grouse in N.E. Scotland. The solid line connecting the years 1970–1977 shows the density of cocks (number per km^2) in spring. I. Pre-laying emigration > 20%. II. Proportion of unmated cocks in the territorial male spring population > 50%. III. Proportion of young in the territorial male spring population < 35%. IV. Post-hatch emigration of adults > 30%. V. Late-winter loss of territorial cocks > 30%. From Watson and Moss (1980).

Watson and Moss (1980) and Watson *et al.* (1984) discussed the immediate demographic causes of population change, i.e., births, deaths, immigration and emigration (Figs. 26, 27). They found that grouse generally tended to breed well in years of increase and more poorly in years of peak and decline, although breeding success was not necessarily linked directly with population change. During decline years emigration occurred, with some loss in winter due to death. In years of low numbers, immigration took place. Changes in spring numbers could be predicted from a measure of breeding success. Using models based entirely on these demographic data, Watson *et*

Fig. 26. Annual fluctuations in the demography of red grouse in N.E. Scotland. (a) $\mathrm{Log_e}$ numbers and spring sex ratio (hens per cock). (b) % of the spring stock that emigrated in spring. (c) % of adults left after spring emigration that emigrated in summer. (d) Numbers of young reared per hen on study area in August. (e) Net % of autumn stock lost over winter, and winter carcases (histograms) as a % of autumn stock (no searching for carcases in winter 1977–78). Cocks ● and ■; hens ○ and □. From Watson *et al.* (1984).

al. (1984) and Rothery *et al.* (1984) obtained reasonable simulations of population trends. They were also able to establish that reductions in brood size alone were not sufficient to cause the observed population decline, but that emigration was also an important factor. Moss *et al.* (1984) also concluded that movement is a major component of population limitation in red grouse. They reiterated that changes in emigration rates and in consequent population densities are caused mostly by changes in spacing behaviour: i.e., during an increase phase birds are relatively tolerant of each other, allowing recruitment of more subordinates and increased density, while in a decline phase birds are less tolerant leading to emigration of subordinates and decreasing density. Moss and Watson (1985) have recently reviewed the importance of spacing behaviour in grouse population dynamics.

Potts *et al.* (1984) developed an alternative model simulating grouse populations which was based on the actions of a parasitic nematode, *Trichostrongylus tenuis* Eberth. They assumed an inverse logistic curve relating mean numbers of *T. tenuis* in adult red grouse to their breeding success, with numbers of worms accumulated per bird varying with the numbers and worm burden of grouse present prior to the breeding season. When stochastic elements representing the effects of weather were introduced their model simulated observed mean population levels. They concluded that *T. tenuis* was important in determining grouse population fluctuations although it may not be the sole cause of them. The studies by Watson *et al.* (1984) did not consider the effects of parasites, although Watson and Moss (1979) suggested that socially-induced stress might predispose some components of the population to increased parasitism. It is possible, therefore, that the models of Watson *et al.* (1984) and Potts *et al.* (1984) are, in effect, taking different routes to the same explanation of the causes of population fluctuations. If demographic and behavioural processes affect the level of parasitism, parasite abundance may then simply be an indirect index of intrinsic population parameters. Of course, the converse may be true if parasitism affects social behaviour instead of *vice versa*.

Clearly there are still gaps in the understanding of the population dynamics of the red grouse, despite the apparent simplicity of the system in which it is found. Nevertheless, the mixture of long-term observation and experimental habitat manipulation in the study of the red grouse has clearly illustrated the complexities of interactions between demographic and habitat factors determining animal population dynamics.

4. *Effects of Grouse on Vegetation*

While it is clear from the foregoing discussion that the condition of the vegetation can affect grouse populations, the reciprocal effect of grouse on vegetation has been less well studied. In general it can be assumed tha grouse are present at such low densities and consume such a small proportion of the available plant material that the direct effects of their herbivory are minimal. However, their strongly selective feeding habits may change the vegetation sufficiently to cause competition between the grouse and another selective grazer, the mountain hare (Moss and Miller, 1976). Grouse may also act as dispersal agents for seeds contained in fruits of species such as *Vaccinium myrtillus* (Wilcock *et al.*, 1984), although seeds of these species can have very low germination rates (Mallik and Gimingham, 1985) and reproduction may be mostly vegetative. There have been no studies conducted to test whether passage through grouse may improve seed germination.

B. Red Deer

Red deer (*Cervus elaphus*) populations have probably existed in Scotland since the end of the Pleistocene in an environment that was much more forested than today. With clearance and the introduction of sheep farming, deer populations declined and, as with grouse, they probably owe their present-day success to the development of shooting as a fashionable sport on Scottish estates. Red deer are now abundant in Scotland, despite increasing sheep stocks, and recent estimates indicate that there are about 260,000 red deer in Scotland. Densities are considerably higher than in the rest of N.W. Europe.

While deer make considerable use of heathland for grazing, they are not purely heath animals and graze and browse on many other types of vegetation. Detailed studies of their behaviour and population dynamics have been carried out on the Island of Rhum in N.W. Scotland and these have recently been reviewed by Clutton-Brock *et al.* (1982). These authors discuss in detail the complex behaviour patterns and social organisation of a deer herd, and we will not discuss these further here. Rather we will concentrate on the interactions between red deer and their habitat, as it usually includes areas of heathland.

1. *Feeding Behaviour*

Studies on the island of Rhum in W. Scotland have indicated that deer have home ranges of about 400 ha for hinds and 500 ha for stags (Lowe, 1966), but this may be much greater (>5000 ha) on more extensive mainland deer "forests" (Staines, 1969). The summer range of deer includes all unfenced areas up to the summit plateaux at above 1000 m elevation, although in winter they are often confined to the lower hill slopes and valley bottoms.

Charles *et al.* (1977) found that in the Rhum study area, grasslands covering one-third of the area sustained 60% of the grazing, while heaths and bogs covering the other two-thirds were much less grazed. Seasonal densities of feeding deer ranged from 0·2 to 0·8 deer ha^{-1} on grasslands, compared with 0 to 0·2 ha^{-1} on heaths and bogs.

Deer show distinct preferences for communities containing the most nutritious foods, and are similar in that respect to Scottish hill sheep (Martin, 1964; Nicholson *et al.*, 1970) and Soay sheep on the island of St Kilda (Milner and Gwynne, 1974). Data from Rhum do, however, indicate marked seasonal variations in amounts of different species eaten (Fig. 28) although differential digestibilities of different plant species may distort the results slightly. The seasonal variations correspond with the movement of deer among community types as each is progressively grazed down over the year: *Molinia* and *Trichophorum* are grazed during early summer while *Calluna* is grazed most in winter when little other forage is available. Similar results have been obtained in N.E. Scotland by Staines (1977). Clutton-Brock *et al.* (1982) found increasing densities of deer on their study area during their investigation but found no consistent change in habitat use with increasing densities.

In comparison with the red grouse, few data are available on the effects of habitat modification on deer populations. Studies by Miles (1971) have shown that deer will preferentially graze recently burnt areas of *Molinia caerulea*-dominated vegetation because of the increased availability of *Molinia* shoots. However, burning in this community type yields only a temporary improvement in grazing. Studies on *Calluna*-dominated communities have indicated that deer graze preferentially in older heather, neglecting pioneer stands (Hewson, 1976; Moss *et al.*, 1981), possibly because taller heather is easier to reach. Moss *et al.* (1981) also found that deer grazed most on *Calluna* in relatively N-rich patches, indicating that they, too, are able to select within apparently uniform vegetation. The picture is complicated, however, by differences in feeding behaviour between stags and hinds. Staines *et al.* (1982) reported that at their study site in N.E. Scotland hinds ate less *Calluna* and more grasses than stags, had ranges lying over relatively richer rocks with access to better quality grasslands, and ingested better quality food (in terms of N per 100 g dry matter). Hinds therefore select better quality food while stags eat a greater amount of poorer quality material.

Observational studies have indicated that regular burning might increase deer populations. An early study by Evans (1890), reported by Miller and Watson (1974), showed an increase in numbers of red deer on Jura, N.W. Scotland, after burning management commenced. The hind population in particular increased by more than 50% in 6 years. Lowe (1969, 1971) also attributed an observed reduction in fecundity in young hinds to the cessation of burning on Rhum. However, neither of these studies had control plots and observed variations could be due to other factors such as weather (Watson, 1971).

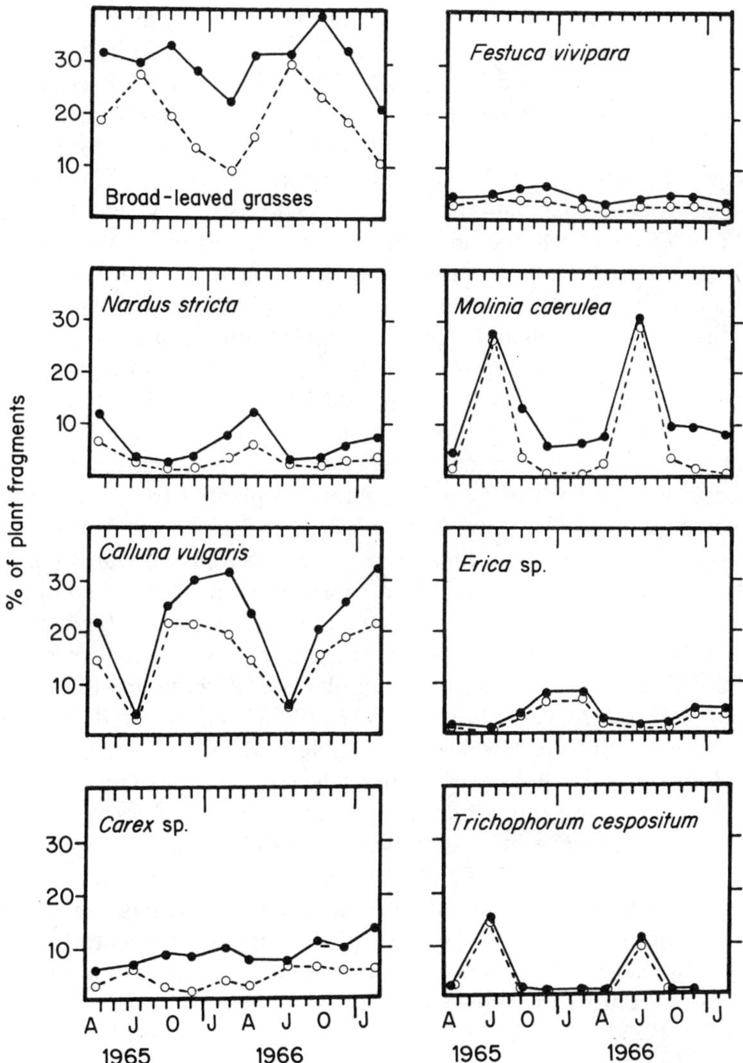

Fig. 28. Percentages of different types of plant fragments in faeces of red deer in the western part of Rhum: –○––○–, percentage of fragments from living plants only; ●–●, fragments from living plus dead plants. From Clutton-Brock *et al.* (1982).

2. *Effects on Vegetation*

Grant *et al.* (1981) studied the response of heath vegetation to the relatively high stocking rates (1·0 to 2·5 "hind equivalents" ha^{-1}) maintained on an experimental deer farming project in N.E. Scotland (Blaxter *et al.*, 1974). They found that the utilisation of *Calluna* in enclosed paddocks could be up to 55–60% of the available material, and cover of *Calluna* was reduced where stocking rates were highest. Older heather was more susceptible to damage, probably because of the reduced ratio of overwintering shoots to wood and root in older plants (Grant *et al.*, 1978). However, since the overall stocking density of red deer in Scotland is around 0·1 ha^{-1} (Mitchell *et al.*, 1977), grazing will generally be less intense. Nevertheless, Nicholson (1971, 1974) has pointed out that in winter deer grazing is concentrated in the lower elevation valley bottoms and may have significant effects on the vegetation.

A particularly important aspect of herbivory by deer is the browsing of young trees and shrubs, as described by Nicholson (1971) and Miller *et al.* (1982). Watson (1977) regarded browsing by red deer as one of the most important checks to the natural regeneration of native trees and shrubs in the eastern Highlands of Scotland. Areas of planted trees must also be protected from deer grazing. Miller *et al.* (1982) found that browsing of saplings of species such as *Pinus sylvestris, Betula pendula* and *Juniperus communis* was most frequent at lower elevations and that tall conspicuous saplings were more liable to be browsed than smaller ones. *Pinus sylvestris* was the most frequently browsed species and browsing usually resulted in death. This matches the known susceptibility of *P. sylvestris* to even light clipping (e.g. references in Mitchell, *et al.*, 1977). *Juniperus* and *Betula* were browsed less and were also less susceptible to damage. Damage to *Pinus sylvestris* saplings by deer clearly inhibits the regeneration of this species and sapling densities are rarely high enough to sustain even normal browsing pressures. Such regeneration is, however, possible on higher ground where browsing is less severe (Miller and Cummins, 1982). Watson (1983) has linked present day pine population structure with regeneration events occurring during periods of low deer numbers in the 18th century. Clearly any attempt at promoting natural regeneration of *P. sylvestris* must include measures to control access by deer.

C. Mountain Hares

The mountain hare (*Lepus timidus*) is perhaps the only mammal truly characteristic of heathland and is found on upland and northern heathlands, usually at densities of under 0·1 ha^{-1}, although densities up to 0·44 ha^{-1} are possible. Like the red grouse, the mountain hare depends almost entirely on *Calluna* which constitutes almost 100% of its diet in winter (Hewson, 1962).

Densities may be related to nutrient levels in the soil and associated availability of nutrients in *Calluna* shoots (Gimingham *et al.*, 1979). The life history and population dynamics of the mountain hare have been studied by Hewson (1965, 1976), Flux (1970), Watson and Hewson (1973) and Watson *et al.* (1973). These studies have indicated that hare mortality is greatest in late winter and early spring, probably due to lack of food of sufficient nutritional quality. This may be caused either by absence of young *Calluna* or its burial by snow.

Hewson (1976) found that in areas where few hares were present, they grazed preferentially on young (pioneer) heather, but where more were present emphasis switched from pioneer heather in the summer and autumn to building and mature heather in winter and spring. Hewson (1976) suggested that the preference for younger heather may be as much for ease of feeding and for observation of the approach of predators as for nutritional reasons. Nevertheless, Miller (1968) found that in summer hares selected areas fertilised with N and in winter areas fertilised with N and P, indicating that nutritive value may be important. Moss *et al.* (1981) found that hares preferred areas high in P and Watson and O'Hare (1979b) found that the number of hares increased on areas of bog treated with fertiliser. Moss and Miller (1976) showed that hare densities were negatively correlated with the over-winter grazing pressure, indicating that a decline in numbers may be related to exhaustion of their preferred food. It is also possible that hares have to compete with grouse for their preferred food (pioneer *Calluna*) in summer and with grouse and deer for the less desirable food available in winter (Moss and Miller, 1976). Miller and Watson (1974) suggested, however, that since hares range more widely than grouse and are not territorial, they are less affected by habitat changes caused by fire and other types of management.

Since hares are often sparse on heathlands, their grazing effect may have been neglected, but studies by Welch and Kemp (1973), Moss *et al.* (1981) and Moss and Hewson (1985) have shown that herbivory by hares may have significant effects on the vegetation. Welch and Kemp (1973) described a distinctive form of *Calluna* characterised by flat, circular bushes, radiating branch structure and high shoot densities caused by hare grazing in small burnt areas. Similar effects are produced by grazing by rabbits (*Oryctolagus cuniculus* L.) which can be common in some heath areas (Gimingham *et al.*, 1979) but which also graze species other than *Calluna* (e.g. Chapuis and Lefeuvre, 1980). Hares removed almost all shoots each year. Welch and Kemp (1973) suggested that grazing by hares would increase lateral branching of regenerating *Calluna* and thus increase the ultimate cover and productivity. This differs from intense grazing by larger herbivores which may remove *Calluna* entirely both by herbivory and trampling damage. Grazing by hares may also maintain stands of chronologically old *Calluna* in a physiologically juvenile state, containing more N, P, K. Mg and soluble carbohydrates than ungrazed *Calluna* of a similar age (Moss *et al.*, 1981; Moss and Hewson, 1985). However, the effects of grazing on shoot nutrient contents are complex and depend on the soil nutrient status and intensity of grazing.

D. Sheep

Hill sheep have been abundant in Scotland since the late 18th century, and are probably the most important form of agricultural production on upland rough grazings. Studies carried out in various parts of Scotland and N. England, especially by the Hill Farming Research Organisation, have yielded much information on the grazing behaviour of sheep, their effects on vegetation and on methods of improving sheep production.

1. *Feeding habits*

Sheep production from hill grazings is generally rather low and is determined mainly by the limits to stocking rates set by the need to provide sheep with a minimum tolerable level of winter nutrition. Sheep, like deer, have large ranges and have access to the various community types present in the upland landscape. Individual sheep tend to remain in a particular area, and the home range may be up to about 50 ha, with up to 50 sheep sharing the same range (Hunter, 1964; Hunter and Milner, 1963). Stocking rates are usually in the range 1·2 to 2·8 sheep ha^{-1}. Sheep show clear preferences for certain community types, especially the relatively rich grass communities such as *Agrostis-Festuca* swards (Hunter, 1962; Arnold, 1964; Rawes and Welch, 1969; Hewson and Wilson, 1979). Although heath types are less preferred, *Calluna* (and other ericaceous species) are important food species, especially in winter. Although considerably less digestible than graminaceous species, it remains green in winter and may be the only food available above lying snow (Kay and Staines, 1981). Thomas and Armstrong (1952) found that sheep graze mainly the current season's shoots, in common with the native herbivores already discussed. The nutritional value of *Calluna* for sheep declines as it grows older, and Grant and Hunter (1968) found that sheep graze younger heather in preference to older. As with hares, however, the selectivity of sheep declined as grazing pressure increased. Milne (1974), on the other hand, found no differences in the voluntary intake and digestibility of shoots from stands of different ages and suggested that stand age did not affect nutritive value for sheep grazing. It is possible that Milne (1974) did not include young enough material (i.e., <3 yr) in his tests, since Grant (1968) noted that sheep tend to congregate on recently burned areas, and burning is thought to improve food for sheep (e.g., Miller and Watson, 1974).

As for deer, distinct seasonal trends in selection of forage species are found. Grazing of preferred species (i.e., grasses) is greatest in summer, with *Calluna* (and on blanket bog, *Eriophorum vaginatum*) forming the mainstay of the sheep's diet over winter (Martin, 1964; Grant *et al.*, 1976). Use of *Calluna* by sheep may, however, be determined by proximity to more nutritious grazings: e.g., Welch (1981) found that sheep used only the part of the moor close to reseeded grassland, despite access to much larger areas.

2. *Effects on vegetation*

The effects of sheep grazing have been studied in detail in several heath community types. Gimingham (1949) discussed the effect of sheep grazing on both *Calluna* morphology and its relationship with *Erica cinerea* (see Section IIIC). From field experiments and clipping experiments to simulate grazing effects, Grant (1968) and Grant and Hunter (1966, 1968) have concluded that grazing can initially reduce production but that if continued it can lead to an increase in overall production. Clipping or heavy grazing creates a cushion-like form of *Calluna* with densely-packed thin branches, without the build-up of lignified branches. This results in an increase in the total number of green shoots and improved production of edible material. Grant and Hunter (1966) suggest that a grazing regime that removes about 60% of the current season's growth would maintain *Calluna* plants more or less indefinitely in a "young" state, as discussed earlier for grazing by hares. However, the effects of grazing depend greatly on its intensity and very high stocking rates could cause heather to decline (Grant, 1971; Grant *et al.*, 1978).

The effects of herbivores include not only the direct grazing effect but also dung deposition and damage to the vegetation through trampling. Welch (1985) has indicated that seeds of several heath species can be dispersed by herbivores, subsequently germinating in dung depositions. Welch (1984b) estimated that grazing by sheep at densities greater than $2{\cdot}7\ ha^{-1}$ on dry heath would cause *Calluna* to decline, and other studies have attributed declines in *Calluna* abundance to increased sheep density (Dalby *et al.*, 1971; Anderson and Yalden, 1981; Bakker, 1978). On blanket bog, too, sheep grazing at low densities can increase *Calluna* cover and shoot production (Rawes and Williams, 1973; Hewson, 1977) but reduce it at high densities or where growth is poor (Welch and Rawes, 1966; Rawes and Hobbs, 1979). However, normal sheep stocking rates are usually much lower than that required to alter *Calluna* abundance or to remove the 60% of the current season's growth suggested by Grant and Hunter (1966). Methods of partial improvement of upland grazings in Scotland to increase stocking rates have been discussed by Eadie and Maxwell (1974) and the Hill Farming Research Organisation (1979).

Welch (1984c) showed that light grazing on dry heath benefited ericaceous species in general, as well as lichens such as *Cladonia portentosa*, while grasses and forbs increased in abundance under heavy grazing. Similar increases in graminaceous species are found with heavy grazing on blanket bog (Rawes and Hobbs, 1979). Even with very low grazing intensities (0·2 sheep ha^{-1} on blanket bog: Rawes and Heal, 1978), exclosure experiments produce marked vegetation changes. For instance, Rawes and Hobbs (1979) found that *Cladonia* lichens increased greatly in abundance in exclosures, a result opposite to that found by Welch (1984c) on dry heath. Rawes (1983) reported the results of long-term exclosures on blanket bog which showed marked increases in ericaceous species. Initially *Empetrum nigrum* became temporarily dominant before being replaced by *Calluna*. Other species such as *Rubus chamaemorus* and *Narthecium ossifragum* also increased greatly and Taylor and Marks (1971) have also reported marked increases in *Rubus chamaemorus* densities following exclusion of sheep. Exclosure of sheep may also result in overall reduction in numbers of species present where one or a few species attain dominance in the absence of grazing (Welch and Rawes, 1964; Ball, 1974). Thus even low intensity grazing may have marked effects on community structure and composition. In this perhaps lies a partial explanation for differences between British heaths where sheep grazing is common and other N.W. European heaths where sheep are often no longer present.

E. Invertebrates

Heathlands possess a relatively rich invertebrate fauna, little of which has been studied in any detail. Studies by Richards (1926) and Nelson (1971) have indicated that both dry and wet heath communities support in the region of 200 invertebrate species. Knowledge of the invertebrate fauna has been reviewed recently by Coulson and Whittaker (1978), Gimingham *et al.* (1979) and Heal (1980), all of whom point out the lack of systematic study and concentration on particular taxonomic groups. Recent studies of invertebrate groups in heathland types have however been reported by Butterfield and Coulson (1983), Coulson and Butterfield (1985) and at the "Colloque sur l'écologie des landes" in Rennes in 1979 (e.g. Gueguen *et al.*, 1980 and other papers in that volume). Barclay-Estrup (1974) and Gimingham (1985) also give information on invertebrates in Scottish heath and relate changes in invertebrate communities to changes in *Calluna* stand characteristics. Invertebrate herbivores associated with *Calluna*-dominated heath include species from many different taxonomic groups, including mites, weevils, thrips, psyllids and lepidopterous larvae. Some species may be present in very large numbers, but, as far as is known, consumption is rarely great enough to cause appreciable damage to the vegetation (Gimingham, 1972). Here we consider only the invertebrate herbivores which have been studied in detail.

1. *Heather beetle*

The species which has perhaps the most noticeable and potentially devastating effects on heath vegetation is the heather beetle, *Lochmaea suturalis*. The larvae and imagos of this insect feed exclusively on *Calluna* leaves, stem apices and bark, and may cause severe damage either in small patches or in extensive areas in years of outbreak (Cameron *et al.*, 1944; Blankwaardt, 1977). Morison (1963) suggested that outbreaks of heather beetle in Scotland are worst after a sequence of 2–3 warm summers, although this is not always the case. Conditions leading to an outbreak may be different elsewhere in N.W. Europe because of the less oceanic conditions (Blankwaardt, 1977; de Smidt, 1977). Blankwaardt (1977) found that outbreaks occurred in Holland every 5–10 years, usually following years of above-average precipitation. The effects of infestations in Holland vary from complete destruction of *Calluna* to hardly any damage, depending on *Lochmaea* density, variations in environmental conditions and age and condition of *Calluna* (de Smidt, 1977). Intensity of beetle attack is also affected by high local mortality of beetles due to an entomopathogenic fungus, *Beauveria bassiani* (Bals) Vuill (Brunsting, 1982). Evidence of heather beetle attack becomes apparent only after periods of hot, dry weather when excessive water loss by *Calluna* is caused by beetle damage. Death of damaged shoots is evident by the following year, and regeneration from basal buds may occur, as happens after fire. Further studies of the effects of *Lochmaea* on *Calluna* physiology are in progress.

Grasses such as *Deschampsia flexuosa* and *Festuca ovina* may invade gaps in the *Calluna* canopy until *Calluna* regains dominance after 5–10 years (de Smidt, 1977), and repeated outbreaks of *Lochmaea* may produce complex vegetation mosaics quite different from those found in Scottish heaths where *Lochmaea* seldom causes heavy damage. Vegetation development may also follow different routes after beetle attack. For instance, Diemont and Heil (1984) recorded evidence for both progressive and cyclical changes in Dutch heathlands previously infested with heather beetle. They suggest that addition of dead plant material following a beetle attack may lead to more nitrogen being mineralised, allowing grasses such as *Festuca ovina* to become dominant at the expense of *Calluna*. Heil and Diemont (1983) were able to produce the same vegetation change by continued addition of N. Brunsting and Heil (1985) suggest that the increased nutrient availability in the soil results in increased nutrient content in *Calluna* shoots, which in turn leads to better growth of beetle larvae. As a consequence beetle populations may increase and more severe infestations may occur.

2. *Strophingia ericae*

Another invertebrate herbivore of *Calluna* which has been studied in some detail is the exopterygote insect, *Strophingia ericae* (Curtis) (Homoptera:

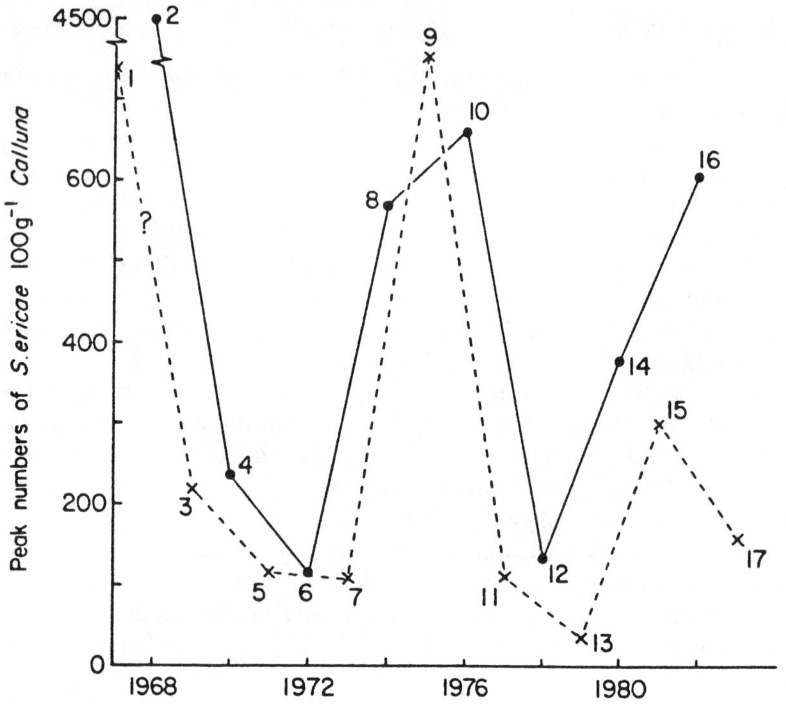

Fig. 29. Peak numbers of psyllid, *Strophingea ericae* (per 100 g dry weight *Calluna*), after recruitment in 2 populations on blanket bog in N. England. Populations A (-●—●-) and B (–X––X–) represent two generations which are out of phase with each other by one year. From Whittaker (1985).

Psylloidea) which feeds on phloem sap (Hodkinson, 1973a; Parkinson and Whittaker, 1975). Hodkinson (1973b) considered this species to be the main invertebrate herbivore of *Calluna* on the blanket bog site he studied, but found that it had no measurable effect on rates of photosynthesis, respiration or growth by *Calluna*, even though densities varied from an average of 2000 m^{-2} up to over 11 000 m^{-2}. Recent population studies of *S. ericae* have revealed population trends which are possibly cyclical, with peaks of abundance about 6 years apart (Fig. 29; Whittaker, 1985). Whittaker (1985) suggests that these fluctuations may be due to interactions between and within cohorts which were competing for the same limiting resource although he does not rule out the possible role of external disturbance. It is interesting to note that the cycle length and timing of peak abundance of *S. ericae* are roughly similar to those found by Watson *et al.* (1984) and Potts *et al.* (1984) for red grouse.

3. *Seed predators*

Seed predation in heathlands has been little studied. The ant species *Tetramorium caespitum* Latrielle has, however, been found to harvest large numbers of seeds of ericaceous and grass species in S. English heath (Brian *et al.*, 1965, 1967). Brian *et al.* (1976, 1977) have followed the effects of fire and post-fire vegetation recovery on ant abundances. They found that fire did not eliminate ant species from the area, but that population sizes increased as the vegetation re-developed. They also suggested that *Tetromorium caespitum* and another species, *Lasius alienus* (Forster) were able to co-exist in the same area of heathland by being adapted to opposite ends of the sequence of post-fire vegetation regeneration. *Lasius alienus* inhabits open vegetation and cuts galleries below bare soil and is most abundant in the immediate post-fire phase, whereas *T. caespitum* can build nest mounds and thus avoid shading by developing vegetation.

Recent studies by Melber (1983) have shown that several species of ground beetles also eat *Calluna* seeds, up to a maximum of 67% of their total diet. The effects of this predation on seed populations are unknown. Recent studies by Kjellsson (1985b) on a species of sedge common in heathland, *Carex pilulifera* L., have, however, shown that predation by ground beetles, *Harpalus fulginosus* Duft, can reduce the seed pool of this species by 65%. Kjellsson (1985a) was also able to demonstrate that the ant *Myrmica ruginodis* Nyl. was an effective dispersal agent for seeds of *Carex pilulifera* and that secondary relocation of seeds by ants is probably of great importance in seedling recruitment. Clearly further research into invertebrate predation and possibly dispersal of heath seeds is required.

V. MANAGEMENT AND CONSERVATION OF HEATHLANDS

Previous sections of this paper have considered the interactions between vegetation, fire and herbivores in heathland. Management has been an integral part of the discussions since virtually all heathlands are managed to some extent. Indeed, since the majority of heathlands are man-made systems, derived from clearance of forests, some form of management is often necessary for their continued existence. Several recent studies have discussed heathland management in detail (e.g. de Smidt, 1979; papers in Farrell, 1983; Miller *et al.*, 1984) and others have discussed the various uses to which heathland is put and the increasing likelihood of land use conflicts in the future (e.g. Wynne-Edwards, 1964; Miller and Watson, 1983; Welch, 1984a). Ball *et al.*, (1982) have also discussed the extent to which upland vegetation, including heathland, is disappearing or changing.

In Scotland, heaths are utilised as rough grazings for sheep and cattle, as 'deer forest' for shooting and as grouse moor. Heath areas are also taken over for reafforestation: usually with the non-native species *Pinus contorta* or *Picea sitchensis*. The use of heath areas for recreation other than game shooting is important, and many people walk, camp and ski in hill areas. The amenity and tourism value of Scotland's unique *Calluna*-clad landscape also cannot be ignored. For conservation purposes, there is a requirement for the setting aside and management of representative areas of different heath types. The goals of conservation may not be best attained by following the same management practices used for production (Gimingham, 1981). Further, there is an increasing requirement for suitable methods of restoration of heathland damaged, for instance, as a result of pipeline laying (e.g., Putwain *et al.*, 1982).

Within the present systems of management, many options are available to the land manager. Gimingham (1981) has reviewed these options, which include burning, various grazing regimes, cutting or mowing, removal of surface organic matter and vegetation, drainage, fertilisation, afforestation and conversion to grass swards or arable land. The use of these practices varies greatly from place to place in N.W. Europe, and Table 2 summarises regional use of heathlands. A further important use of the wetter heaths in N.W. Scotland and Ireland is peat extraction for fuel. Here we summarize the effects of alterations in the main forms of management discussed in previous sections: i.e., grazing and fire.

Table 2

Variation in the agricultural use of heathland and associated fields. From Gimingham and de Smidt (1983).

	Burning	Cutting sods	Grazing	Cutting heath	Permanent fields	Shifting fields
North (Denmark, S. Sweden)	X	X	(X)			X
West (Scotland, Wales, England)	X		X		X	
Middle (Germany, Netherlands, Belgium)	(X)	X	X		X	
South (France, Spain, Portugal)			(X)	X		X

A. Grazing and Burning

As grazing land, heathlands are remarkably unproductive. Miller and Watson (1974) estimated that usually less than 10% of the annual primary production of a *Calluna* moor is eaten by herbivores. Table 3 gives estimates for secondary production in N.E. Scotland, indicating very low outputs of saleable product. Nevertheless, such systems remain viable because of the profit to be gained from recreational shooting – such profits may be greater than those obtained by biologically more efficient methods of production such as sheep farming or forestry, at least in the short term and in the absence of capital input. Traditional forms of management for deer, sheep and grouse can be mixed with amenity uses fairly readily, but usually grazing and forestry are mutually exclusive. The balance between these opposing land uses will probably be determined by economics. When game shooting enjoys a high degree of popularity there may be little incentive to reafforest tracts of heathland. However, the demand for more home-grown timber products is increasing, and there is a trend towards large-scale expansion of forestry on upland heaths in N. Britain.

The heathland system is deceptively simple in that all the herbivorous species present utilise almost the same resource at some time during the year. Red grouse and hares both selectively graze young shoots of young heather. Sheep also graze recently burned heather, although they and red deer utilise mainly older heather in the winter and other species in the summer. Despite the low herbivore production relative to *Calluna* primary production, studies quoted in Section IV have shown that strongly selective

Table 3

Some notional figures for the cropping of herbivores from the heather moorland ecosystem in the east-central Highlands of Scotland. From Miller and Watson (1983).

Crop	Average density in spring (no. km^{-2})	Biomass (kg km^{-2})	Production of young (no. km^{-2})	Annual crop taken by man (kg km^{-2})	Main sources of data
Sheep	50	2250	40	1000	Hill Farming Research Organisation (1979)
Red deer	10	685	2	110	Mitchell, Staines, and Welch (1977)
Mountain hare	16	43	50	0	Flux (1970); Watson *et al.* (1973)
Red Grouse	65	41	89	25	Jenkins, Watson, and Miller (1963, 1967)

grazing by the herbivores probably limits that production and maintains populations at what appear to be rather low densities. The problem is not the amount of primary production but availability of adequate nutrients.

One herbivore species may affect the food supply available to another, and, for instance, it may be difficult to manage a moor for sheep and grouse simultaneously. Other indirect interactions may also be important: for instance the sheep tick (*Ixodes ricinus* (Linn.)) and its associated louping ill virus are known to cause grouse mortality (Duncan *et al.*, 1978). Some interactions may be rather subtle: for instance, Hewson (1984) suggests that moor-burning for sheep management on wet heath which leads to the reduction of *Calluna* and its replacement by graminaceous species may reduce numbers of hares and grouse. These species are the favoured prey of foxes and eagles, and if their numbers are reduced the risk of predation on lambs may be increased.

Potential effects of variations in fire and grazing regimes in dry heath are shown in Fig. 30. Burning and grazing generally act to alter the vegetation in similar ways. Increased burning frequency or grazing intensity cause a change from *Calluna* to graminaceous dominance, and reduced burning or grazing results ultimately in scrub or woodland. Such changes may also take place on wetter heath. The management regime adopted must be related to the vegetation type present and to the management goal. For instance, frequent fire on blanket bog shifts the vegetation from *Calluna* to *Eriophorum* or *Molinia* dominance. This may improve the habitat for sheep grazing but not for grouse. Rawes and Hobbs (1979) suggested that fires may not be necessary on blanket bog and that the optimum management for blanket bog may be light grazing by sheep. The same would not be true on drier heath. Here, fire is an inexpensive management tool which can be used effectively to improve grazing for sheep or to improve grouse stocks. If carried out efficiently, fire does not seem to damage the heathland system as it is at present: instead it perpetuates the heathland plant community, which consists of a set of "opportunistic" or disturbance-adapted species (Hobbs *et al.*, 1984). However, the long-term effects of burning on nutrient status need to be reviewed, and the importance of fire frequency in determining community composition must be stressed. The possibility of accidental fire is also important, especially in areas where public pressure is increasing. It is possible that, in some cases, occasional accidental fires have a much greater

Fig. 30. Changes in *Calluna*-dominated dwarf shrub heaths as a result of variation in intensity of grazing and burning. Present vegetation shown in double circles; probable and possible stages in the vegetation succession are shown by single and broken circles respectively. Rate of change, where known, is shown above the arrows and management intensity below the arrows; sheep numbers are ha^{-1}. Differing soil conditions are noted in the left-hand margin. From Miles *et al.* (1978).

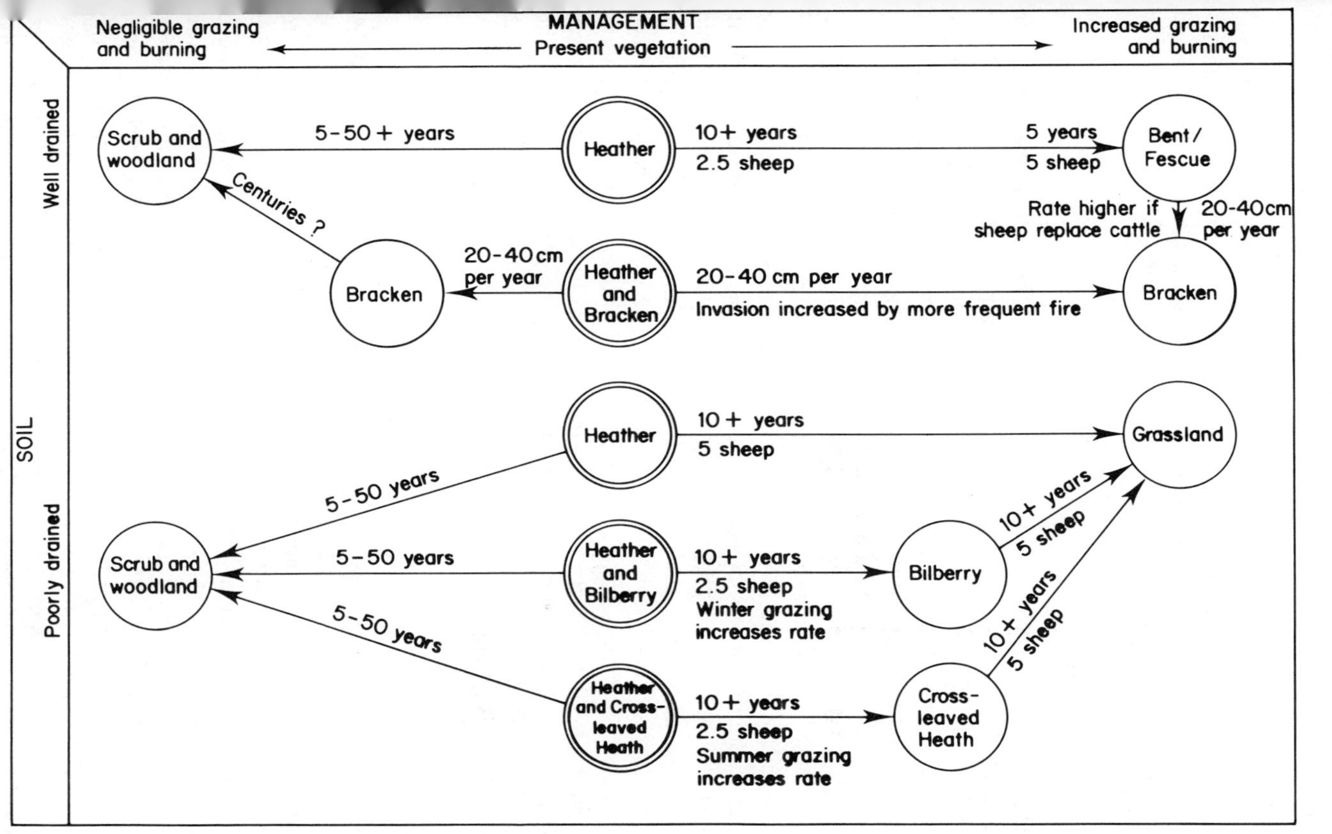
MANAGEMENT
Negligible grazing and burning
Present vegetation
Increased grazing and burning
SOIL
Well drained
Poorly drained
Scrub and woodland
5-50 + years
Heather
10+ years
2.5 sheep
5 years
5 sheep
Bent / Fescue
Rate higher if sheep replace cattle
20-40cm per year
Centuries ?
Bracken
20-40cm per year
Heather and Bracken
20-40 cm per year
Invasion increased by more frequent fire
Bracken
Heather
10 + years
5 sheep
Grassland
5-50 years
Scrub and woodland
5-50 years
Heather and Bilberry
10 + years
2.5 sheep
Winter grazing increases rate
Bilberry
10 + years
5 sheep
5-50 years
Heather and Cross-leaved Heath
10 + years
2.5 sheep
Summer grazing increases rate
Cross-leaved Heath
10 + years
5 sheep

impact on the heathland system than regular management fires. A further important problem facing land managers is the presence of large areas of old *Calluna* which have not been burned recently due to reduced manpower on estates. These stands do not regenerate well following burning (Hobbs and Gimingham, 1984b), and it may be that some alternative form of management is required, e.g. mowing or "burn-beating" as described by Froment (1981), although the effectiveness of this is uncertain.

B. Recreation

Amenity and recreational use also has potential impacts on the vegetation and fauna and may require different forms of management (e.g., Yates, 1972; Daniels, 1983). Erosion of footpaths, the formation of new paths and the widening of old ones are all problems caused by increased public pressure. Liddle and Chitty (1981) have also shown that horse-riding causes distinct localised changes in lowland heath vegetation. Developments such as skiing also have potentially disruptive effects on vegetation and fauna, and these have been discussed by Bayfield (1970) and Watson (1979, 1985).

In addition to these recreational developments, estate management for the traditional sports of game shooting have become more disruptive in recent years. Visiting game shooters now expect to be driven to within shooting distance of their quarry and this has meant the construction of vehicle tracks over heath which was previously accessible only by foot. Such tracks are nowadays bulldozed, often with a view to economy rather than to minimizing the impact on the surrounding heath. Scars caused by track construction are recolonized by plants, but the process is extremely slow and at first only mosses such as *Polytrichum piliferum* and *P. juniperinum* can establish (Bayfield *et al.*, 1984; V. J. Hobbs, 1984).

C. Conservation

Gimingham (1982) points out that heathland communities can be maintained by the continuation of traditional management practices, but that some alteration in the normal management regime may be necessary to promote biotic diversity rather than increased herbivore production. For instance, the fire regime might be altered to either more frequent burning to allow herbaceous species, pioneer mosses and lichens to become more abundant or to less frequent fires to allow the formation of uneven-aged heterogeneous stands. Or a mixture of these might be required. In some cases, cessation of burning altogether might be appropriate to allow the development of mosaics of heath, scrub and woodland. Reduced frequency of burning might allow an increase in *Juniperus communis*, which has been virtually eliminated from many Scottish heaths due to its lack of fire-surviving mechanisms (Mallik and Gimingham, 1985).

Grazing may also be an important management tool for conservation: e.g., Bakker *et al.* (1983) report experiments started in Holland to test the effects of reintroduction of sheep on the regrowth of heathland. Kottman *et al.* (1985) have indicated that sheep grazing may provide a more economical means of heath management than other practices such as mechanical cutting. Again, changes in management may have subtle effects on the rest of the heath system. For instance Tye (1980) attributes the decline in populations of wheatears (*Oenanthe oenanthe* L.) on Breckland heath partially to reductions in grazing pressure by rabbits and sheep. Without adequate grazing the short turf that the wheatears prefer is not maintained. On the other hand, too intense browsing pressure may prevent regeneration of heath plant species (e.g., Tubbs, 1974).

Fertilization may also have significant effects on the abundance of heath fauna (Watson and O'Hare, 1979b) as well as on the heath vegetation (Helsper *et al.* 1983). Brunsting and Heil (1985) suggest that the cessation of the old practice of cutting sods, coupled with increased nutrient inputs (from rain water and agriculture) have led to an eutrophication of Dutch heathlands which has set them on a spiral of increased heather beetle attacks (see Section IVE) and conversion to grassland. In this case, reductions in the levels of nutrient inputs and a return to traditional management may be the only way to prevent the eventual disappearance of the heath. Werger *et al.* (1985) have recently discussed the traditional methods of sod-cutting.

Berendse (1985) also points to increased nutrient availability as being the key to the conversion of wet heaths from *Erica tetralix* to *Molinia caerulea* dominance. He argues that restoration of grazing by sheep may not be sufficient to return the heath to *Erica* dominance because this will not lower the nutrient availability. Clearly, increased nutrient availability within the usually nutrient-poor heath system is a problem the N.W. European heaths share with heaths in other parts of the world (e.g., Heddle and Specht, 1975; Specht *et al.*, 1977).

An increasing problem in the management of lowland heaths in S. England and elsewhere is the spread of unwanted species such as *Pteridium aquilinum* and various scrub species. Experiments on the use of cutting or herbicides as control measures are presently underway (e.g., Marrs, 1983, 1985; Lowday, 1983; Bostock, 1983).

Considerable attention has been paid to the problem of fragmentation of heathland habitats, especially in the south of England. Studies by Moore (1962) and more recently by Webb (1980) and Webb and Haskins (1980) have illustrated the decline of heathlands in Dorset over the past 200 years (Fig. 31). Webb and Haskins (1980) found that the rate of loss of heath was less than 0·5% per year up to 1934, but between then and 1973 it increased to as much as 3% per annum. Although this rate of loss has since declined again, it is still high enough to cause concern for the long-term future of heath in that area.

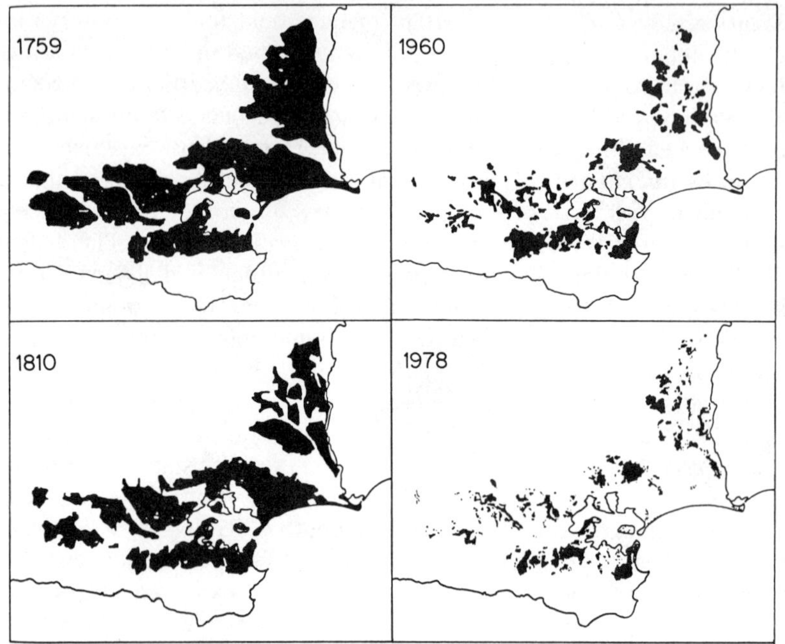

Fig. 31. Reduction in area and fragmentation of Dorset heathlands over the past 200 years. From Webb (1980).

Webb and Hopkins (1984), Hopkins and Webb (1984) and Webb *et al.* (1984) have investigated various aspects of the invertebrate faunas on fragmented heathlands. They found that species diversities of several invertebrate groups were negatively correlated with area of fragment and positively correlated with the extent to which the heath was isolated. Thus small, isolated heathlands tended to have a richer invertebrate fauna than either large heathlands or those in close proximity to other heaths. This was due mostly to the invasion of smaller heath areas by species from surrounding communities. Spider diversity, on the other hand, showed no correlation with fragment size or isolation. However, when a subset of spider species defined as 'heath species' was taken, these species were found to contribute a larger proportion to the total spider fauna on large heaths than on smaller areas. Within this set of heath species, those species with the poorest powers of dispersal were confined to larger heaths. Webb *et al.* (1984) also found that invertebrate diversity could be related to the structural diversity of the surrounding vegetation. Such findings have important implications for nature conservation and reserve selection. Especially important in this respect is the conservation of a *heath* fauna which is relatively low in diversity rather than a more diverse fauna made up of species not characteristic of heathlands (Webb *et al.*, 1984).

Similar studies are required in other areas to examine the relationships between heaths and surrounding areas. Problems of fragmentation are perhaps less severe in N. Britain where large areas of heathland still remain, but conservation problems are still present, especially with increasing conflicts of land use. The maintenance of heath ecosystems depends on a thorough understanding of the interactions between their various components and with surrounding systems. Changes in heath systems can occur rapidly, e.g. after fire, but in other cases changes may take place over much longer time periods. Thus short-term studies can help explain only part of the functioning of heath systems. This review has illustrated the value of long-term studies in unravelling the complexities of plant and animal dynamics.

VI. CONCLUSIONS

In this review we have concentrated almost exclusively on heathlands dominated by *Calluna* found in N.W. Europe and N. Britain in particular. Although this system has a relatively limited distribution, there is a depth of information on these heathlands which is lacking for heath and shrub communities elsewhere in the world. In this final section we outline the main components and interactions in the heath systems of N.W. Europe as a basis for comparison with other areas.

Groves (1981) emphasised that the ecological feature most strongly controlling the distribution of heathlands and related shrublands throughout the world is their occurrence on acid, nutrient-poor soils. The importance of this low nutrient status to the functioning of the *Calluna*-dominated system has been emphasised throughout this review. Nutrient status affects plant community composition, with poorer soils dominated almost exclusively by *Calluna* and richer soils supporting a more varied community with greater numbers of associated species.

We hypothesise that the progressive locking-up of nutrients by *Calluna* during its growth cycle is the key factor determining the dynamics of many components of the heath system. As *Calluna* ages, most of its biomass is in the form of woody material, litter decomposition is slow, and nutrient availability thus declines as more nutrients are locked up in dead material. The interactions between *Calluna* and its associated species are well known (e.g., Gimingham, 1983), but the mechanisms by which *Calluna* excludes other species as it matures have not been studied in any detail. It has generally been assumed that shading by the *Calluna* canopy is the major factor preventing the growth of other species. However recent work, especially from Holland, has suggested that nutrient relations may be of great importance in the interactions between *Calluna* and other species (e.g., Heil and Diemont, 1983). Associated species may therefore be denied resources as nutrients are locked up in the *Calluna* canopy. Anything which disrupts this process is likely to release resources and allow other species to grow. Large or small-scale disturbances, such as fire or heather beetle attack, or heavy grazing will act in this way. In all cases the woody canopy is removed and nutrients released through mineralisation or in the form of ash or faeces. Species with higher nutrient requirements are then able to grow in the more open community.

Superimposed on the constraints placed on the heath system by its low nutrient status therefore are various forms of disturbance and the responses of the biota to these. The heathland plant community is essentially a collection of opportunistic species, able to survive in a system subject to frequent disturbances. Species responses to disturbances vary greatly, and these variations produce the patterns of vegetation development observed. This applies to both large-scale disturbances such as fire and the more localised disturbances caused by heather beetle attacks. Alterations in management regime such as increased or decreased fire frequency or grazing intensity favour different sets of species in a predictable manner. More intense or frequent disturbance favours herbaceous species, while ericaceous species are more prevalent where disturbance is less intense or frequent.

The heathland community depends on disturbance for its continued survival and without fire and/or grazing would gradually change into woodland. We doubt that heaths with phasic development would persist as uneven-aged stands of *Calluna* where tree seeds were available and grazing was restricted. Grazing by deer and sheep is undoubtedly one of the main factors preventing the change of heath to woodland in N. Britain. Management practices aimed at maintaining heathland involve disruption of the phasic development of *Calluna* and subsequent development to woodland. This may not be true in wetter heaths, where phasic growth of *Calluna* does not take place.

Management for herbivore production also generally involves disruption of the ageing process of *Calluna*. Nutrient availability is again a key factor affecting herbivores, which are known to concentrate on areas of more nutrient rich vegetation and on more nutritious plants or plant parts. Creation of nutrient rich patches by fire is one means of management for improved herbivore production. Intense grazing itself can alter the plant community either by changing the structure and increasing the nutrient content of *Calluna* or by producing a shift in the community composition. Herbivore populations, while affected by the structure and nutrient status of the vegetation, are also to some extent internally regulated. Studies of the red grouse are still unravelling the complexity of factors influencing population densities.

A clear understanding of the interactions between vegetation, herbivores and disturbance including fire is central to management for both production and conservation. We suggest that future research effort should be directed at further elucidating these interactions, with emphasis on nutrient relations and life histories in relation to disturbance. We need to re-examine the effects of fire on nutrient status and to investigate the role of nutrients in the interactions between *Calluna* and other plant species throughout its life cycle. Detailed studies of the functional role of invertebrates in heath systems are revealing much of importance, and further studies are required which concentrate on seed predation and dispersal. Finally, much could be gained from detailed comparisons of heath systems controlled on the one hand by large-scale disturbance such as fire and on the other by small-scale disturbance caused by heather beetle. Studies to date indicate that similar processes may be operating in both systems, and an integrated approach to the effects of disturbance in heathlands could yield results of considerable theoretical and practical interest.

VII. SUMMARY

1. Studies on heathlands in N.W. Europe have provided a remarkably detailed picture of interactions between the heath vegetation, the major herbivores and disturbance by fire.

2. Low nutrient status and disturbances such as fire and herbivory are identified as the main factors controlling the dynamics of heath systems.

3. The classical idea of the "*Calluna* cycle" should be interpreted as one of a larger set of possible pathways for heath vegetation development.

4. Production, shoot nutrient content and species composition are all greatly affected by the age of the dominant, *Calluna vulgaris*. Stand age also affects the outcome of fire. Variations in the regenerative capacity of the dominant and associated species in stands of different ages determine the post-fire vegetation response.

5. Effects of disturbance on the vegetation vary with disturbance intensity and scale. Light-moderate grazing or fires at certain frequencies may maintain heathland communities in a relatively productive state and with *Calluna* as the dominant. More intense grazing or frequent fires may lead to the replacement of *Calluna* by graminaceous species. Phytophagous insects have similar effects, but on a smaller scale.

6. Herbivore populations that are strongly dependent on *Calluna* (e.g., the red grouse) respond to alteration in the vegetation productivity or nutritive value. However, red grouse numbers are also determined by intrinsic population processes.

7. For conservation purposes, traditional management practices may require to be continued or restored, although variations may be necessary to promote biotic diversity.

ACKNOWLEDGEMENTS

We thank Drs E. D. Ford, J. Grace, V. J. Hobbs, C. J. Legg and G. R Miller for their constructive comments on the manuscript, and C. Taplin fo typing it all.

REFERENCES

Airlie (Earl of). (1971). Making full use of an upland estate. *Landowning in Scotlan* **142**, 3–6.

Allen, S. E. (1964). Chemical aspects of heather burning. *J. appl. Ecol.* **1**, 347–367

Allen, S. E., Evans, C. C. and Grimshaw, H. M. (1969). The distribution of minera nutrients in soil after heather burning. *Oikos* **20**, 16–25.

Anderson, P. and Yalden, D. W. (1981). Increased sheep numbers and the loss o heather moorland in the Peak District, England. *Biol. Conserv.* **20**, 195–213.

Arnold, G. W. (1964). Factors within plant associations affecting the behaviour an performance of grazing animals. *In* "Grazing in Terrestrial and Marine Environ ments" (Ed. D. J. Crisp), pp. 133–154. Blackwell, Oxford.

Bakker, J. P. (1978). Some experiments on heathland conservation and regenera tion. *Phytocoenosis* **7**, 351–370.

Bakker, J. P., de Bie, S., Dallinga, J. H., Tjaden, P. and de Vries, Y. (1983) Sheep-grazing as a management tool for heathland conservation and regeneratio in the Netherlands. *J. appl. Ecol.* **20**, 541–560.

Ball, D. F., Dale, J., Sheail, J. and Heal O. W. (1982). "Vegetation Change i Upland Landscapes." Institute of Terrestrial Ecology. Cambridge.

Ball, M. E. (1974). Floristic changes on grasslands and heaths on the Isle of Rhu after a reduction or exclusion of grazing. *J. Env. Manag.* **2**, 299–318.

Ballester, A., Albo, J. M. and Vieitez E. (1977). The allelopathic potential of *Eric scoparia* L. *Oecologia* **30**, 55–61.

Ballester, A., Arines, J. and Vieitez, E. (1972). Compuestos fenólicos en suelos d brezal. *Anal. Edafol. Agrobiol* **31**, 359–366.

Bannister, P. (1964a). Stomatal responses of heath plants to water deficits. *J. Eco* **52**, 151–158.

Bannister, P. (1964b). The water relations of certain heath plants with reference t their ecological amplitude. I. Introduction, germination and establishment. *J Ecol.* **52**, 423–432.

Bannister, P. (1964c). The water relations of certain heath plants with reference t their ecological amplitude. II. Field studies. *J. Ecol.* **52**, 468–481.

Bannister, P. (1964d). The water relations of certain heath plants with reference t their ecological amplitude. III. Experimental studies; General conclusions. *J Ecol.* **52**, 499–510.

Bannister, P. (1976). "Introduction to Physiological Plant Ecology" Blackwell Oxford.

Barclay-Estrup, P. (1970). The description and interpretation of cyclical processes i a heath community. II. Changes in biomass and shoot production during th *Calluna* cycle. *J. Ecol.* **58**, 243–249.

Barclay-Estrup, P. (1971). The description and interpretation of cyclical processes in a heath community. III. Microclimate in relation to the *Calluna* cycle. *J. Ecol.* **59**, 143–166.

Barclay-Estrup, P. (1974). Arthropod populations in a heathland as related to cyclical changes in the vegetation. *Entomol. Mon. Mag.* **109**, 79–84.

Barclay-Estrup, P. and Gimingham, C. H. (1969). The description and interpretation of cyclical processes in a heath community. I. Vegetation change in relation to the *Calluna* cycle. *J. Ecol.* **57**, 737–758.

Barclay-Estrup, P. and Gimingham, C. H. (1975). Seed-shedding in heather (*Calluna vulgaris* (L.) Hull). *Trans. Bot. Soc. Edinb.* **42**, 275–278.

Bayfield, N. G. (1970). Some effects of walking and skiing on vegetation at Cairngorm. *In* "The Scientific Management of Animal and Plant Communities for Conservation" (Eds E. Duffey and A. S. Watt), pp. 469–485. Blackwell, Oxford.

Bayfield, N. G. (1984). The dynamics of heather (*Calluna vulgaris*) stripes in the Cairngorm Mountains, Scotland. *J. Ecol.* **72**, 515–527.

Bayfield, N. G., Urquhart, U. H. and Rothery, P. (1984). Colonisation of bulldozed track verges in the Cairngorm Mountains, Scotland. *J. appl. Ecol.* **21**, 343–354.

Berendse, F. (1985). The effect of grazing on the outcome of competition between plant species with different nutrient requirements. *Oikos* **44**, 35–39.

Berendse, F. and Aerts, R. (1984). Competition between *Erica tetralix* L. and *Molinia caerulea* (L.) Moench as affected by the availability of nutrients. *Acta Oecologia Oecol. Plant* **5**, 3–14.

Blankwaardt. H. F. H. (1977). Het optreden van plagen van de heidekever (*Lochmaea suturalis* Thomson) in Nederland sedert 1915. *Entomol. Berichten.* **37**, 33–40.

Blaxter, K. L., Kay, R. N. B., Sharman, G. A. M., Cunningham, J. M. M. and Hamilton, W. J. (1974). "Farming the Red Deer" Rowett Research Institute and Hill Farming Research Organisation Report No. 1. Department of Agriculture and Fisheries for Scotland, Edinburgh.

Bøcher, T. W. and Jørgensen, C. A. (1972). Jyske dvoergbuskheder. Experimentelle undersøgelser af forskellige kulturindgrebs indflydelse pa vegetationen. *Biol. Skr. Dan. Vid. Selsk.* **19**, 1–55.

Bostock, J. (1983). Countryside Commission experiments on Cannock Chase. *In* "Heathland Management" (Ed. L. Farrell), pp. 78–87. Nature Conservancy Council, Shrewsbury.

Braathe, P. (1950). Granas veksthemning på lyngmark. *Tidsskr. Skogbr.* **58**, 42–44.

Brian, M., Abbott, A., Pearson, B. and Wardlaw, J. (1977). "Ant Research 1954–76" Institute of Terrestrial Ecology, Wareham.

Brian, M. V., Elmes, G. and Kelly, A. F. (1967). Populations of the ant *Tetramorium caespitum* Latrielle. *J. Anim. Ecol.* **36**, 337–342.

Brian, M. V., Hibble, J. and Stradling, D. J. (1965). Ant pattern and density in a southern English heath. *J. Anim. Ecol.* **34**, 545–555.

Brian, M. V., Mountford, M. D., Abbott, A. and Vincent, S. (1976). The changes in ant species distribution during ten years post-fire regeneration of a heath. *J. Anim. Ecol.* **45**, 115–133.

Brown, R. T. and Mikola, P. (1974). The influence of fruticose soil lichens on the mycorrhizae and seedling growth of forest trees. *Acta for. Fenn.* **141**, 1–23.

Brunsting, A. M. H. (1982). The influence of the dynamics of a population of herbivorous beetles on the development of vegetational patterns in a heathland system. *In* "Proc. 5th Int. Symp. on Insect-Plant Relationships" (Eds J. H. Visser and A. H. Minks), pp. 215–224. Center for Agricultural Publishing and Documentation, Wageningen.

Brunsting, A. M. H. and Heil, G. W. (1985). The role of nutrients in the interactions between a herbivorous beetle and some competing plant species in heathlands. *Oikos* **44**, 23–46.

Butterfield, J. and Coulson, J. C. (1983). The carabid communities on peat and upland grasslands in northern England. *Holarctic Ecol.* **6**, 163–174.

Cameron, A. E., McHardy, J. W. and Bennett, A. N. (1944). "The Heather Beetle (*Lochmaea suturalis*): its Biology and Control." British Field Sports Society, Petworth.

Carballeira, A. and Cuervo, A. (1980). Seasonal variation in allelopathic potential of soils from *Erica australis* L. heathland. *Acta Oecologia Oecol. Plant.*, **1**, 345–353.

Carlisle, A. (1977). The impact of man on the native pinewoods of Scotland. *In* "Native Pinewoods of Scotland" (Eds R. G. H. Bunce and J. N. R. Jeffers), pp. 70–77. Institute of Terrestrial Ecology, Cambridge.

Caughley, G. and Lawton, J. H. (1981). Plant-herbivore systems. In "Theoretical Ecology, Principles and Applications, 2nd edn" (Ed. R. M. May), pp. 132–166. Sinauer, Sunderland. Massachusetts.

Charles, W. N., McCowan, D. and East, K. (1977). Selection of upland swards by red deer (*Cervus elaphus* L.) on Rhum. *J. appl. Ecol.* **14**, 55–64.

Chapman, S. B. (1967). Nutrient budgets for a dry heath ecosystem in the south of England. *J. Ecol.* **55**, 677–689.

Chapman, S. B., Hibble, J. and Rafarel, C. R. (1975a). Net aerial production by *Calluna vulgaris* on lowland heath in Britain. *J. Ecol.* **63**, 233–253.

Chapman, S. B., Hibble J. and Rafarel, C. R. (1975b). Litter accumulation under *Calluna vulgaris* on a lowland heath in Britain. *J. Ecol.* **63**, 259–271.

Chapman, S. B. and Webb, N. R. (1978). The productivity of a *Calluna* heathland in southern England. *In* "Production Ecology of British Moors and Montane Grasslands" (Eds O. W. Heal and D. F. Perkins), pp. 247–262. Springer, Berlin.

Chapuis, J. L. and Lefeuvre, J. C. (1980). Evolution saisonniere du regime alimentaire du lapin de garenne. *Oryctolagus cuniculatus* (L.) sur une lande bretonne. *Bull. Ecol.* **11**, 587–597.

Clément, B., Forgeard, F. and Touffet, J. (1980). Importance de la vegetation muscinale dans les premiers stades de recolonisation des landes après incendie. *Bull. Ecol.* **11**, 359–364.

Clément, B. and Touffet, J. (1981). Vegetation dynamics in Brittany heathland after fire. *Vegetatio* **46**, 157–166.

Clutton-Brock, T. H., Guinness, F. E. and Albon, S. D. (1982). "Red Deer Behaviour and Ecology of Two Sexes" University of Chicago Press, Chicago.

Clymo, R. S. (1978). A model of peat bog growth. *In* "Production Ecology of British Moors and Montane Grasslands" (Eds O. W. Heal and D. F. Perkins), pp. 187–223. Springer, Berlin.

Coppins, B. J. and Shimwell, D. W. (1971). Variations in cryptogam complement and biomass in dry *Calluna* heath of different ages. *Oikos* **22**, 204–209.

Coulson, J. C. and Butterfield, J. E. L. (1985). The invertebrate communities of peat and upland grasslands in the north of England and some conservation implications. *Biol. Conserv.* **34**, 197–225.

Coulson, J. C. and Whittaker, J. B. (1978). Ecology of moorland animals. *In* "Production Ecology of British Moors and Montane Grasslands" (Eds O. W. Heal and D. F. Perkins), pp. 52–93. Springer, Berlin.

Currall, J. E. P. (1981). Some effects of management by fire on wet heath vegetation in western Scotland. PhD Thesis, University of Aberdeen.

Daggitt, S. (1981). The carbon dioxide exchange of *Sphagnum capillifolium* (Enrh.) Hedw. growing in a blanket mire habitat. PhD Thesis, University of Leeds.

Dalby, M., Fidler, J. H., Fidler, A. and Duncan, J. E. (1971). The vegetative changes on Ilkley Moor. *Naturalist* **97**, 49–56.

Daniels, J. L. (1983). "Heathland Management in Amenity Areas" Countryside Commission, Cheltenham.

Delany, M. J. (1953). Studies on the microclimate of *Calluna* heathland. *J. Anim. Ecol.* **22**, 227–239.

Deleuil, G. (1950). Mise en evidence de substances toxiques pour les therophytes dans les associations du Rosmarino-Ericon. *Compt. Rend. Acad. Sci. Paris* **230**, 1362–1364.

Diemont, W. H. and Heil, G. W. (1984). Some long-term observations on cyclical and seral processes in Dutch heathlands. *Biol. Conserv.* **30**, 283–290.

Dimbleby, G. W. (1962). "The Development of British Heathlands and their Soils" Oxford Forestry Memoirs 23. Clarendon Press.

Duncan, J. S., Reid, H. W., Moss, R., Phillips, J. D. and Watson, A. (1978). Ticks, louping ill and red grouse on moors in Speyside, Scotland, *J. Wildl. Mgmt.* **42**, 500–505.

Eadie, J. and Maxwell, T. J. (1974). Operational Research in Agriculture. *In* "Symposium on the study of Agricultural Systems" University of Reading.

Elliot, R. J. (1953). The effects of burning on heather moors of the southern Pennines. PhD Thesis, University of Sheffield.

Evans, C. C. and Allen, S. E. (1971). Nutrient losses in smoke produced during heather burning. *Oikos* **22**, 149–154.

Evans, H. (1890). "Some Account of Jura Red Deer" Private Publication. Derby.

Farrell, L. (1983). "Heathland Management" Nature Conservancy Council, Shrewsbury.

Fenton, E. W. (1949). Vegetation changes in hill grazings with particular reference to heather (*Calluna vulgaris*). *J. Brit. Grassland Soc.* **4**, 95–103.

Flux, J. E. C. (1970). Life history of the mountain hare (*Lepus timidus scoticus*) in north-east Scotland. *J. Zool. Lond.* **161**, 75–123.

Forgeard, F. and Chapuis, J. L. (1984). Impact du lapin de garenne, *Oryctolagus cuniculatus* (L.) sur la végétation des pelouses incendiées de Paimpont (Ille-et-Vilaine). *Acta Oecologica Oecol. Gener.* **5**, 215–228.

Forgeard, F. and Touffet, J. (1980). La recolonisation des landes et des pelouses dans la region de Paimpont. Evolution de la végétation au cours de trois années suivant l'incendie. *Bull. Ecol.* **11**, 349–358.

Forrest, G. I. (1971). Structure and production of North Pennine blanket bog vegetation. *J. Ecol.* **59**, 453–479.

Forrest, G. I. and Smith, R. A. H. (1975). The productivity of a range of blanket bog types in the northern Pennines. *J. Ecol.* **63**, 173–202.

Froment, A. (1981). Conservation of *Calluna* – *Vaccinietum* heathland in the Belgian Ardennes: an experimental approach. *Vegetatio* **47**, 193–200.

Gimingham, C. H. (1949). The effects of grazing on the balance between *Erica cinerea* L. and *Calluna vulgaris* (L.) Hull in upland heath, and their morphological responses. *J. Ecol.* **37**, 100–119.

Gimingham, C. H. (1960). Biological flora of the British Isles. *Calluna vulgaris* (L.) Hull. *J. Ecol.* **48**, 455–483.

Gimingham, C. H. (1972). "Ecology of Heathlands" Chapman and Hall, London.

Gimingham, C. H. (1975). "An Introduction to Heathland Ecology" Oliver and Boyd, Edinburgh.

Gimingham, C. H. (1977). The status of pinewoods in British ecosystems. *In* "Native Pinewoods of Scotland" (Eds R. G. H. Bunce and J. N. R. Jeffers), pp. 1–4. Institute of Terrestrial Ecology. Cambridge.

Gimingham, C. H. (1978). *Calluna* and its associated species: some aspects of coexistence in communities. *Vegetatio* **36**, 179–186.

Gimingham, C. H. (1981). Conservation: European heathlands. *In* "Heathlands and Related Shrublands. B. Analytical Studies" (Ed. R. L. Specht), pp. 249–259. Elsevier, Amsterdam.

Gimingham, C. H. (1982). Plant strategies and ecosystem processes in managed vegetation: some implications for conservation. *In* "Evolution and Environment" (Eds V. J. A. Novak and J. Mlíkovský), pp. 1015–1033. ČSAU, Prague.

Gimingham, C. H. (1985). Age-related interactions between *Calluna vulgaris* and phytophagous insects. *Oikos* **44**, 12–16.

Gimingham, C. H., Chapman, S. B. and Webb, N. R. (1979). European heathlands. *In* "Heathlands and Related Shrublands. A. Descriptive Studies" (Ed. R. L. Specht), pp. 365–413. Elsevier, Amsterdam.

Gimingham, C. H. and de Smidt, J. T. (1983). Heaths as natural and semi-natural vegetation. *In* "Man's Impact on Vegetation" (Eds M. J. A. Werger and I. Ikusima), pp. 185–199. Junk, The Hague.

Glass, A. D. M. (1976). The allelopathic potential of phenolic acids associated with the rhizosphere of *Pteridium aquilinum*. *Can. J. Bot.* **54**, 2440–2444.

Gliessman, F. L. S. (1976). Allelopathy in a broad spectrum of environments as illustrated by bracken. *Bot. J. Linn. Soc.* **73**, 95–104.

Gong, W. K. (1976). Birch regeneration in heathland vegetation. PhD Thesis, University of Aberdeen.

Gore, A. J. P. and Olson, J. S. (1967). Preliminary models for accumulation of organic matter in an *Eriophorum/Calluna* ecosystem. *Aquilo, Ser. Botanica* **6**, 297–313.

Grace, J. and Marks, T. C. (1978). Physiological aspects of bog production at Moor House. *In* "Production Ecology of British Moors and Montane Grasslands" (Eds O. W. Heal and D. F. Perkins), pp. 37–51. Springer, Berlin.

Grace, J. and Woolhouse, H. W. (1970). A physiological and mathematical study of the growth and productivity of a *Calluna – Sphagnum* Community. I. Net photosynthesis of *Calluna vulgaris* (L.) Hull. *J. appl. Ecol.* **7**, 363–381.

Grace, J. and Woolhouse, H. W. (1973a). A physiological and mathematical study of the growth and productivity of a *Calluna – Sphagnum* community. II. Light interception and photosynthesis in *Calluna*. *J. appl. Ecol.* **10**, 63–76.

Grace, J. and Woolhouse, H. W. (1973b). A physiological and mathematical study of the growth and productivity of a *Calluna – Sphagnum* community. III Distribution of photosynthate in *Calluna vulgaris* L. Hull. *J. appl. Ecol.* **10**, 77–91

Grace, J. and Woolhouse, H. W. (1974). A physiological and mathematical study of the growth and productivity of a *Calluna – Sphagnum* community. IV. A model of growing *Calluna*. *J. appl. Ecol.* **11**, 281–295.

Grant, S. A. (1968). Heather regeneration following burning: a survey. *J. Brit Grassld. Soc.* **23**, 26–33.

Grant, S. A. (1971). Interactions of grazing and burning on heather moors. 2. Effect on primary production and level of utilisation. *J. Brit. Grassld. Soc.* **26**, 173–181.

Grant, S. A., Barthram, G. T., Lamb, W. I. C. and Milne, J. A. (1978). Effects of season and level of grazing on the utilisation of heather by sheep. 1. Responses of the sward. *J. Brit. Grassld. Soc.* **33**, 289–300.

Grant, S. A., Hamilton, W. J. and Souter, C. (1981). The responses of heather-dominated vegetation in north-east Scotland to grazing by red deer. *J. Ecol.* **69**, 189–204.

Grant, S. A. and Hunter, R. F. (1962). Ecotypic differentiation of *Calluna vulgaris* (L.) Hull in relation to altitude. *New Phytol.* **61**, 44–55.

Grant, S. A. and Hunter, R. F. (1966). The effects of frequency and season of clipping on the morphology, productivity and chemical composition of *Calluna vulgaris* (L.) Hull. *New Phytol.* **65**, 125–133.

Grant, S. A. and Hunter, R. F. (1968). Interactions of grazing and burning on heather moors and their implications in heather management. *J. Brit. Grassld. Soc.* **23**, 285–293.

Grant, S. A., Lamb, W. I. C., Kerr, C. D. and Bolton, G. R. (1976). The utilisation of blanket bog vegetation by grazing sheep. *J. appl. Ecol.* **13**, 857–869.

Groves, R. H. (1981). Heathland soils and their fertility status. *In* "Heathlands and Related Shrublands. B. Analytical Studies" (Ed. R. L. Specht), pp. 143–150. Elsevier, Amsterdam.

Grubb, P. J., Green, H. E. and Merrifield, R. C. J. (1969). The ecology of chalk heath: its relevance to the calcicole-calcifuge and soil acidification problems. *J. Ecol.* **57**, 175–212.

Grubb, P. J. and Suter, M. B. (1971). The mechanism of acidification by *Calluna* and *Ulex* and the significance for conservation. *In* "The Scientific Management of Animal and Plant Communities for Conservation". (Eds E. Duffey and A. S. Watt), pp. 115–133. Blackwell, Oxford.

Gueguen, A., Lefeuvre, J. C., Forgeard, F. and Touffet, J. (1980). Analyse comparée de la dynamique de la restauration du peuplement d'Orthoptères et du peuplement végétal dans une zone brûlée de lande. *Bull. Ecol.* **11**, 747–763.

Handley, W. R. C. (1963). Mycorrhizal associations and *Calluna* heathland afforestation. *For. Comm. Bull.* **36**. HMSO, London.

Hansen, K. (1964). Studies on the regeneration of heath vegetation after burning-off. *Bot. Tidsskr.* **60**, 1–41.

Hansen, K. (1969). Edaphic conditions of Danish heath vegetation and the response to burning-off. *Bot. Tidsskr.* **64**, 121–140.

Hansen, K. (1976). Ecological studies in Danish heath vegetation. *Dansk. Bot. Arkiv.* **41**, 1–118.

Hawksworth, D. L., James, P. W. and Coppins, B. J. (1980). Checklist of British lichen-forming, lichenicolous and allied fungi. *Lichenologist* **12**, 1–115.

Heal, O. W. (1980). Fauna of heathlands in the United Kingdom. *Bull. Ecol.* **11**, 413–420.

Heal, O. W. and Perkins, D. F. (1978). "The Ecology of Some British Moors and Montane Grasslands". Springer, Berlin.

Heddle, E. M. and Specht, R. L. (1975). Dark Island Heath (Ninety-Mile plain, South Australia). VIII. The effect of fertilisers on composition and growth, 1950–1972. *Aust. J. Bot.* **23**, 151–164.

Heil, G. W. and Diemont, W. H. (1983). Raised nutrient levels change heathland into grassland. *Vegetatio* **53**, 113–120.

Helsper, H. P. G., Glenn-Lewin, D. and Werger, M. J. A. (1983). Early regeneration of *Calluna* heathland under various fertilization treatments. *Oecologia* **58**, 208–214.

Helsper, H. P. G. and Klerken, G. A. M. (1984). Germination of *Calluna vulgaris* (L.) Hull in vitro under different pH-conditions. *Acta. bot. Neerl.* **33**, 347–353.

Hewson, R. (1962). Food and feeding habits of the mountain hare *Lepus timidus scoticus* Hilzheimer. *Proc. Zool. Soc. Lond.* **139**, 515–526.

Hewson, R. (1965). Population changes in the mountain hare *Lepus timidus* L. *J. Anim. Ecol.* **34**, 587–600.

Hewson, R. (1976). Grazing by mountain hares *Lepus timidus* L., red deer *Cervus elaphus* L. and red grouse *Lagopus l. scoticus* on heather moorland in North-East Scotland. *J. appl. Ecol.* **13**, 657–666.

Hewson, R. (1977). The effect on heather *Calluna vulgaris* of excluding sheep from moorland in north-east England. *Naturalist* **102**, 133–136.
Hewson, R. (1984). Scavenging and predation upon sheep and lambs in west Scotland. *J. appl. Ecol.* **21**, 843–865.
Hewson, R. and Wilson, C. J. (1979). Home range and movements of Scottish blackface sheep in Lochaber, North-West Scotland. *J. appl. Ecol.* **16**, 743–751.
Hill Farming Research Organization. (1979). "Science and Hill Farming" Hill Farming Research Organization, Penicuik.
Hobbs, R. J. (1981). Post-fire succession in heathland communities. PhD Thesis. University of Aberdeen.
Hobbs, R. J. (1983). Markov models in the study of post-fire succession in heathland communities. *Vegetatio* **56**, 17–30.
Hobbs, R. J. (1984a). Possible chemical interactions among heathland plants. *Oikos* **43**, 23–29.
Hobbs, R. J. (1984b). Length of burning rotation and community composition in high-level *Calluna-Eriphorum* bog in N. England. *Vegetatio* **57**, 129–136.
Hobbs, R. J. (1985). The persistence of *Cladonia* patches in closed heathland stands. *Lichenologist* **17**, 103–109.
Hobbs, R. J., Currall, J. E. P. and Gimingham, C. H. (1984). The use of "thermo-color" pyrometers in the study of heath fire behaviour. *J. Ecol.* **72**, 241–250.
Hobbs, R. J. and Gimingham, C. H. (1980). Some effects of fire and grazing on heath vegetation. *Bull. Ecol.* **11**, 709–715.
Hobbs, R. J. and Gimingham, C. H. (1984a). Studies on fire in Scottish heathland communities. I. Fire characteristics. *J. Ecol.* **72**, 223–240.
Hobbs, R. J. and Gimingham, C. H. (1984b). Studies on fire in Scottish heathland communities. II. Post-fire vegetation development. *J. Ecol.* **72**, 585–610.
Hobbs, R. J. and Legg, C. J. (1983). Markov models and initial floristic composition in heathland vegetation dynamics. *Vegetatio* **56**, 31–43.
Hobbs, R. J., Mallik, A. U. and Gimingham, C. H. (1984). Studies on fire in Scottish heathland communities. III. Vital attributes of the species. *J. Ecol.* **72**, 963–976.
Hobbs, V. J. (1984). The structure and dynamics of *Polytrichum piliferum* communities. PhD Thesis. University of Aberdeen.
Hodkinson, I. D. (1973a). The biology of *Strophingia ericae* (Curtis) (Homoptera, Psylloidea) with notes on its primary parasite *Tetrastichus actis* (Walker) (Hym., Eulophidae). *Norsk. Entomol. Tidsskrift* **20**, 237–243.
Hodkinson, I. D. (1973b). The population dynamics and host plant interactions of *Strophingia ericae* (Curt.) (Homoptera: Psylloidea). *J. Anim. Ecol.* **42**, 565–583.
Hopkins, P. J. and Webb, N. R. (1984). The composition of the beetle and spider faunas on fragmented heathlands. *J. appl. Ecol.* **21**, 935–946.
de Hullu, E. and Gimingham, C. H. (1984). Germination and establishment of seedlings in different phases of the *Calluna* life cycle in a Scottish heathland. *Vegetatio* **58**, 115–121.
Hunter, R. F. (1962). Hill sheep and their pasture: a study of sheep grazing in south-east Scotland. *J. Ecol.* **50**, 651–680.
Hunter, R. F. (1964). Home range behaviour in hill sheep. *In* "Grazing in Terrestrial and Marine Environments." (Ed. D. J. Crisp), pp. 155–171. Blackwell, Oxford.
Hunter, R. F. and Milner, C. (1963). The behaviour of individual, related and groups of South Country Cheviot sheep. *Anim. Behav.* **11**, 507–513.
Imeson, A. C. (1971). Heather burning and soil erosion on the North Yorkshire Moors. *J. appl. Ecol.* **8**, 537–542.
Jalal, M. A. F. and Read, D. J. (1983a). The organic acid composition of *Calluna*

heathland soil with special reference to phyto- and fungitoxicity. I. Isolation and identification of organic acids. *Plant and Soil* **70**, 257–272.

Jalal, M. A. F. and Read, D. J. (1983b). The organic acid composition of *Calluna* heathland soil with special reference to phyto- and fungitoxicity. II. Monthly quantitative determination of the organic acid content of *Calluna* and spruce dominated soils. *Plant and Soil.* **70**, 273–286.

Jarvis, P.G. (1964). Interference by *Deschampsia flexuosa* L. Trin. *Oikos* **15**, 56–78.

Jenkins, D., Watson, A. and Miller, G. R. (1963). Population studies on red grouse, *Lagopus lagopus scoticus* (Lath.) in north-east Scotland. *J. Anim. Ecol.* **32**, 317–376.

Jenkins, D., Watson, A. and Miller, G. R. (1967). Population fluctuations in the red grouse (*Lagopus lagopus scoticus*). *J. Anim. Ecol.* **36**, 97–122.

Jones, H. E. and Gore, A. J. P. (1978). A simulation of production and decay in blanket bog. *In* "Production Ecology of British Moors and Montane Grasslands" (Eds O. W. Heal and D. F. Perkins), pp. 160–186. Springer, Berlin.

De Jong, T. J. and Klinkhamer, P. G. L. (1983). A simulation model for the effects of burning on the phosphorous and nitrogen cycle of a heathland ecosystem. *Ecol. Model.* **19**, 263–284.

Kashimura, T. (1985). The distribution of some heathland plant species along a microtopographic gradient at Dinnet, Scotland, and their dehydration resistances. *Vegetatio* **20**, 57–65.

Kay, R. N. B. and Staines, B. W. (1981). The nutrition of red deer (*Cervus elaphus*). *Nutrition Abstracts Review, B.* **57**, 601–622.

Kayll, A. J. and Gimingham, C. H. (1965). Vegetative regeneration of *Calluna vulgaris* after fire. *J. Ecol.* **53**, 729–734.

Kayll, A. J. (1966). Some characteristics of heath fires in N. E. Scotland. *J. appl. Ecol.* **3**, 29–40.

Keatinge, T. H. (1975). Plant community dynamics in wet heathland. *J. Ecol.* **63**, 163–172.

Kenworthy, J. B. (1963). Temperatures in heather burning. *Nature, Lond.* **200**, 1226.

Kenworthy, J. B. (1964). A study of the changes in plant and soil nutrients associated with moorburning and grazing. PhD Thesis, University of St. Andrews.

Kinako, P. D. S. (1975). Effects of heathland fires on the microhabitat and regeneration of vegetation. PhD Thesis, University of Aberdeen.

Kinako, P. D. S. and Gimingham, C. H. (1980). Heather burning and soil erosion on upland heath in Scotland, U.K. *J. Env. Manag.* **10**, 277–284.

Kjellsson, G. (1985a). Seed fate in a population of *Carex pilulifera* L. I. Seed dispersal and ant-seed mutualism. *Oecologia* (Berlin) **67**, 416–523.

Kjellsson, G. (1985b). Seed fate in a population of *Carex pilulifera* L. II. Seed predation and its consequences for dispersal and seed bank. *Oecologia* (Berlin) **67**, 424–429.

Kottman, H. J., Schwoeppe, W., Willers, T. and Wittig, R. (1985). Heath conservation by sheep grazing: a cost–benefit analysis. *Biol. Conserv.* **31**, 67–74.

Lance, A. N. (1978a). Survival and recruitment success of individual young cock red grouse *Lagopus L. scoticus* tracked by radio telemetry. *Ibis* **120**, 369–379.

Lance, A. N. (1978b). Territories and the food plant of individual red grouse. II. Territory size compared with an index of nutrient supply in heather. *J. Anim. Ecol.* **47**, 307–313.

Legg, C. J. (1978). Succession and homeostasis in heathland vegetation. PhD Thesis, University of Aberdeen.

Legg, C. J. (1980). A Markovian approach to the study of heath vegetation dynamics. *Bull. Ecol.* **11**, 393–404.

Liddle, M. J. and Chitty, L. D. (1981). The nutrient budget of horse tracks on an English lowland heath. *J. appl. Ecol.* **18**, 841–848.

Lippe, E., de Smidt, J. T. and Glenn-Lewin, D. C. (1985). Markov models and succession: a test from a heathland in the Netherlands. *J. Ecol.* **73**, 775–791.

Lovat, Lord. (1911). Heather burning. *In* "The Grouse in Health and in Disease" (Ed. A. S. Leslie), pp. 392–412. Smith, Elder, London.

Lowday, J. (1983). Bracken control on lowland heaths. *In* "Heathland Management" (Ed. L. Farrell), pp. 68–77. Nature Conservancy Council, Shrewsbury.

Lowe, V. P. W. (1966). Observations on the dispersal of red deer on Rhum. *In* "Play, Exploration and Territory in Mammals" (Eds P. A. Jewell and C. Loizos), pp. 211–228. Academic Press, London.

Lowe, V. P. W. (1969). Population dynamics of red deer (*Cervus elaphus* L.) on Rhum. *J. Anim. Ecol.* **38**, 425–457.

Lowe, V. P. W. (1971). Some effects of a change in estate management on a deer population. *In* "The Scientific Management of Plant and Animal Communities for Conservation" (Eds E. Duffey and A. S. Watt), pp. 437–456. Blackwell, Oxford.

Mackenzie, J. M. D. (1952). Fluctuations in the number of British Tetronids. *J. Anim. Ecol.* **21**, 128–153.

McVean, D. N. and Lockie, J. D. (1969). "Ecology and Land Use in Upland Scotland" University Press, Edinburgh.

Mallik, A. U. (1982). Post-fire microhabitat and plant regeneration in heathland. PhD Thesis. University of Aberdeen.

Mallik, A. U. and Gimingham, C. H. (1983). Regeneration of heathland plants following burning. *Vegetatio* **53**, 45–58.

Mallik, A. U. and Gimingham, C. H. (1985). Ecological effects of heather burning. II. Effects on seed germination and vegetative regeneration. *J. Ecol.* **73**, 633–644.

Mallik, A. U., Gimingham, C. H. and Rahman, A. A. (1984). Ecological effects of heather burning. I. Water infiltration, moisture retention and porosity of surface soil. *J. Ecol.* **72**, 767–776.

Mallik, A. U., Hobbs, R. J. and Legg, C. J. (1984). Seed dynamics in *Calluna-Arctostaphylos* heath in north-eastern Scotland. *J. Ecol.* **72**, 855–871.

Mallik, A. U. and Rahman, A. A. (1985). Soil water repellency in regularly burned *Calluna* heathlands: comparison of three measuring techniques. *J. Env. Manag.* **20**, 207–218.

Maltby, E. (1980). The impact of severe fire on *Calluna* moorland in the North York Moors. *Bull. Ecol.* **11**, 683–708.

Maltby, E. and Legg, C. J. (1983). Revegetation of fossil patterned ground exposed by severe fire on the North York Moors. *In* "Permafrost: fourth International Conference, Proceedings" pp. 792–797. National Academy Press, Washington D.C.

Mantilla, J. L. G., Arines, J. and Vieitez, E. (1975). Actividad biologica sobre el crecimiento y germinacion de extractos de *Calluna vulgaris* (L.) Hull. *Anal. Edafol. Agrobiol.* **34**, 789–795.

Marrs, R. (1983). Scrub control on lowland heaths. *In* "Heathland Management" (Ed. L. Farrell), pp. 59–67. Nature Conservancy Council, Shrewsbury.

Marrs, R. H. (1985). The effects of potential bracken and scrub control herbicides on lowland *Calluna* and grass heath communities in East Anglia, U.K. *Biol. Conserv.* **32**, 13–32.

Marrs, R. H. and Bannister, P. (1978). The adaptation of *Calluna vulgaris* (L.) Hull to contrasting soil types. *New Phytol.* **81**, 753–761.

Martin, D. J. (1964). Analysis of sheep diet utilising plant epidermal fragments in faeces samples. *In* "Grazing in Terrestrial and Marine Environments" (Ed. D. J. Crisp), pp. 173–188. Blackwell, Oxford.

Melber, A. (1983). *Calluna* – Samen als Nahrungsquelle fur Laufkafer in einer nordwestdeutschen Sandheide (Col: Carabidae). *Zool. Jb. Syst.* **110**, 87–95.

Middleton, A. D. (1934). Periodic fluctuations in British game populations. *J. Anim. Ecol.* **3**, 231–249.

Miles, J. (1971). Burning *Molinia*-dominant vegetation for grazing by red deer. *J. Brit. Grassld. Soc.* **26**, 247–250.

Miles, J. (1974a). Experimental establishment of new species from seed in Callunetum in north-east Scotland. *J. Ecol.* **62**, 527–551.

Miles, J. (1974b). Effects of experimental interference with stand structure on establishment of seedlings in Callunetum. *J. Ecol.* **62**, 675–687.

Miles, J. (1975). Performance after six growing seasons of new species established from seed in Callunetum in north-east Scotland. *J. Ecol.* **63**, 891–901.

Miles, J. (1981a). Problems in heathland and grassland dynamics. *Vegetatio* **46**, 61–74.

Miles, J. (1981b). "Effect of Birch on Moorlands" Institute of Terrestrial Ecology, Cambridge.

Miles, J. and Kinnaird, J. W. (1979). The establishment and regeneration of Birch, Juniper and Scots Pine in the Scottish highlands. *Scot. For.* **33**, 102–117.

Miles, J., Welch, D. and Chapman, S. B. (1978). Vegetation and management in the uplands. *In* "Upland Land Use in England and Wales" (Ed. O. W. Heal), pp. 77–95. Countryside Commission Publication CCP111, Cheltenham.

Miles, J. and Young, W. F. (1980). The effects on heathland and moorland soils in Scotland and Northern England following colonisation by Birch (*Betula* spp). *Bull. Ecol.* **11**, 233–242.

Miller, G. R. (1964). Land use in the Scottish Highlands VII. The management of heather moors. *Adv. Sci.* (Lond.) **21**, 163–169.

Miller, G. R. (1968). Evidence for selective feeding on fertilised plots by red grouse, hares and rabbits. *J. Wildl. Mgmt.* **32**, 849–853.

Miller, G. R. (1979). Quantity and quality of annual production of shoots and flowers by *Calluna vulgaris* in north-east Scotland. *J. Ecol.* **67**, 109–129.

Miller, G. R. (1980). The burning of heather moorland for red grouse. *Bull. Ecol.* **11**, 725–733.

Miller, G. R. and Cummins, R. P. (1982). Regeneration of scots pine *Pinus sylvestris* at a natural tree-line in the Cairngorm Mountains. *Holarctic Ecol.* **5**, 27–34.

Miller, G. R., Jenkins, D. and Watson, A. (1966). Heather performance and red grouse populations. 1. Visual estimates of heather performance. *J. appl. Ecol.* **3**, 313–326.

Miller, G. R., Kinnaird, J. W., Cummins, R. P. (1982). Liability of saplings to browsing on a red deer range in the Scottish highlands. *J. appl. Ecol.* **19**, 941–951.

Miller, G. R. and Miles, A. M. (1969). Productivity and management of heather. *In* "Grouse Research in Scotland" 13th Progress Report, pp. 31–45. Nature Conservancy, Edinburgh.

Miller, G. R. and Miles J. (1970). Regeneration of heather (*Calluna vulgaris* (L.) Hull) at different ages and seasons in north-west Scotland. *J. appl. Ecol.* **7**, 51–60.

Miller, G. R., Miles, J. and Heal, O. W. (1984). "Moorland Management: a Study of Exmoor" Institute of Terrestrial Ecology, Cambridge.

Miller, G. R. and Watson, A. (1974). Some effects of fire on vertebrate herbivores in the Scottish Highlands. *Proc. Annual Tall Timbers Fire Ecology Conf.* **12**, 39–64.

Miller, G. R. and Watson, A. (1978a). Heather productivity and its relevance to the regulation of red grouse populations. *In* "Production Ecology of some British Moors and Montane Grasslands" (Eds O. W. Heal and D. F. Perkins), pp. 277–285. Springer, Berlin.

Miller, G. R. and Watson, A. (1978b). Territories and the food plant of individual red grouse. I. Territory size, number of mates and brood size compared with the abundance, production and diversity of heather. *J. Anim. Ecol.* **47**, 293–305.

Miller, G. R. and Watson, A. (1983). Heather moorland in northern Britain. *In* "Conservation in Pespective" (Eds A. Warren and F. B. Goldsmith), pp. 101–117. Wiley, London.

Miller, G. R., Watson, A. and Jenkins, D. J. (1970). Responses of red grouse populations to experimental improvement of their food. *In* "Animal Populations in Relation to their Food Resources" (Ed. A. Watson), pp. 323–335. Blackwell, Oxford.

Milne, J. A. (1974). The effects of season and age of stand on the nutritive value of heather (*Calluna vulgaris* L. Hull) to sheep. *J. Agric. Sci.* **83**, 281–288.

Milner, C. and Gwynne, D. (1974). The Soay sheep and their food supply. *In* "Island Survivors: The Ecology of Soay Sheep" (Eds P. A. Jewell, C. Milner and J. Morton Boyd), pp. 273–325. Athlone Press, London.

Mitchell, B., Staines, B. W. and Welch, D. (1977). "Ecology of Red Deer: a Research Review Relevant to their Management in Scotland" Institute of Terrestrial Ecology. Cambridge.

Mohamed, B. F. and Gimingham, C. H. (1970). The morphology of vegetative regeneration in *Calluna vulgaris. New Phytol.* **69**, 743–750.

Moore, N. W. (1962). The heaths of Dorset and their conservation. *J. Ecol.* **50**, 369–391.

Moran, P. A. P. (1952). The statistical analysis of game bag records. *J. Anim. Ecol.* **3**, 154–158.

Morison, G. D. (1963). "The Heather Beetle (*Lochmaea suturalis* Thomson)" North of Scotland College of Agriculture, Aberdeen.

Moss, R. (1969). A comparison of red grouse (*Lagopus L. scoticus*) stocks with the production and nutritive value of heather (*Calluna vulgaris*). *J. Anim. Ecol.* **38**, 103–122.

Moss, R. (1972). Food selection by red grouse [*Lagopus Lagopus scoticus* (Lath.)] in relation to chemical composition. *J. Anim. Ecol.* **41**, 411–428.

Moss, R. (1977). The digestion of heather by red grouse during the spring. *Condor* **79**, 471–477.

Moss, R. and Hewson, R. (1985). Effects on heather of heavy grazing by mountain hares. *Holarctic Ecol.* **8**, 280–284.

Moss, R. and Miller, G. R. (1976). Production, dieback and grazing of heather (*Calluna vulgaris*) in relation to the numbers of red grouse (*Lagopus L. scoticus*) and mountain hares (*Lepus timidus*) in north east Scotland. *J. appl. Ecol.* **13**, 369–377.

Moss, R., Miller, G. R. and Allen, S. E. (1972). The selection of heather by captive red grouse in relation to the age of the plant. *J. appl. Ecol.* **9**, 771–782.

Moss, R. and Watson, A. (1980). Inherent changes in the aggressive behaviour of a fluctuating red grouse *Lagopus lagopus scoticus* population. *Ardea* **68**, 113–119.

Moss, R. and Watson, A. (1985). Adaptive value of spacing behaviour in population cycles of red grouse and other animals. *In* "Behavioural Ecology" (Eds R. M. Sibly and R. H. Smith), pp. 275–294. Blackwell, Oxford.

Moss, R., Watson, A., and Parr, R. (1975). Maternal nutrition and breeding success in red grouse (*Lagopus lagopus scoticus*). *J. Anim. Ecol.* **44**, 233–244.

Moss, R., Watson, A. and Rothery, P. (1984). Inherent changes in the body size, variability and behaviour of a fluctuating red grouse (*Lagopus lagopus scoticus*) population. *J. Anim. Ecol.* **53**, 171–190.

Moss, R., Welch, D. and Rothery, P. (1981). Effects of grazing by mountain hares and red deer on the production and chemical composition of heather. *J. appl. Ecol.* **18**, 487–496.

Muirburn Working Party (1977). "A Guide to Good Muirburn Practice" HMSO, Edinburgh.

Nelson, J. M. (1971). The invertebrates of an area of Pennine Moorland within the Moor House Nature Reserve in Northern England. *Trans. Soc. Brit. Entomol.* **19**, 173–235.

Nicholson, I. A. (1971). Some effects of animal grazing and browsing on vegetation. *Trans. Bot. Soc. Edinb.* **41**, 85–94.

Nicholson, I. A. (1974). Red deer range and the problems of carrying capacity in the Scottish Highlands. *Mammal Review* **4**, 103–118.

Nicholson, I. A. Paterson, I. S. and Currie, A. (1970). A study of vegetational dynamics: selection by sheep and cattle in *Nardus* pasture. *In* "Animal Populations in Relation to their Food Resources" (Ed. A. Watson), pp. 129–143. Blackwell, Oxford.

Parkinson, J. D. and Whittaker, J. B. (1975). A study of two physiological races of the heather psyllid, *Strophingia ericae* (Curtis) (Homoptera: Psylloidea). *Biol. J. Linn. Soc.* **7**, 73–81.

Pearman, P. J. (1959). The influence of heather (*Calluna vulgaris* L.) on root growth in tree seedlings. *J. Oxford Univ. For. Soc.* **5**, 28–33.

Picozzi, N. (1968). Grouse bags in relation to the management and geology of heather moors. *J. appl. Ecol.* **5**, 483–488.

Potts, G. R., Tapper, S. C. and Hudson, P. J. (1984). Population fluctuations in red grouse: analysis of bag records and a simulation model. *J. Anim. Ecol.* **53**, 21–36.

Putwain, P. D., Gillham, D. A. and Holliday, R. J. (1982). Restoration of heather moorland and lowland heathland, with special reference to pipelines. *Environ. Conserv.* **9**, 225–235.

Radley, J. (1965). Significance of major moorland fires. *Nature* (*London*) **205**, 1254–1259.

Ramaut, J. L. and Corvisier, M. (1975). Effets inhibiteurs des extraits de *Cladonia impexa* Harm., *C. gracilis* (L.) Willd. et *Cornicularia muricata* (Ach.) Ach. sur la germination des graines de *Pinus sylvestris* L. *Oecol. Plant.* **10**, 295–299.

Rawes, M. (1983). Changes in two high altitude blanket bogs after the cessation of sheep grazing. *J. Ecol.* **71**, 219–235.

Rawes, M. and Heal, O. W. (1978). The blanket bog as part of a Pennine moorland. *In* "The Ecology of Some British Moors and Montane Grasslands". (Eds O. W. Heal and D. F. Perkins), pp. 224–243. Springer, Berlin.

Rawes, M. and Hobbs, R. (1979). Management of semi-natural blanket bog in the northern Pennines. *J. Ecol.* **67**, 789–807.

Rawes, M. and Welch, D. (1969). Upland productivity of vegetation and sheep at Moor House National Nature Reserve, Westmorland, England. *Oikos Suppl.* **11**, 72 pp.

Rawes, M. and Williams, R. (1973). Production and utilisation of *Calluna* and *Eriophorum*. *Potassium Inst. Colloq. Proc.* **3**, 115–119.

Read, D. J. and Jalal, M. A. F. (1980). The physiological basis of interaction between *Calluna vulgaris*, forest trees, and other plant species. *Proc. Conf. Weed Control in Forestry*. Univ. of Nottingham. pp. 21–32.

Reader, R. J. (1984). Comparison of the annual flowering schedules for Scottish heathland and mediterranean type shrublands *Oikos* **43**, 1–8.

Reader, R. J., Mallik, A. U., Hobbs, R. J. and Gimingham, C. H. (1983). Shoot regeneration after fire or freezing temperatures and its relation to plant life form for some heathland species. *Vegetatio* **55**, 181–189.

Richards, O. W. (1926). Studies on the ecology of English Heaths. III. Animal communities of the felling and burn succession at Oxshott heath, Surrey. *J. Ecol.* **14**, 244–281.

Robertson, K. P. and Woolhouse, H. W. (1984a). Studies of the seasonal course of carbon uptake of *Eriophorum vaginatum* in a moorland habitat. I. Leaf production and senescence. *J. Ecol.* **72**, 423–435.

Robertson, K. P. and Woolhouse, H. W. (1984b). Studies of the seasonal course of carbon uptake of *Eriophorum vaginatum* in a moorland habitat. II. The seasonal course of photosynthesis. *J. Ecol.* **72**, 685–700.

Robertson, R. A. (1957). Heather management. *Scott. Agric.* **37**, 126–129.

Robertson, R. A. and Davies, G. E. (1965). Quantities of plant nutrients in heather ecosystems. *J. appl. Ecol.* **2**, 211–219.

Robinson, R. K. (1971). Importance of soil toxicity in relation to the stability of plant communities. *In* "The Scientific Management of Animal and Plant Communities for Conservation" (Eds E. Duffey and A. S. Watt), pp. 105–113. Blackwell, Oxford.

Robinson, R. K. (1972). The production by roots of *Calluna vulgaris* of a factor inhibitory to growth of some mycorrhizal fungi. *J. Ecol.* **60**, 219–224.

Roff, W. J. (1964). An analysis of competition between *Calluna vulgaris* and *Festuca ovina*. PhD Thesis, University of Cambridge.

Rothery, P., Moss, R. and Watson, A. (1984). General properties of predictive population models in red grouse (*Lagopus lagopus scoticus*). *Oecologia* (Berlin) **62**, 382–386.

Roze, F. and Forgeard, F. (1982). Evolution de la mineralisation de l'azote dans les sols des landes incendiées et non incendiées de la region de Paimpont (Bretagne, France). *Acta. Oecol., Oecol. Plant.* **3**, 249–268.

Savory, C. J. (1974). The feeding ecology of red grouse in N.E. Scotland. PhD Thesis, University of Aberdeen.

Savory, C. J. (1978). Food consumption by red grouse in relation to the age and productivity of heather. *J. Anim. Ecol.* **47**, 269–282.

Smidt, J. T. de. (1977). Interaction of *Calluna vulgaris* and the heather beetle (*Lochmaea suturalis*). *In* "Vegetation und Fauna" (Ed. R. Tüxen), pp. 179–186. Cramer, Vaduz.

Smidt, J. T. de. (1979). Origin and destruction of northwest European heath vegetation. *In* "Werden und Vergehen von Pflanzengesellschaften" (Ed. O. Wilmanns and R. Tuxen), pp. 411–435. Cramer, Vaduz.

Smith, A. J. E. (1978). "The Moss Flora of Britain and Ireland". Cambridge University Press. Cambridge.

Specht, R. L. (1979). Heathlands and related shrublands of the world. *In* "Heathlands and Related Shrublands of the World. A. Descriptive Studies" (Ed. R. L. Specht), pp. 1–18. Elsevier. Amsterdam.

Specht, R. L., Conner, D. J. and Clifford, H. T. (1977). The heath-savannah problem: the effect of fertiliser on sand-heath vegetation of North Stradbroke Island, Queensland. *Aust. J. Ecol.* **2**, 179–186.

Staines, B. W. (1969). Our knowledge of deer behaviour and its possible effects on deer husbandry. *In* "The Husbanding of Red Deer" (Eds. M. M. Bannerman and K. L. Blaxter), pp. 29–31. Rowett Research Institute, Aberdeen.

Staines, B. W. (1977). Factors affecting the seasonal distributions of red deer (*Cervus elaphus*) in Glen Dye, north-east Scotland. *Ann. appl. Biol.* **87**, 495–512.

Staines, B. W., Crisp, J. M. and Parish, T. (1982). Differences in the quality of food eaten by red deer (*Cervus elaphus*) stags and hinds in winter. *J. appl. Ecol.* **19**, 65–77.

Tansley, A. G. (1939). "The British Islands and their Vegetation" Cambridge University Press. Cambridge.

Taylor, K. and Marks, T. C. (1971). The influence of burning and grazing on the growth and development of *Rubus chamaemorus* L. in *Calluna-Enophorum* bog. *In* "The Scientific Management of Animal and Plant Communities for Conservation" (Eds E. Duffey and A. S. Watt), pp. 153–166. Blackwell, Oxford.

Thomas, B. (1934). The composition of common heather. *J. agric. Sci., Camb.* **24**, 151–155.

Thomas, B. (1937). The composition and feeding value of heather at different periods of the year. *J. Minist. Agric. Fish.* **43**, 1050–1055.

Thomas, B. and Armstrong, D. G. (1952). The nutritive value of common heather (*Calluna vulgaris*). 1. The preparation of samples of *Calluna vulgaris* for analytical purposes and for digestibility studies. *J. agric. Sci., Camb.* **42**, 461–464.

Thomas, B. and Dougall, H. W. (1947). Yield of edible material from common heather. *Scott. Agric.* **27**, 35–38.

Tivy, J. (1973). "The Organic Resources of Scotland: their Nature and Evaluation". Oliver and Boyd, Edinburgh.

Torkildsen, G. B. (1950). On årsakene til granens dårlige gjenvekst i einstapelestand. *Blyttia* **8**, 160–164.

Trepp, W. (1961). Die Planteform des Heidelbeer-Fichtenwaldes der Alpen (Picetum subalpinum myrtilletosum). *Schweiz. Z. Forstw.* **112**, 337–350.

Tubbs, C. R. (1974). Heathland management in the New Forest, Hampshire, England. *Biol. Conserv.* **6**, 303–306.

Tutin, T. G., Heywood, V. H., Burgess, N. A., Valentine, D. H., Walters, S. M. and Webb, D. A. (1964–80). "Flora Europaea", Vols I–V. Cambridge University Press, London.

Tye, A. (1980). The breeding biology and population size of the wheatear (*Oenanthe oenanthe*) on the Breckland of East Anglia, with implications for its conservation. *Bull. Ecol.* **11**, 559–569.

Tyler, G., Gullstrand, C., Holmquist, K. and Kjellstrand, A. (1973). Primary production and distribution of organic matter and metal elements in two heath ecosystems. *J. Ecol.* **61**, 251–268.

Veinstein, E. and Tolpysheva, T. Y. (1975). On the influence of lichen extracts on higher plants. *Bot. Zh.* **60**, 1004–1011.

Wallace, R. (1917). "Heather and Moor Burning for Grouse and Sheep" Oliver and Boyd, Edinburgh.

Wallén, B. (1980). Structure and dynamics of *Calluna vulgaris* on sand dunes in south Sweden. *Oikos* **35**, 20–30.

Wallén, B. (1983). Translocation of ^{14}C in adventitious rooting *Calluna vulgaris* on peat. *Oikos* **40**, 241–248.

Watson, A. (1971). Climate and the antler-shedding and performance of red deer in north-east Scotland. *J. appl. Ecol.* **8**, 53–67.

Watson, A. (1977). Wildlife potential in the Cairngorms region. *Scott. Birds* **9**, 245–262.

Watson, A. (1979). Bird and mammal numbers in relation to human impact at ski lifts on Scottish hills. *J. appl. Ecol.* **16**, 753–764.

Watson, A. (1983). Eighteenth century deer numbers and pine regeneration near Braemar, Scotland. *Biol. Conserv.* **25**, 289–305.

Watson, A. (1985). Soil erosion and vegetation damage near ski lifts at Cairn Gorm, Scotland. *Biol. Conserv.* **33**, 363–381.

Watson, A. and Hewson, R. (1973). Population densities of mountain hares (*Lepus timidus*) on western Scottish and Irish moors and on Scottish hills. *J. Zool.* **170**, 151–159.

Watson, A., Hewson, R., Jenkins, D. and Parr, R. (1973). Population densities of mountain hares compared with red grouse on Scottish moors. *Oikos* **24**, 225–230.

Watson, A. and Miller, G. R. (1971). Territory size and aggression in a fluctuating red grouse population. *J. Anim. Ecol.* **40**, 367–383.

Watson, A. and Miller, G. R. (1976). "Grouse Management" The Game Conservancy. Booklet 12. Fordingbridge.

Watson, A. Miller, G. R. and Green, F. H. W. (1966). Winter browning of heather (*Calluna vulgaris*) and other moorland plants. *Trans. Bot. Soc. Edinb.* **40**, 195–203.

Watson, A. and Moss, R. (1972). A current model of population dynamics in red grouse. *Proc. int. orn. Congr.* **15**, 134–149.

Watson, A. and Moss, R. (1979). Population cycles in the Tetronidae. *Ornis Fennica* **56**, 87–109.

Watson, A. and Moss R. (1980). Advances in our understanding of the population dynamics of red grouse from a recent fluctuation in numbers. *Ardea* **68**, 103–111.

Watson, A., Moss, R. and Parr, R. (1984). Effects of food enrichment on numbers and spacing behaviour of red grouse. *J. Anim. Ecol.* **53**, 663–678.

Watson, A., Moss, R., Phillips, J. and Parr, R. (1977). The effect of fertilisers on red grouse stocks on Scottish moors grazed by sheep, cattle and deer. *In* "Ecologie du Petit Gibier" (Eds P. Pesson and M. G. Birkan), pp. 193–212. Bordas and Gauthier-Villars. Paris.

Watson, A., Moss, R., Rothery, P. and Parr, R. (1984). Demographic causes and predictive models of population fluctuations in red grouse. *J. Anim. Ecol.* **53**, 639–662.

Watson, A. and O'Hare, P. J. (1979a). Red grouse populations on experimentally treated and untreated Irish bog. *J. appl. Ecol.* **16**, 433–452.

Watson, A. and O'Hare, P. J. (1979b). Bird and mammal numbers on untreated and experimentally treated Irish bog. *Oikos* **33**, 97–105.

Watt, A. S. (1947). Pattern and process in the plant community. *J. Ecol.* **35**, 1–22.

Watt, A. S. (1955). Bracken versus heather: a study in plant sociology. *J. Ecol.* **43**, 490–506.

Watt, A. S. (1964). The community and the individual. *J. Ecol.* **52**, (Suppl.), 203–211.

Webb, N. R. (1980). The Dorset heathlands: present status and conservation. *Bull. Ecol.* **11**, 659–664.

Webb, N. R. Clarke, R. T. and Nicholas, J. T. (1984). Invertebrate diversity on fragmented *Calluna* heathland: effects of surrounding vegetation. *J. Biogeog.* **11**, 41–46.

Webb, N. R. and Haskins, L. E. (1980). An ecological survey of heathlands in the Poole Basin, England in 1978. *Biol. Conserv.* **15**, 281–296.

Webb, N. R. and Hopkins, P. J. (1984). Invertebrate diversity on fragmented *Calluna* heathland. *J. appl. Ecol.* **21**, 921–933.

Welch, D. (1981). Diurnal movements of Scottish Blackface sheep between improved grassland and a heather hill in north-east Scotland. *J. Zool.* **194**, 267–271.

Welch, D. (1984a). Studies in the grazing of heather moorland in north-east

Scotland. I. Site descriptions and patterns of utilization. *J. appl. Ecol.* **21**, 179–195.

Welch, D. (1984b). Studies in the grazing of heather moorland in north-east Scotland. II. Response of heather. *J. appl. Ecol.* **21**, 197–207.

Welch, D. (1984c). Studies in grazing of heather moorland in north-east Scotland. III. Floristics. *J. appl. Ecol.* **21**, 209–225.

Welch, D. (1985). Studies in the grazing of heather moorland in north-east Scotland. IV. Seed dispersal and plant establishment in dung. *J. appl. Ecol.* **22**, 461–472.

Welch, D. and Kemp, E. (1973). A Callunetum subjected to intensive grazing by mountain hares. *Trans. Bot. Soc. Edinb.* **42**, 89–99.

Welch, D. and Rawes, M. (1964). The early effects of excluding sheep from high-level grasslands in the North Pennines. *J. appl. Ecol.* **1**, 281–300.

Welch, D. and Rawes, M. (1966). The intensity of sheep grazing on high-level blanket bog in upper Teasdale. *Irish J. Agric. Res.* **5**, 185–196.

Werger, M. J. A., Prentice, I. C. and Helsper, H. P. H. (1985). The effect of sod-cutting to different depths on *Calluna* heathland regeneration. *J. Env. Manag.* **20**, 187–188.

White, J. (1980). Demographic factors in populations of plants. *In* "Demography and Evolution in Plant Populations" (Ed. O. T. Solbrig), pp. 21–48. Blackwell, Oxford.

Whittaker, E. (1961). Temperatures in heath fires. *J. Ecol.* **49**, 709–715.

Whittaker, E. and Gimingham, C. H. (1962). The effects of fire on regeneration of *Calluna vulgaris* (L.) Hull from seed. *J. Ecol.* **50**, 815–822.

Whittaker, J. B. (1985). Population cycles over a 16-year period in an upland race of *Strophingia ericae* (Homoptera: Psylloidea) on *Calluna vulgaris. J. Anim. Ecol.* **54**, 311–321.

Whittow, J. B. (1977). "Geology and Scenery in Scotland" Penguin. Harmondsworth.

Wilcock, C. C., Robeson, C. K. and MacLean, C. A. (1984). On fruit eating by Scottish red grouse. *Biol. J. Linn. Soc.* **23**, 331–341.

Williams, J. (1985). Statistical analysis of fluctuations in red grouse bag data. *Oecologia* **65**, 269–272.

Winter, A. G. (1961). New physiological and biological aspects in the interrelationships between higher plants. *In* "Mechanisms in Biological Competition". (Ed. F. L. Milthorpe), *Symp. Soc. Exp. Biol.* **15**, 229–244.

Woolhouse, H. W. and Kwolek, A. V. A. (1981). Seasonal growth and flowering rhythms in European heathlands. *In* "Heathlands and Related Shrublands. B. Analytical Studies" (Ed. R. L. Specht), pp. 29–38. Elsevier. Amsterdam.

Wynne-Edwards, V. C. (1964). Land use in the Scottish Highlands. X. Multipurpose land use and conservation in the Highlands. *Adv. Sci.* (London) **21**, 177–183.

Yates, E. M. (1972). The management of heathlands for amenity purposes in south-east England. *Geographia Polonica* **24**, 227–240.

Yoda, K., Kira, T., Ogawa, H., Hozumi, K. (1963). Intraspecific competition among higher plants XI. Self-thinning in overcrowded pure stands under cultivated and natural conditions. *J. Biol. Osaka City Univ.* **14**, 107–129.

Developments in Ecophysiological Research on Soil Invertebrates

E. N. G. JOOSSE and H. A. VERHOEF

I. Introduction 175
II. Coastal Sand Dunes and Heathlands 177
A. Drought 177
B. Cold and Heat 190
C. Starvation 200
D. Mineral Shortage 206
E. Toxic Compounds 209
IV. Polluted Environments 213
A. Air Pollution 214
B. Heavy Metals 216
IV. Summary and Conclusions 233
References 234

I. INTRODUCTION

In ecology there is a growing appreciation of ecophysiological research. This reductionistic approach in ecology assigns a central position to the individual organism in relation to its environment. In general a reductionistic approach concentrates particularly on the chosen subject of study: the molecule, the cell or the individual, and does not consider the higher levels of organization (Levins and Lewontin, 1980). In ecology a reductionistic approach involves the risks concentrating on organisms, to the exclusion of the higher organizations, the population and the community. One has to realize that starting from the individual organism, ecophysiological knowledge has also to provide insight in the functioning of populations and communities.

Molecular knowledge is not often applied in higher organization levels. Recent environmental problems, however, resulting from chemical contamination by industry and pesticides demand an extension of the applica-

ADVANCES IN ECOLOGICAL RESEARCH Vol. 16
ISBN 0-12-013916-2

tion of chemical knowledge in ecology to form the discipline of ecotoxicology. Ecological test-systems have to be developed, to find out and predict the consequences of contamination, not only for individuals, but also for the functioning of populations and communities.

Ecophysiology is slowly emerging from the direction of physiology (Phillips, 1975; Bligh *et al.*, 1976; Barker Jørgensen, 1983), and only recently from an ecological viewpoint (i.e., Townsend and Calow, 1981; Vannier, 1983; Joosse, 1983; Sibly and Calow, 1986). Ecophysiology can help to gain insight into the flexibility of individuals and the way the variability of populations enables them to adapt to a changing environment whether caused by natural or by human disturbances.

Any environmental factor can develop into a stress factor. Water and food may be in short supply, temperature can be too high or too low and heavy metals can exert a toxic influence. Stress resistance can be a measure for the ability of the organism to "avoid" changes in its physiology and can also be a measure of the ability to "tolerate" irreversible injurious consequences of the changes in the environment. Injurious consequences of stress can possibly be repaired, but only by an expenditure of energy, and even avoidance mechanisms are energy demanding. Where energy is allocated to resistance mechanisms, it may affect growth and reproduction.

The mechanisms required to avoid or to tolerate a stress may be very different. Stress can be avoided by developing dormant stages in less mobile and sessile organisms, or by dispersal in insects, birds and bats. "Tolerance" mechanisms are based on a changed, but not disturbed metabolism and may operate via quantitative (e.g. osmoregulation) or qualitative strategies (e.g., detoxication of toxic substances with specific proteins), or may for instance be performed by storing toxic substances in definite cells or organs (compartmentation resistance, Ernst, 1983a).

One of the objectives of ecophysiological studies is to detect and classify the diversity of physiological and behavioural adaptations to different stress factors, occurring in the environment and acting on the organisms. Research in plant ecology has shown that plant species also possess a diversity of adaptation strategies to tolerate or avoid the many stress factors in their environment (Levitt, 1980; Ernst, 1983a).

Apart from the physiological effects on individuals, we consider parameters which enable us to express the consequences of environmental changes for populations. The response of a population to environmental stress is indicated by its rate of increase, which is in fact expressed in the classic logistic growth equation. This rate of increase (r) is essentially dependent on fertility and mortality, which in turn depend on individual growth and reproduction. Some attempts have been made to explain life history patterns as outcomes of single selective pressures in plants (Grime 1979) and in animals (Vannier, 1983; Block, 1982).

In the present paper the advance in our knowledge concerning some physiological and behavioural adaptations and the relation to population parameters in soil animals is discussed. To trace the multiple adaptation to a natural complex of factors, the mechanisms and strategies of invertebrates will be related to two specific extreme habitats namely: coastal sand dunes and heathlands (Section II) and polluted environments (Section III). This review cannot be exhaustive, and contains only examples concerning selected, mainly saprophagous species. It tries to identify causal mechanisms in the evolution of resistance and to link these properties with the resulting life history patterns in soil animals.

II. COASTAL SAND DUNES AND HEATHLANDS

Coastal sand dune and heathland ecosystems are characterized by extreme changes of microclimatic factors. Considerable spatial and temporal differences in temperature can occur in the nearly bare sand during the day in summer, whereas between the vegetation and the soil beneath the surface the conditions are found to be more homogeneous. The water supply in the soil is in general very low as a result of the low water capacity of the sand, of the high evaporation especially during the summer and of the low organic content of the soil. Coastal sand dune ecosystems are especially characterized by a deficiency in the nutrients N and P. The very well adapted tree species in dune areas, the evergreen gymnosperms, have only about half the foliar N (1–2%) of deciduous angiosperms (Mattson, 1980). Moreover, during the growing season concentrations of N and other minerals decline sharply until senescence. It is thus to be expected that the litter layer, the food supply of saprophagous soil animals, exhibits a shortage in N and other mineral elements.

A special characteristic of ecosystems with periodic drought is the presence of toxic compounds in plants. Leaves have generally high concentrations of organic polymeric substances such as phenols and lignin, which are thought to reduce the digestibility of the plant material for invertebrate animals.

The following stress factors will be focused on: A. Drought; B. Cold and heat; C. Starvation; D. Mineral shortage; E. Toxic compounds.

A. Drought

As mentioned before, drought is a general phenomenon in sand dunes and heathlands and soil invertebrates must have developed resistance mechanisms in order to survive.

The effect water shortage has on animals is mainly an indirect one; it works through body water content. If it is reduced below certain critical limits by exposure to desiccating conditions, the animal dies. In general, water is lost by diffusion from the respiratory and general body surfaces, by secretion of fluids involved in digestion, defense, pheromones, reproduction, oviposition, defaecation and excretion. These losses can be balanced by water gained by imbibition through the ingestion of hydrated foods or drinking, by the production of metabolic water and by active or passive uptake from the ambient air or aqueous environment (Arlian and Veselica, 1979).

The cumulative function of loss and gain: the body water content, is not a fixed value. In arthropods, the group attention is focused on in this and the following paragraph, it may differ dependent on the taxonomic group and ranges from 45% (fresh weight) for some Coleoptera, to 80–90% for Lepidoptera (Edney, 1977; Rapoport and Tschapek, 1967). Further, it may vary within an individual as a function of age, developmental stage, sex, physiological condition, diet, season (see Arlian and Veselica, 1979).

To find out if a certain water content of an animal is the result of water balance mechanisms, it is necessary to integrate avenues of water loss and gain, to determine their relative importance, the means available for control of each of these and the limits within which regulation is possible. Since for soil arthropods such integrative work on water balance is lacking, the examples concern Orthoptera and Coleoptera, and serve to illustrate the need for similar research on soil arthropods.

The first example concerns the water balance of an adult, non-flying locust, *Locusta migratoria* (see Fig. 1). The largest component of water gain is with the food (:fresh grass = A). When water intake via food is reduced by offering dry food (B), a natural feature in dry sand dunes, both faecal and spiracular loss are reduced. If water is available, locusts drink quickly to restore balance.

The water balance of the grasshopper *Oedipoda* (Fig. 1), shows that the two above-mentioned conditions of wet and dry food for the locust are similar, in the grasshopper, with regard to the situation during the afternoon and at night respectively, for the same animal: *Oedipoda* undergoes a daily cycle of water gain during the afternoon and of water loss at night (Sell and Houlihan, 1985).

In the third example, two Coleoptera species, *Eleodes armata* and *Cryptoglossa verrucosa*, are compared as to their water balance (Fig. 2). Both species are active in the Mojave desert during periods when ambient temperatures fall within the range 10–30°C (Cooper, 1983). If temperature exceeds 30°C or drops below 10°C, these beetles use burrows with equable temperature conditions underground. Water influx, efflux and metabolic water gain were measured in the field by using doubly labelled water. During rainfall in winter the main influx of water in *E. armata* is by drinking. In the

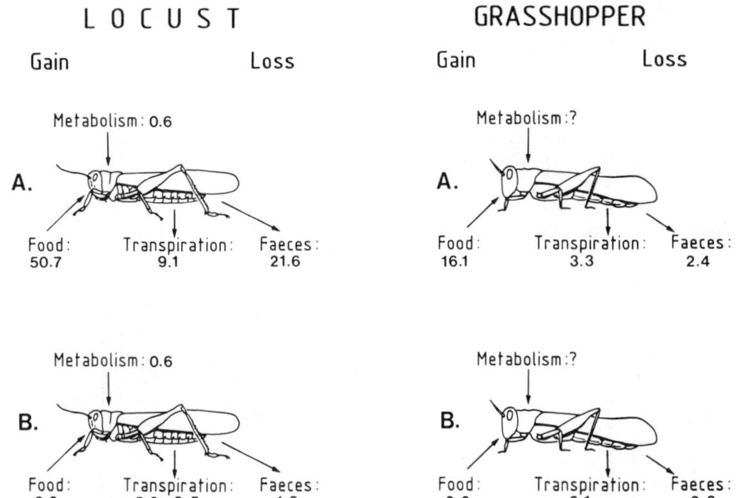

Fig. 1. Water balances of a locust and a grasshopper (mg H_2O/animal/h). Locust A: animal feeding on fresh grass; B: animal feeding on dry food. Grasshopper A: animal during the afternoon; B: animal at night. Locust: redrawn from Edney (1977); grasshopper: redrawn from Sell and Houlihan (1985).

absence of rain the main influx is with the food. The proportion of faecal water loss decreases in the latter case. In spring a smaller proportion is derived from rain water, whereas during summer nearly all the water comes from the plant eaten by this herbivorous species during crepuscular-nocturnal activity periods (Cooper, 1985).

Insectivorous *C. verrucosa* too, gets most of its water from its (dead) food. During winter and spring, however, this species goes into some kind of quiescence and then all the water influx is metabolic water. Total transpirational water loss rates are lower in *C. verrucosa* than in *E. armata* (see Table 1). Apparently, adult *C. verrucosa* is more xeric adapted than adult *E. armata*. As a result of its greater fecundity, however (more, small eggs and a greater hatching success), *E. armata* populations may be better adapted to the extremes of the desert than *C. verrucosa* populations.

In the foregoing examples different cooperative adaptation mechanisms are present, involving physiology, behaviour, and life history. These adaptation mechanisms are categorized as "avoidance mechanisms" and concern individuals as well as populations. The second category, "tolerance mechanisms", mainly concerns the cellular and molecular level (Levitt, 1980; Alexandrov, 1977). In the following, examples of the different avoidance and tolerance mechanisms for dealing with drought will be presented, and the lack of knowledge about soil arthropods will be apparent.

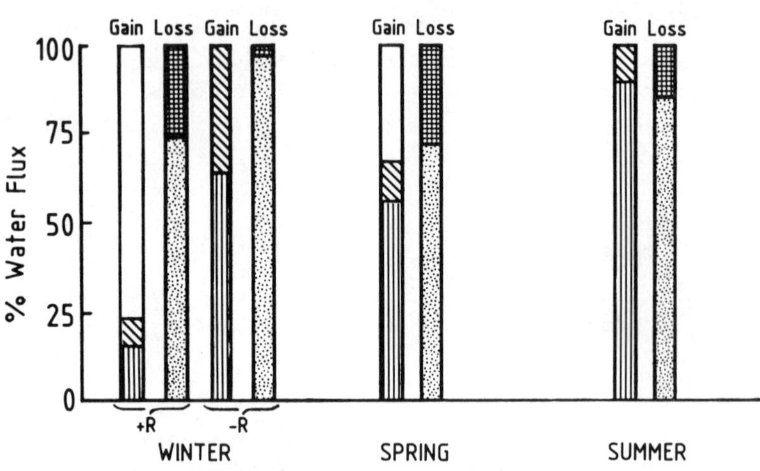

CRYPTOGLOSSA VERRUCOSA

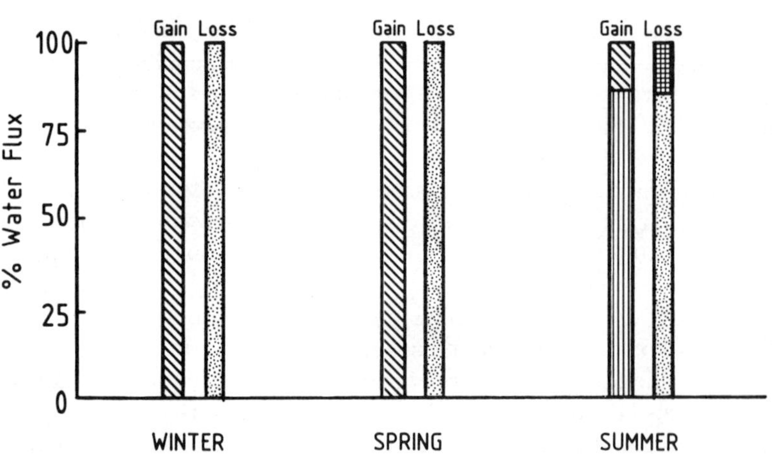

Fig. 2. Proportions of water gain and loss for the Coleoptera *Eleodes armata* an *Cryptoglossa verrucosa* during the different seasons. Gain: food (▥), metabolism (▧), drinking (□). Loss: net transpiration (⁙), faeces and urine (▦). +R: wit rain; −R: without rain. After Cooper (1985).

1. *Physiological Avoidance Mechanisms*

The water activity (a_w) of the haemolymph, the extracellular fluid in arthropods, ranges from 0·995 to 0·998 (i.e., 300 to 600 mOsm kg^{-1}). This means that, unless the water vapour activity (a_v) of the ambient air is saturated, the activity gradient favours the net loss of water from the animal to the atmosphere by simple diffusion. This can be curtailed by diminishing the permeability of the cuticular surface, i.e., by increasing the cuticular resistance. The cuticular structure, lipid layers and the epidermal layer play a role. A body of data is available on this subject (Table 1). Concerning

Table 1
Water loss in some land arthropods and plants.

Arthropod groups	Transpiration rate ($\mu g\,cm^{-2}\,h^{-1}\,mm\,Hg^{-1}$)	Resistance ($s\,cm^{-1}$)
Isopoda		
Philoscia muscorum	180	18·9
Porcellio scaber	110	31·0
Myriapoda		
Lithobius sp.	270	12·6
Glomeris marginata	200	17·0
Insecta		
Collembola		
Onychiurus fimatus	813	4·2
Tomocerus minor	581	5·9
Podura aquatica	510	6·7
Tetrodontophora bielanensis	385	8·9
Orchesella villosa	160	21·3
Orchesella cincta	143	23·9
Entomobrya nivalis	35	97·5
Seira domestica	3	1137·7
Coleoptera		
Eleodes armata	17·2	198·0
Cryptoglossa verrucosa	8·4	406·0
Arachnida		
Scorpionidea		
Hadrurus arizonensis	1·2	2786·0

Plant groups		Transpiration rate ($\mu g\,cm^{-2}\,h^{-1}\,mm\,Hg^{-1}$)	Resistance ($s\,cm^{-1}$)
Stomata open	Mesophytes	1706–340	2–10
	Xerophytes	683–171	5–20
Stomata closed	Mesophytes	171–43	20–80
	Xerophytes	34–17	100–200

Arthropod data from Edney (1977) and Verhoef and Witteveen (1980); Plant data from Edney (1977).

Collembola, the representative species arranged from the drought-sensitive *Onychiurus fimatus* up to the drought-resistant *Seira domestica*, show a clear relation between their resistance value and the water conditions of their microhabitat (Verhoef and Witteveen, 1980; Vannier, 1983). However, there are no important differences between drought-resistant, temperate specimens and the above-mentioned desert beetles.

The resistance value can change, dependent on the developmental stage: often gravid females have a low resistance, with the consequence that eggs are laid in humid places. It can also be influenced by air humidity and even controlled by hormones (Willmer, 1980). Comparison of some plant data, with those for arthropods shows their relatively low cuticular resistance, and the great effect of the position of the stomata. This is related to the problems which arise from an impermeable cuticle, as this affects not only water vapour, but also gaseous exchange. One feature found in arthropods is confinement of the place of gaseous exchange to small areas, the remainder of the cuticle being impermeable. Examples of these small permeable areas are: ventral tube vesicles, pseudotracheae, book lungs, tracheal systems. Each of these systems can be closed to some degree. An example of the effect of the opening of the tracheal system of the cricket *Acheta domesticus* on CO_2 and water loss is given in Fig. 3. In Collembola, the surface of the

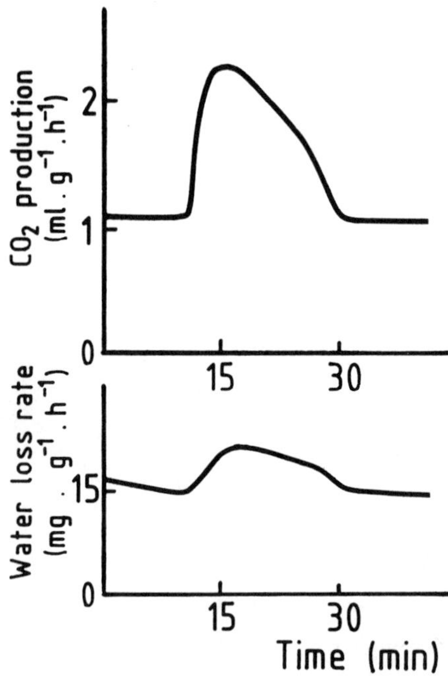

Fig. 3. Simultaneous measurement of water loss and CO_2 production in a cricke (alive, with blocked anus and mouth) at 40°C. After Hadley and Quinlan (1982).

ventral tube vesicles, which have a respiratory as well as an osmoregulatory function, is kept moist by urine from the labial nephridia via an external transport system. The morphology in this system shows in the different species a development from an open groove, via a groove partially covered by two lateral flanges, to a tubular structure. As in those species there is a parallel development to drier habitats, the covering of this groove is probably a morphological adaptation to dryness. This system, which is analogous to the water conducting system in isopods (Hoese, 1981), is considered to be a special adaptation to the terrestrial habitat in atracheate soil arthropods (Verhoef *et al.*, 1983).

2. *Behavioural Avoidance Mechanisms*

In animals avoiding stress we can distinguish cryptobiontic behaviour, and a diel activity pattern. Cryptobiontic reactions to stress are e.g., the hiding away in small spaces such as crevices, cracks, holes and burrows in rock, sand or soil, in wood, in and on plants (ranging from leaf-miners and gall-makers to animals living in flowers, buds and on leaves) and in litter layers. Figure 4 gives an example of the humidity conditions at different depths within the burrow of a soil-nesting wasp (*Cerceris arenaria*).

They can also creep away in self-made structures such as webs, or cases

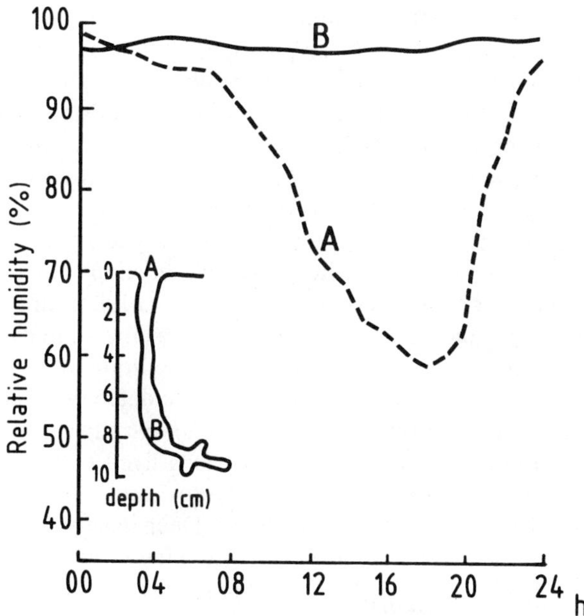

Fig. 4. Humidity conditions at different depths within the burrow of a soil-nesting wasp through a summer day. After Willmer (1982).

(Chauvin *et al.*, 1979). Aggregations are formed by many soil arthropods such as Isopoda, Acarina and Collembola by means of orthokinetic reactions to water and/or pheromones (Vannier, 1970; Joosse and Groen, 1970; Verhoef and Nagelkerke, 1977; Verhoef *et al.*, 1977). The progressive adaptation to terrestrial life as found within the group of Collembola is attended by an increasing importance of these aggregation pheromones (Verhoef, 1984). A peculiar form of behaviour, namely rolling-up, is common to some terrestrial isopods and many millipedes. It may protect pleopods of armadillid isopods from drying (see Crawford, 1981).

Temporary resting in stable, near-saturated environments has its drawbacks: there is the danger of increased pathogenic attack and the animals experience lower temperatures and lower light intensities. This may also account for the diel activity pattern. By shifting towards a crepuscular or nocturnal activity pattern animals can avoid drought, but are confronted with the above-mentioned drawbacks.

3. *Life History Patterns*

The ecological and evolutionary success of arthropods in extreme biotopes, such as deserts or exposed mountains, depends strongly on how they will time their reproduction, structure their development and adapt their resource utilization (Crawford, 1981; Leinaas and Fjellberg, 1985). These three parameters are often closely related to the timing of precipitation and resource availability. This relation is found in long-lived herbivores and detritivores, such as millipedes, cryptostigmatid mites, Coleoptera (Crawford, 1981). Short-lived animals often demonstrate synchronization with temporal resources by rapid breeding, followed by dormancy or dispersal, and rapid resumption of activity when the appropriate situation returns. Likewise, the collembolan *Folsomides* quickly becomes active after months in dry soil upon the arrival of rain (Greenslade, 1981).

In more temperate environments prolonged dryness leads to quiescence in Collembola (Joosse, 1983). After a primary reaction of increased feeding and metabolic activity (see Verdier and Vannier, 1984), drought-resistant species enter a permanent "pre-ecdysis" state, as the animals do not reach the critical volume at which stretch-induced moulting takes place. In that state, both transpiration rate and metabolic rate decrease (Verhoef and Li 1983). The utilization of relatively long, physiologically less active stages to survive dry periods can also be found in long-lived detritivores. Besides diapause, nearly lifeless stages of crypto- and anhydrobiosis exist (Greenslade, 1981) up to simple quiescence. Deep inactive stages, however might prevent a rapid reaction on the return of favourite conditions.

4. *Tolerance Mechanisms*

Water loss tolerances vary considerably between arthropods, and may range from 17% to 89% of the species' normal water content (Arlian and Veselica

1979). This is, however, dependent on the speed of water loss, as was shown for Collembola (Verhoef, in prep.). Fast dehydrated (0% R.H.) drought-sensitive species such as *Tomocerus minor* tolerate water loss up to 22% of their normal water content (Vegter, 1985). Slower dehydrated (36% and 96% R.H.) specimens, however, tolerate a loss of 47%, a value, common for drought-tolerant species like *Orchesella cincta* and *O. flavescens* (Verhoef, in prep.) Besides the maintenance of a critical water content, simultaneous osmotic regulation is required to ensure that internal water concentration remains within tolerable limits. In hydrated arthropods values are found between 300 and 600 mOsm (Edney, 1977). The two main water compartments are the haemolymph and the tissues. Between these compartments water may be moved but each may change volume and osmoregulate independently. Although the study of the distribution of water between the compartments and cell volume regulation in arthropods is still in its infancy, recent research on locusts, beetles, cockroaches and springtails has shown that the haemolymph acts as a reservoir from which water may be withdrawn to prevent large reductions of cellular water and cell volume (Machin, 1981; Hyatt and Marshall, 1985; Verhoef, in prep.). This process is called water compartmentalization (Machin, 1981). In the drought-tolerant tenebrionid beetle *Onymacris* tissue water is partially regulated at the expense of the haemolymph (Fig. 5). This sharing of water is brought about by the mobilization of osmotically active solutes in hydrated animals and by storage

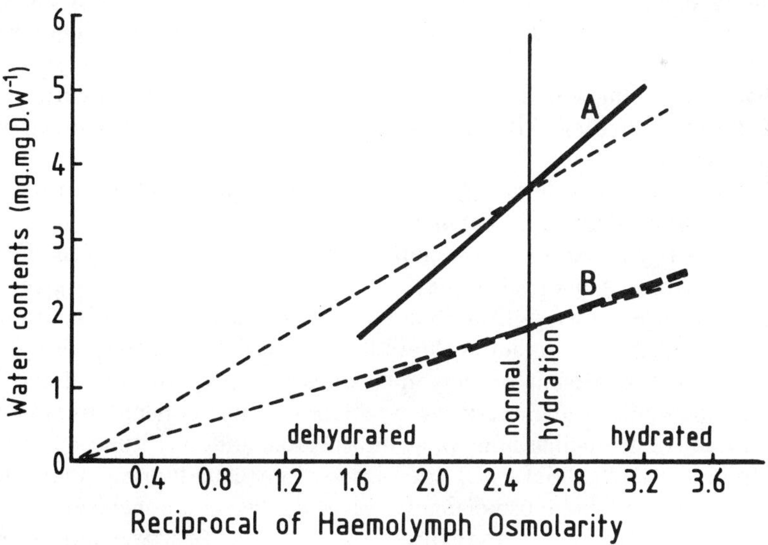

Fig. 5. The relationships between total (A) and intracellular (B) water content and the reciprocal of haemolymph osmolarity in the tenebrionid beetle *Onymacris*. After Machin (1981).

and osmotic inactivation in dehydrated animals. The solutes involved are amino acids, trehalose, sodium and chloride. Another example of water compartmentalization has recently been found for cockroaches (Hyatt and Marshall, 1985): after dehydration the haemolymph volume lost 50% of its original volume as against 25% of tissue water. Haemolymph osmolality and sodium, potassium and chloride concentrations in haemolymph and tissue water are all regulated within narrow limits. Sodium and potassium ions are sequestered within the fat body as insoluble urates during dehydration. Upon rehydration these ions are again mobilized from the fat tissue and released into the haemolymph. Only small percentages of sodium and potassium, removed from the haemolymph, are excreted (4 and 11%, respectively). Chloride ions appear to be associated with the cuticle during times of water deprivation.

Selective removal of organic solutes, such as amino acids during dehydration is supposed to take place in some lepidopteran larvae, although in *Pieris* larvae a relative accumulation of amino acids occurs in the blood, due to a proportionately greater sequestration of ions (Willmer, 1982). According to Woodring and Blakeney (1980), free amino acids such as proline, play an important role in haemolymph osmoregulation.

Literature on the osmoregulation of terrestrial arthropods deals mainly with insects (Edney, 1977; Willmer, 1982). Recently, this work has been supplemented with work on isopods (Price and Holdich, 1980), myriapods and arachnids (Riddle, 1985) and Collembola (Verhoef, in prep.). During prolonged aerial dehydration the osmolality of *O. cincta* rises according to the expected rise without osmoregulation (see Fig. 6). This does not depend on the speed of dehydration (0%, 36%, 96% RH). In *T. minor* the speed of dehydration is important: in fast dehydrated animals the rise accords with the expected rise. In slowly dehydrated animals there is a higher rise than the expected value.

So, in both species there seems to be no osmoregulation (Verhoef, 1981). This is supported by our research on the functioning of the kidney system in Collembola. The relation between the haemolymph of several Collembola species, hydrated and dehydrated, and the urine osmolality is given in Fig. 7. The urine is hypo- to isoosmotic with the haemolymph. The urine in Isopoda and that in Opiliones shows hyperosmotic urine values too. This means that at increasing haemolymph values absorption of water in the gut or rectum would contribute to the regulation of haemolymph osmolality. Separation of the total osmotic value into an ionic component and an organic component (see Verhoef *et al.*, 1983) shows ionic regulation in both species. Figure 8 gives the total osmolality (:T) and the two components, ionic (:I) and organic (:O) fraction, for hydrated, dehydrated (20% weight loss) and rehydrated animals. The changes in total osmolality are in accordance with Fig. 6. In *O. cincta*, however, there is a clear ionic regulation. Furthermore,

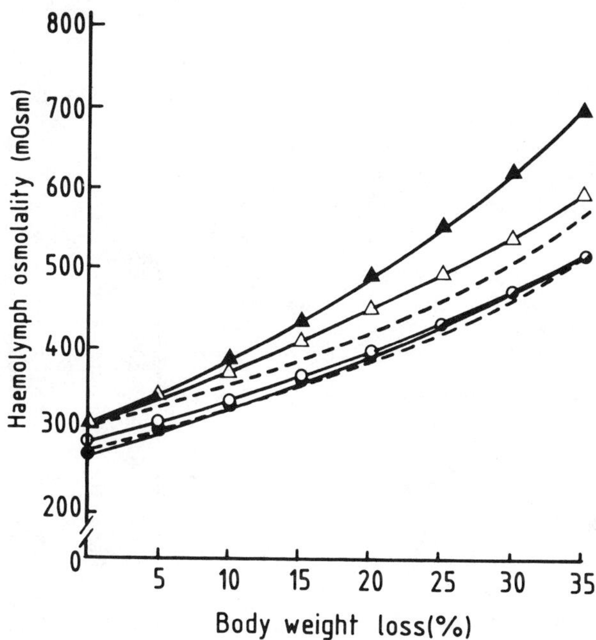

Fig. 6. The effects of water loss on the haemolymph osmolality of the Collembola *Tomocerus minor* (at 96% RH: ▲—▲ and 36% RH: △—△) and *Orchesella cincta* (at 96% RH: ●—● and 36% RH: ○—○). Expected values for *T. minor* (upper ----) and for *O. cincta* (lower ---). From Verhoef, unpublished.

after rehydration with de-ionized water the ion concentration nearly returns to its original value (Fig. 8a). This points to both immobilization and excretion of ions. Preliminary data on the ionic component of urine after rehydration sustain this idea (Verhoef, unpubl. data). In *T. minor* ionic regulation depends on the speed of dehydration: fast dehydrated animals show no regulation (Fig. 8b). Slowly dehydrated animals, however, are able o regulate (Fig. 8c). The regulation of the ionic component, together with he absence of regulation of the total osmotic value as in these two collembolan species, has also been found in the drought-sensitive isopod *Oniscus asellus*, which is capable of a remarkable degree of sodium control, but incapable of effective osmoregulation (Price and Holdich, 1980). The ise in the organic fraction in both Collembola species might be caused by xtra production of organic material, by a failure of the kidney system to excrete it, or might be explained on the assumption of water compartmenalization. A high osmolality might facilitate the uptake of water through the uticle of the ventral tube vesicles.

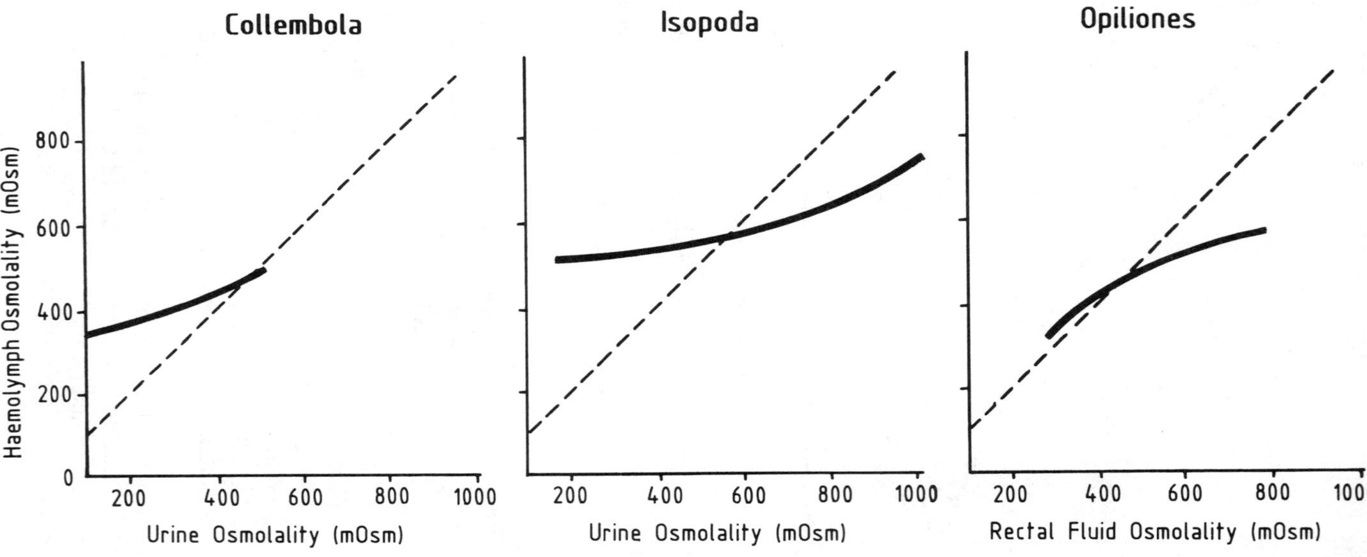

Fig. 7. The relationships between the osmolality of the haemolymph and the urine or rectal fluid of Collembola, Isopoda and Opiliones. Data of Collembola (Verhoef, unpublished); data of Isopoda (derived from Lindquist and Fitzgerald, 1976); data of Opiliones (after Riddle, 1985).

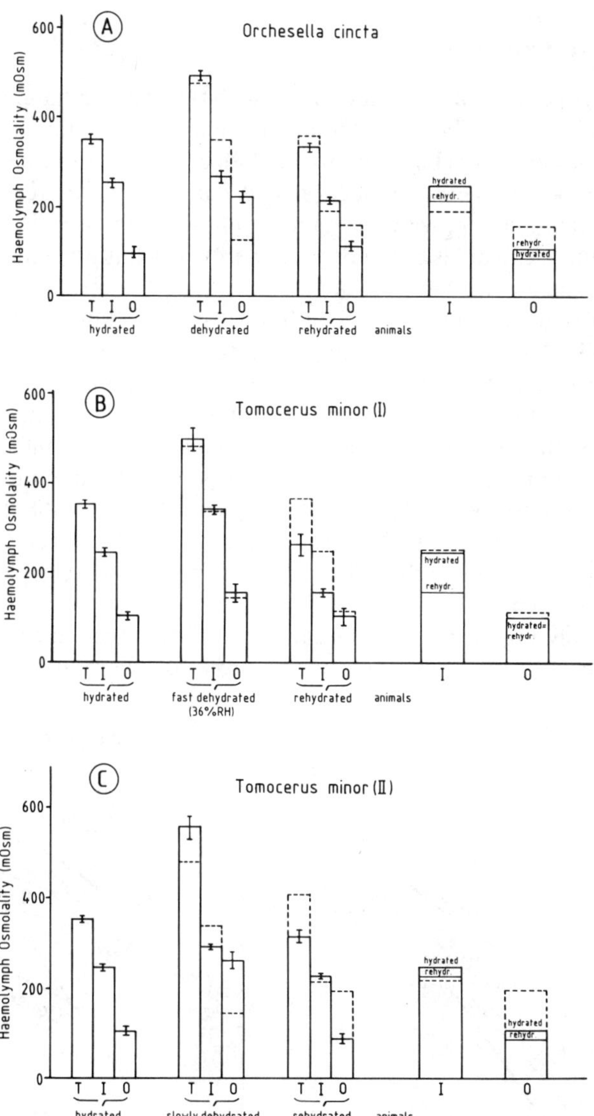

Fig. 8. The haemolymph osmolalities for hydrated, dehydrated and rehydrated Collembola. Total osmolality (T) and the two components, the ionic (I) and organic (O) fraction, are shown for *Orchesella cincta* (A) and fast dehydrated *Tomocerus minor* (B) and slowly dehydrated *Tomocerus minor* (C). The expected values (---) and the means ± S.E.M. are presented. On the right for each species a summary is given of the ionic (I) and the organic (O) fraction for hydrated animals and the actual and expected (---) values for rehydrated animals. From Verhoef, unpublished.

B. Cold and Heat

In an arthropod which is in equilibrium with the ambient temperatures, the net exchange of heat between the animal and its environment is zero, and thus the gain of heat is balanced by the loss of heat. Loss of heat occurs by long-wave radiation, by conduction and convection, whereas gain takes place via solar and long-wave radiation and by metabolism (Bursell, 1970). This "metabolic heat" is derived from energy which is not captured in high energy phosphate linkage.

The fluctuating temperature conditions of sand dunes and heathlands might cause large changes in the body temperature of the inhabitants. Maintenance of body temperature within certain limits is desirable, because physiological and biochemical mechanisms may be deranged at either extreme of the temperature range. Enzymic reactions, muscle activity, oxygen consumption, all have different Q_{10}s so that temperature changes can disturb interdependent metabolic reactions (Willmer, 1982).

In Fig. 9 examples of body temperature control are given for different arthropod larvae and adults. The adults seem to be more capable of body temperature control than the larvae.

Excessively high temperatures can cause disturbance in the integrity of membrane function, denaturation of proteins, thermal inactivation of

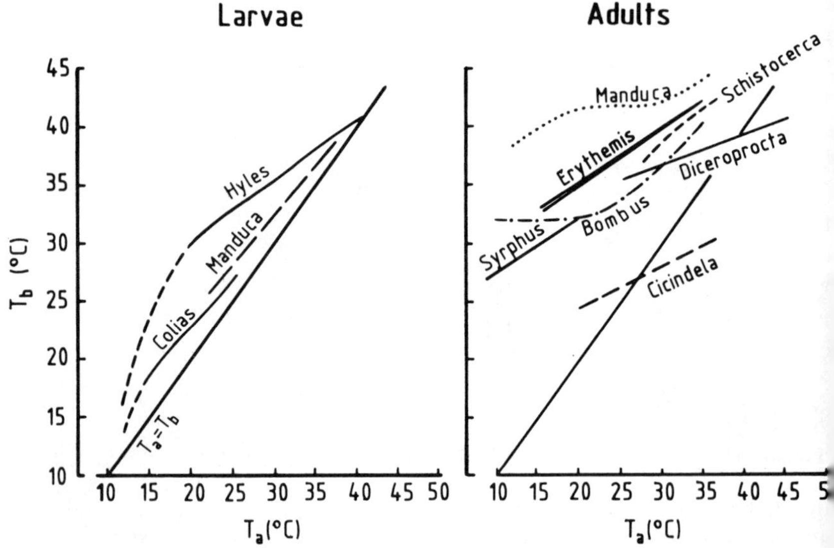

Fig. 9. Regulation of body temperature (T_b) in several arthropods, comparing T with the ambient temperature (T_a). Data are redrawn from Willmer (1982) excep those of adult *Manduca* and *Bombus* (Heinrich, 1974).

enzymes, exhaustion of food reserves, build-up of end products of metabolism, and changes in the permeability of epidermal or cuticular layers, causing death from desiccation (see amongst others Gilby, 1980).

Excessively low temperatures may cause structural changes of membranes, build-up of harmful substrates normally detoxified, freezing injury which involves solute-concentration effects, cell-volume changes, lipid-phase transitions, protein destabilization etc. (amongst others Baust and Rojas, 1985). To avoid these dangers body temperature regulation can take place by either changes of metabolic heat production, or alterations of heat exchange with the environment.

1. *Physiological Avoidance Mechanisms*

Heat production, evaporative cooling, insulation and colour may all play a role in the physiological avoidance of temperature extremes.

Thermoregulation by the production of heat is important in flying arthropods. The exclusive source of metabolic heat of thermoregulatory significance is flight muscle, although walking may elevate body temperature slightly. Prolonged flight in air temperatures above 38°C is impossible for locusts, because their thorax rapidly reaches a lethal 45°C. (Flight is therefore interrupted by frequent periods of rest while the insect cools down.) It is known that large arthropods such as bees, beetles and butterflies are able to warm-up by "shivering", in which heat is produced by contracting the flight muscles against each other. This behaviour is accompanied by a rapid rise in thoracic temperature, but usually by little change in abdominal temperature, on account of the poor insulation of the abdomen (May, 1979). It is an energetically expensive mechanism. In non-flying arthropods metabolic heat is quantitatively unimportant (Heinrich, 1974) and is balanced by evaporative cooling.

Evaporative cooling is probably not an important component of the thermal regulation in arthropods, owing to their relatively small size and consequently high surface/volume ratio. It does occur though in arthropods with large water supplies, like animals that feed on plant juices or succulent vegetation (such as aphids) (May, 1979). In soil arthropods, like isopods, evaporative cooling is confined to brief periods.

Body temperature can also be strongly influenced by insulation. In general, the creation of a microclimate by means of an external insulating layer around the animal is often used in arthropods. Bristles, hairs, scales, the subelytral cavity of beetles play a role in reducing the rates of temperature change. The effects of pubescence are shown in Fig. 10. Hairs and scales as found in many Collembola species, may also have an insulating effect.

The effects of colour (:surface absorptivity) have been somewhat controversial. It has been stated that visible colour has little effect on body temperature because of the greater importance of short wave infrared in

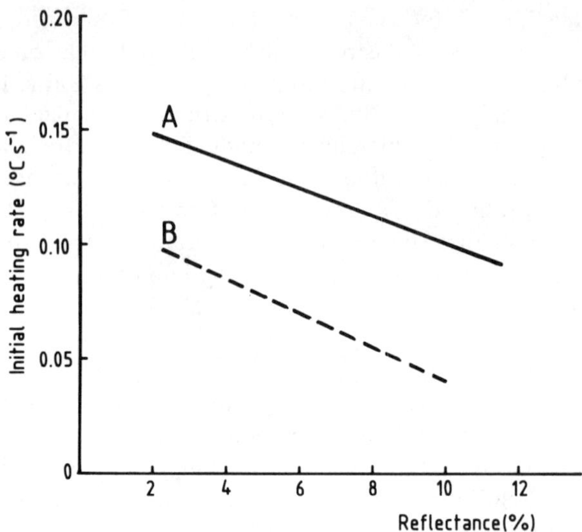

Fig. 10. The relationship between reflectance (:colour) and heating rate for temper-ate arthropods: normally pigmented (:A) and pubescent arthropods (:B). Afte Willmer (1982).

solar radiation. Important body temperature differences between black anc white beetles and their different diel activity give clear arguments for th importance of colour as a thermal strategy (Edney, 1971). Furthermore there are examples of individual arthropods (such as grasshoppers) changin colour in response to changing ambient temperatures (May, 1979). The dar colour of many Collembola species from exposed habitats might function a a defense mechanism against U.V.

2. *Behavioural Avoidance Mechanisms*

Behavioural responses are often more closely adapted to environmenta conditions than physiological responses. Three major groups of response can be distinguished: posture, microhabitat selection, diel activity cycle.

Posture has a very direct effect on body temperature as has been found i grasshoppers, locusts, beetles and spiders (Willmer, 1982). The orientatio of the body to the sun is used to control body temperature at both extreme of the temperature range. Many soil arthropods in bright sunshine at hig ambient temperatures "stilt" , i.e., they extend their legs and raise the body out of the hot boundary layer of air within a few millimetres of th ground and are perhaps exposed to higher wind velocities. This was found i locusts, beetles and micro-arthropods like collembolans (May, 1979).

Probably the most common mechanism for body temperature control i arthropods is the short-term selection of thermally optimal microclimates. Microhabitat selection comprises sun-shade alternation and basking (spiders, tiger beetles, locusts and collembolans), aggregation (by which the effective thermal size is changed; caterpillars and locusts,) vertical movements (using the vegetation; caterpillars, locusts, beetles and collembolans) and burrowing (amongst others tiger beetles, dung beetles) (Willmer, 1982; Bowden *et al.*, 1976).

Temperature extremes, often together with humidity, appear to mould the diel activity pattern in grasshoppers, beetles and ants. In summer their activities are often carried out during early morning and in the evening, whereas in cool periods there is a shift towards noon (amongst others Dreisig, 1980). Such a flexible shift has also been found in the laboratory for the collembolan *Orchesella cincta*. This species is supposed to be less heat-sensitive than the strictly nocturnal *Tomocerus minor* (Joosse, unpublished).

3. *Life History Patterns*

As stated before, some arthropods are able to time their reproduction, to structure their development and to adapt their resource utilization, in order to optimize their survival in seasonal environments. Timing mechanisms such as reproductive diapause, egg diapause and simple quiescence are used to survive adverse periods.

Examples of summer quiescence have been described for temperate Collembola. This involves a developmental arrest during a period of high temperatures and low water and food availability (Testerink, 1983) and can be considered to be a flexible response to resource variability on a short time scale. After the adverse period, synchronization of reproductive activity occurs (Joosse and Testerink, 1977a), which gives the drought- and heat-sensitive hatchlings a good chance to survive (Vegter, 1985). This phenomenon is found especially in surface living species, characterized by unstable microhabitat conditions. In deeper living soil species, in an environment with a moderate seasonality, it is absent (Verhoef and Li, 1983).

Diapause occurs in most of the major arthropod groups in a wide variety of climates as an adaptation to relatively long-term variability and is associated with different developmental stages (Masaki, 1980). In some groups a certain stage is favoured for both summer and winter diapause such as the adult stage of Coleoptera and the pupal stage of Lepidoptera and Diptera.

Summer diapause is induced by long photoperiod and high temperatures and is prevented and/or terminated by short photoperiod and low temperatures. As selection factors responsible for the evolution of summer diapause biotic factors like food supply and predation may, however, be of more importance than heat and drought (Masaki, 1980).

In winter diapause too, photoperiod and temperature play a role in induction and termination. For temperate Collembola short photoperiod induces and high temperature terminates winter diapause (van der Woude, in prep.). As these abiotic factors are also thought to be a trigger of the development of cold tolerance, a relationship between these two mechanisms could be supposed. Several studies, however, have shown that cold tolerance and winter diapause are two separated mechanisms, occurring in the same season (Young and Block, 1980; van der Woude, 1986). Based on an extensive study of the comparative demography of forest floor Collembola populations, van Straalen (1985b) concludes that for temperate Collembola winter is a more adverse period than summer. Mortality factors operating during winter have a stronger selective effect on the population than those operating during summer (see Vegter, 1985). Diapause occurs in Collembola in all developmental stages (van der Woude, in prep.; Leinaas, 1983). Egg diapause has been found in the univoltine species *Tomocerus longicornis* and *Anurida maritima*. In the intertidal *A. maritima* the eggs are laid in summer; the adults die in autumn, probably of starvation (Witteveen *et al.*, in prep.). Diapause is terminated in autumn by temperatures below +5°C. Egg development is suppressed due to the low temperatures of autumn and winter until the next spring (see Fig. 11). Simultaneous egg and adult diapause, together with juvenile quiescence have been found for

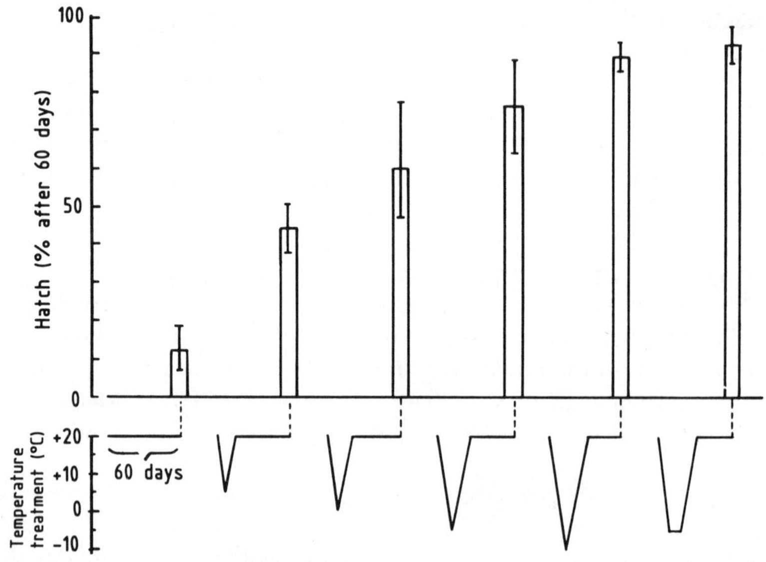

Fig. 11. The effects of different temperature treatments on the hatch of the Collembola *Anurida maritima*. From Witteveen, Verhoef and Huipen (in prep.).

Lepidocyrtus lignorum (Leinaas, 1983), which may be explained by a greater cold tolerance of juveniles compared to adults (amongst others Block, 1982). In the temperate *O. cincta* and *T. minor* no such difference was found. Here both stages are in diapause. The intensity of the dormant stage is higher for the more exposed living *O. cincta* than for the deeper living *T. minor*. This might prevent premature termination in *O. cincta* (van der Woude, in prep.).

4. *Tolerance Mechanisms*

(a) Cold tolerance. Active animal life is limited to a narrow range of temperatures from a few degrees below the freezing point of pure water to approximately 50°C. The critical limits of the temperature range depend strongly on the cooling *c.q.* heating rate and exposure time. In this section a restriction will be made to the effects of subzero temperatures leaving out adaptations to chilling. Arthropods meeting freezing temperatures appear to have a wide range of methods for survival in these extreme situations. Generally a distinction is made between the attainment of cold hardiness via a freeze-tolerant route or a freeze-susceptible route (Block, 1982): in short, freeze-susceptible animals die if frozen, whereas freeze-tolerant animals survive the formation of extracellular and possibly intracellular ice. They use different "composite mechanisms" of tolerance which are considered to be two distinct ecophysiological strategies, and till recently all overwintering arthropods studied fell into either one or the other category; therefore they will be discussed separately:

(a)1 Freeze susceptibility. Freeze-susceptible animals possess or develop the ability to supercool extensively. They maintain their body fluids in the liquid phase below the melting point, by the production of low molecular weight antifreeze agents which lower the freezing and supercooling points. Such compounds include polyhydric alcohols (such as glycerol, sorbitol), free amino acids (such as alanine, proline) and sugars. Their effects depend upon their concentration, which may rise to multimolar levels, varying between and within species (Baust and Rojas, 1985). In bark beetles even toxic ethylene glycol, at a concentration of more than 2 M, has been found (Gehrken, 1984).

In addition to these antifreeze agents, macro molecular weight haemolymph proteins (thermal hysteresis antifreeze proteins or THPs) have been found to function in providing subzero temperature tolerance. Antifreeze proteins, in some cases with extremely high levels of cysteine demonstrate thermal hysteresis (the difference between the freezing and melting points of an aqueous solution). They have advantages over polyols as they function by means of a non-colligative mechanism (Duman and Horwath, 1983). Polyols, at concentrations required to significantly lower the freezing and supercooling points, drastically increase the osmotic pressure of the body fluids and can therefore be toxic, especially at temperatures above freezing when the arthropods become active. Further, polyols are readily lost at higher ambient temperatures, whereas THPs are produced in early autumn and are maintained until late spring (Duman and Horwath, 1983).

Removal and/or masking of nucleating agents from the haemolymph, gut, and other tissues in winter may also extend supercooling (Block, 1982; Zachariassen, 1982).

Freeze susceptibility is often found in micro-arthropods, like Collembola and Acarina, the small volume of which will have a reduced ice nucleator content compared to larger poikilotherms. There is also a reduced probability of inoculative freezing in micro-arthropods with sclerotized exoskeletons. Furthermore, THPs might protect a supercooled arthropod against nucleative as well as inoculative freezing (Zachariassen and Husby, 1982). In Collembola which live in the extreme habitats of the polar and alpine zones, lowest mean supercooling points ranging from −17·2 to −32·4°C have been found. In some cases glycerol, manitol and trehalose have been found (Sømme, 1982). In an extensive study on the temperate Collembola *O. cincta* and *T. minor*, van der Woude (1986) has found lowest mean supercooling points of −14·4°C and −11·0°C, respectively, for adults in winter. For juveniles −15·3°C and −13·6°C were measured. The lower lethal temperatures (see Fig. 12) and the supercooling points of both species decrease in early autumn and rise in spring. As the total osmotic value of the

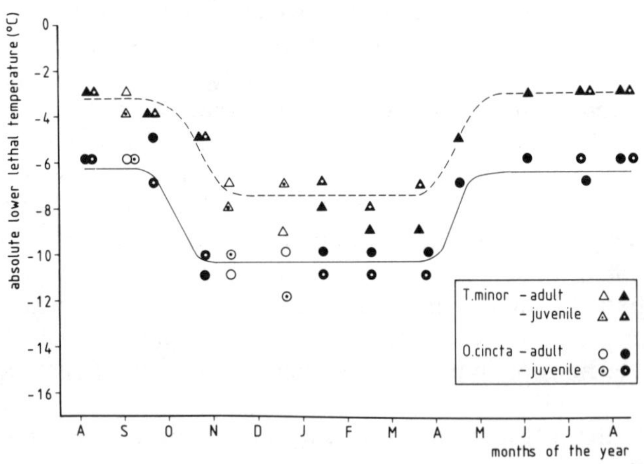

Fig. 12. The seasonal change of absolute lower lethal temperatures during a year for *Orchesella cincta* (juveniles and adults) and for *Tomocerus minor* (juveniles and adults). Open symbols refer to 1981 and filled symbols to 1982. From van der Woude and Verhoef (in press).

haemolymph is constant throughout the year (about 350 mOsm), with small fluctuations caused by osmotic changes during a moulting instar (Verhoef, 1981), the decrease of the supercooling points of these Collembola is assumed to be caused by a non-osmotic factor like a thermal hysteresis protein. This is also stated for collembolan species living in severe environments on glaciers such as *Isotoma hiemalis* and *Entomobrya nivalis*. In species living in habitats with buffered temperatures thermal hysteresis was absent (Zettel, 1984).

A similar trend can be seen in the two above-mentioned temperate collembolan species: *O. cincta* is the cold-hardiest of the two, and lives exposed, partially in trees and partially in the upper litter layer, whereas *T. minor* lives in the deeper litter layers. *O. cincta* is also more drought-tolerant than *T. minor* (Verhoef and Witteveen, 1980); this exemplifies the correlation between cold hardiness and drought tolerance as often found in animals (e.g. Leinaas and Fjellberg, 1985) and plants (Levitt, 1980). The role of THPs in water retention and their presence in dry summers in the drought-tolerant *E. nivalis* (Zettel, 1984) underlines this correlation. Seasonal change in the ability to supercool has been reported for a number of overwintering arthropods (Sømme, 1982; van der Woude, 1986). For the temperate Collembola belonging to the "autumn-dynamic" type of seasonal change in supercooling points (see Sømme, 1982), photoperiod and low temperatures are considered to be the main triggers for induction.

Diet changes and even water availability may also modify such effects (Baust and Rojas, 1985). The effect of feeding activity on the supercooling capacity of an animal has often been demonstrated (Whitmore *et al.*, 1985). Correlations found between empty ("cleared") guts and high supercooling capacity and full guts and low supercooling capacity, leading to a division into a group with high and a group with low supercooling points (Block, 1982), the effect of starvation, the clearing of guts in preparation for overwintering, all point to the original hypothesis of Salt (1961) i.e., that any animal maintaining a full gut is susceptible to gut nucleation, and thus to freezing at elevated subzero temperatures. Although recent studies have provided challenges to the "gut content" idea (see review Baust and Rojas, 1985), the two collembolan species *O. cincta* and *T. minor* show a clear relation between feeding and supercooling point. Because of their frequent moulting at all times of the year there is a fixed part of the population with an empty gut due to the preparation of a new gut epithelium. If the moulting cycle is not disturbed (e.g., by quiescence) 50–60% of a population of *O. cincta* and 55–65% of *T. minor* have filled guts. All through the year a group with high and a group with low values of both supercooling points and lower lethal temperatures is present, correlated with the degree of filling of the gut. Gut evacuation at about zero °C has been found in *O. cincta* as an adaptation mechanism by the disappearance of the, vulnerable, group of animals with high values. In *T. minor* this mechanism is absent. This species acts on the lowering of temperatures by avoidance behaviour: verticle migration in the soil (van der Woude and Verhoef, in press). The same phenomenon is also found in other temperate Collembola (Anderson and Healey, 1972).

(a)2 Freeze tolerance. The adaptations involved in arthropod freeze tolerance are complicated and not completely understood. Polyols, especially glycerol and sorbitol, help to prevent freeze damage as cryoprotective agents. It has not yet been shown clearly how polyols act in the production of freezing tolerance. Tissue changes at the membrane level or molecular level are likely to be of critical importance to the development of freezing tolerance. The important shifts in the major cations (Ca, Na, Mg and K) of the tenebrionid beetle (*Upis ceramboides*) reflect major changes at the molecular level. Of possible interest is the effect of calcium on protoplasmatic viscosity and membrane permeability (Miller, 1982). Most of the animals employing this strategy do not supercool extensively prior to freezing. Recently, several arthropods have been shown to actively synthesize haemolymph ice nucleating agents which minimize supercooling, initiate ice formation in the extracellular fluid at relatively high subzero temperatures and thus prevent lethal intracellular ice formation. The agents appear to be proteinaceous (Zachariassen and Hammel, 1976; Duman and Horwath, 1983). Examples of freeze-tolerant arthropods with ice-nucleating factors are several Coleoptera species, Lepidoptera larvae, Hymenoptera species and gallfly larvae (Block, 1982; Miller, 1982).

In many of these freeze-tolerant arthropods the water content is considerably reduced. Much of this water might be held in a metabolically "bound" state. Layers of "bound" water around intracellular macromolecular structures might serve to limit intracellular freezing and protect molecules, such as proteins, from denaturation (Storey *et al.*, 1981).

There are exceptions to the rule of the high supercooling points in freeze-tolerant arthropods. Several freeze-tolerant larvae and adult arthropods (beetles, parasitic wasps and Hymenoptera) combine low supercooling points with the presence of glycerol or other cryoprotective substances (Miller, 1982).

The two "strategies" appear to have much in common, polyols, sugars and THPs function in both groups. Changes from freeze-tolerant to freeze-susceptible between different developmental stages (larva and adult, respectively) and switching from freeze tolerance to freeze susceptibility from one year to the other have been found in the beetle larvae *Dendroides canadensis* and *Cucujus clavipes*. Their lower lethal temperatures were similar from year to year, regardless of the strategy employed (Horwath and Duman, 1984). The reasons behind this change from a freeze-tolerant to a freeze-susceptible mode are still unknown. The classical distinction between the achievement of cold hardiness via a freeze-tolerant or a freeze-susceptible route may not be so rigid (Horwath and Duman, 1984); the same phenomenon may be present in many other arthropods.

(b) Heat tolerance. Arthropods, like other animals, can tolerate heat up to temperatures of about 50°C, which is up to 10°C lower than summer temperatures occurring at the surface of temperate dune ecosystems (e.g., Ernst and van Andel, 1985). As a measure of upper temperature sensitivity the Critical Thermal Maximum (CTM), is often used (Hutchinson, 1961). CTM values have been determined for only a few arthropods, including meloid beetles (Cohen and Pinto, 1977), termites, sawfly larvae, cockroaches (see Appel *et al.* 1983; Cohen and Cohen, 1981; Vannier and Ghabbour, 1983) and also Collembola (Thibaud, 1968, 1977a, b). CTM values for the different cockroaches studied range from 51·4°C to 47·6°C and show a good correlation with habitat temperatures (Appel *et al.*, 1983). The same has been found in meloid beetles (Cohen and Pinto, 1977) and Collembola species (see Table 2).

In temperate Collembola (CTM = 35–45°C), troglobic species appear to have CTMs 5°C lower than those of epigeic ones (Thibaud, 1977b). In these arthropods inverse relationships can be found between transpiration rate and heat tolerance, especially when transpiration has been measured at higher temperatures (see Vannier and Verdier, 1981).

As the thermal damage of cellular functions, measured in animals, plants and micro-organisms, is often due to thermal denaturation of proteins, it is generally assumed that thermostability of cells is an indicator of resistance of cellular proteins to the denaturing action of heat (Alexandrov, 1977; Levitt, 1980). Research on the thermostability of animal tissues and cells has mainly

Table 2

Upper lethal temperatures of Collembola species from thermo-stable (t.s.) and thermo-unstable (t.u.) habitats.

Species	Upper lethal temperature	Habitat
Folsomia quadrioculata	35°C	t.s.
Isotoma notabilis	38°C	t.s.
Isotoma violacea	35°C	t.s.
Onychiurus armatus	35°C	t.s.
Entomobrya marginata	46°C	t.u.
Entomobrya nivalis	45°C	t.u.
Hypogastrura viatica	38°C	t.u.
Lepidocyrtus cyaneus	48°C	t.u.
Podura aquatica	40°C	t.u.
Vertagopus westerlundi	50°C	t.u.
Xenylla maritima	48°C	t.u.

Data from Thibaud, 1977a; Leinaas and Fjellberg, 1985.

been done with molluscs, reptiles, amphibians, fishes and birds (Alexandrov, 1977) and there is a paucity in arthropod literature. The proposed mechanisms of thermotolerance (Alexandrov, 1977; Levitt, 1980) comprise thermostability of proteins, lipid properties and thermostability of nucleic acids.

An increased thermostability of proteins might be caused by structural changes in protein molecules leading to conformational flexibility, increase in anti-denaturing substances (e.g., cations like Ca), and increased resynthesis of proteins (repair). Recently, the heat shock proteins (hsps) system has been discovered. These proteins are synthesized in a few minutes soon after the temperature stress and are supposed to be involved in the process of the acquisition of thermal resistance. This system seems to be a common characteristic as it has been found in bacteria, yeasts, insects, sea urchins, birds and mammals, with a high variability of the molecular weight of the hsps among the different groups (Stephanou *et al.*, 1983). In *Drosophila* the survival of two strains, one sensitive (S_1) and another resistant (R_1) to heat shock, has been found to be correlated with the regulation of hsps synthesis, the R_1 strain synthesizing more hsps compared with the S_1 strain, which points to the adaptive significance of these proteins (Alahiotis and Stephanou, 1982).

Relations between lipid properties and high temperatures like the replacement of unsaturated by saturated fatty acids have been found in plants, amphibians, fishes and also in fly larvae (see Prosser, 1973).

Only a few publications exist concerning the relation between the properties of nucleic acid and high temperature and they are exclusively limited to thermostable micro-organisms. This relation might, however, also apply to animals.

C. Starvation

Given the low humus content of the sandy soils and the low N and P content, the question arises whether animals living in the dune soil environment suffer from limited energy and minerals and eventually exhibit adaptations to it. Attention is focused here mainly on springtails (Collembola) and woodlice (Isopoda).

Studies on the food preferences of Collembola by examination of the gut contents of animals collected in the field, have nearly all led to the conclusion that most Collembola are unspecialized feeders (Petersen, 1971; McMillan, 1975; Vegter, 1983). Also in laboratory cultures most Collembola feed on a wide variety of food: fungal hyphae, spores, bacteria, pollen grains, decaying plant and animal material and unicellular algae. In many cases the diet of springtails is found to be essentially what is available (Poole, 1959; McMillan and Healey, 1971; Bödvarsson, 1970; Gilmore and Raffensperger, 1970;

Petersen, 1980; Vegter, 1983). This low degree of food specialization could be associated with an excess of food available for saprophagous species (Anderson and Healey, 1972). However, a varied diet would also seem to be a selective advantage when food becomes scarce at certain times (Emlen, 1973). Food generalism might thus also evolve as a result of food scarcity or nutrient imbalance or an unpredictable food supply.

Joosse and Testerink (1977a) demonstrated that in forests on sandy soils food occasionally becomes scarce for the collembolan *Orchesella cincta*, especially when its habitat becomes very dry during summer. The question was raised, whether the observed starvation during dry periods is caused by an absence or by an inaccessibility of food and it was thought that we meet here an active arrest of feeding, although food is present. This has also been found in some other insects (Murray, 1968). For Collembola the arrest of feeding under those circumstances is interpreted as an adaptation to drought. This immediate response to adverse humidity conditions has been described as a quiescence or oligopause (Testerink, 1983). Gut contents in particular appear to be disadvantageous under dry conditions, because the transpiration rate in fed animals is found to be significantly higher than in starved ones (Vannier and Verhoef, 1978). The ability to survive dry periods is especially apparent in species which demonstrate the mechanism of evacuating the gut contents (Verhoef and van Selm, 1983).

It is not very clear if woodlice (Isopoda) species also suffer from food shortage under natural conditions. They are usually found in nature with food in their digestive tract. It is remarkable however, that species like *Oniscus asellus* can tolerate very lengthy fasts. In laboratory conditions at 20°C, survival was measured beyond 7 weeks (Hartenstein, 1964), and the large cells of the hepatopancreas which undergo obvious alterations after food deprivation, remain capable of regeneration after several weeks of starving (Storch, 1982). This marked ability to starve suggests the occurrence of starvation periods under natural conditions, but the question remains still largely unresolved (Al-Dabbagh and Block, 1981). A characteristic feature in general appears to be a very high death rate of new born young, but no attention has been given to nutrient deficiency. Brody and Lawlor (1984) described fluctuating food availability for *Armadillidium vulgare*, due to temperature and humidity changes. Just like Collembola these animals remain below the surface layers of the leaf litter and do not feed during periods of drought.

1. *Physiological Avoidance Mechanisms*

It is to be expected that metabolic adaptations exist as a mechanism to survive starvation periods, for fasting certainly has consequences for energy demanding activities. It was found that in Collembola the food reserves, lipids and glycogen, were depleted during these periods (Testerink, 1981). It

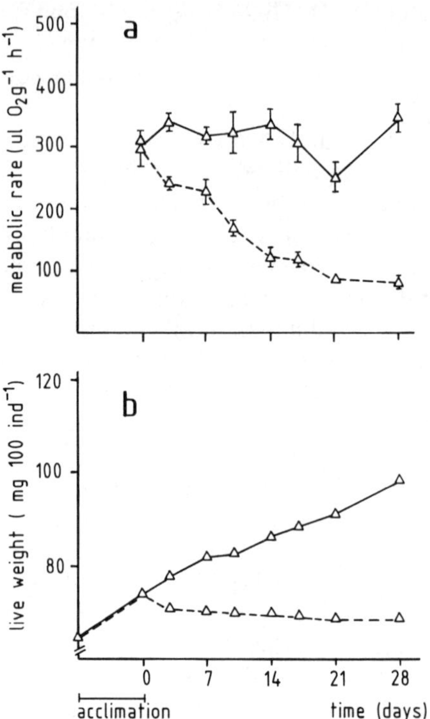

Fig. 13. Effect of starvation on metabolic rate and growth in *Orchesella cincta*: △—△ fed animals; △---△ starved animals. After Testerink (1983).

has been demonstrated by Testerink (1983) that after one week of starvation the decrease in the metabolic rate of springtails (as measured by respiration of inactive individuals) was more than 40%. This greatly reduced the rate of energy expenditure (Fig. 13).

As found in the laboratory the quantity of food also has a direct effect on energy metabolism, measured as oxygen consumption, of the isopod *Armadillidium vulgare* (Hubbell, 1971). The rate of oxygen consumption of *Porcellio scaber* however, was unaffected by 2–4 days starvation (Wieser, 1972; Newell *et al.*, 1974) and several other authors found for this species that the oxygen consumption rate is independent of the depletion of nutritional reserves (Husain and Alikhan, 1974).

It remains a question, whether food could be responsible for some of the observed differences in metabolic rate. *Philoscia muscorum* (isopod) showed respiration rates in winter (measured at a given temperature) which are significantly higher than in spring or autumn. This is probably due to the animals acclimatizing by raising their metabolic rates higher than normal at

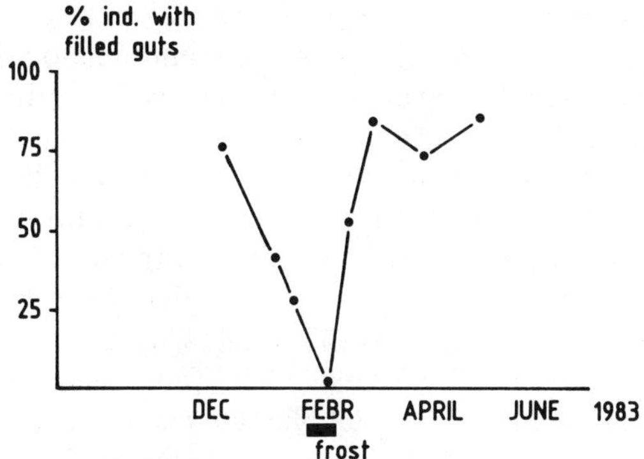

Fig. 14. Percentage of individuals of *Orchesella cincta* with filled guts during a season. After van der Woude (1986).

lower temperatures (Hassall, 1983). The respiration rate is also higher in summer than in spring and autumn and this is attributed to the increased physiological activity associated with the development of gonads as was found also by Phillipson and Watson (1965) for *Oniscus asellus*.

2. *Life History Patterns*

In field populations of Collembola starvation occurs rather frequently, not only during dry periods in summer but also during low temperatures in winter (Testerink, 1981; van der Woude, 1986) (Fig. 14). The effect of starvation on energy-demanding activities is great. Most important for the population is an arrested reproduction for some time, which only restarts when the conditions are favourable again (Joosse and Testerink, 1977a; Joosse, 1981). Since the collembolan life-cycle is characterized by alternating feeding and reproductive instars (Fig. 15), separated by an ecdysis (de With and Joosse, 1971; Joosse, 1981), all the individuals synchronously enter a reproductive phase after the arrest. The synchronization can be traced in the field for several instars (Testerink, 1981). The ability to adjust the reproduction cycle to favourable times by physiological means increases the chance for the animals of reproductive success in a temporarily uncertain and spatially heterogeneous environment (Giesel, 1976; Stearns, 1976).

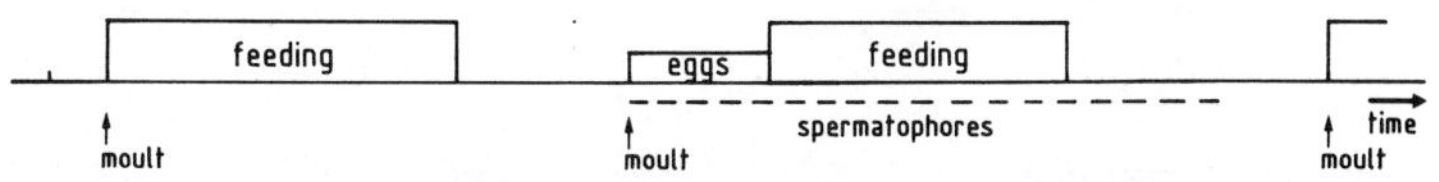

Fig. 15. Feeding and reproductive instars in Collembola. After Joosse (1981).

This not only prevents wasting of eggs but also saves reserves to permit higher endurance of starvation and a prolonged reproduction phase, which was shown in periodically starved animals (Joosse and Testerink, 1977b).

It has been demonstrated by a demographic field study that a species such as *Orchesella cincta* which exhibits such a trait, has relatively high fertility compared with other species (van Straalen, 1983). This high fertility, however, is not only caused by this reproductive timing mechanism, but mainly by a more rapid succession of the reproductive instars during its whole lifetime. The species in fact has an overall high metabolism compared to other collembolan species (Testerink, 1982) (Table 3), and thus a higher energy requirement. Such a high metabolic rate enables the animals to be more active, to moult and reproduce more frequently (Joosse and Veltkamp, 1970) and thus to have a higher fertility. But the higher mobility of *O. cincta* and comparable species (Joosse, 1971) allows also a more efficient exploration of the habitat and adults of these species can even be found occasionally on plants and in trees (Bowden *et al.*, 1976; Bauer, 1979; van der Woude and Verhoef, 1986). With this way of living the individuals have, however, to cope with a higher predatory pressure; higher than inactive species. They encounter many predators (Ernsting and Joosse, 1974) and although they apparently avoid predation by an efficient use of their escape mechanism in the form of a spring tail (Ernsting *et al.*, 1977, Ernsting and Jansen, 1978), mortality in field populations appears to be relatively high in this species, apparently related to the predatory pressure (van Straalen 1985a, Ernsting and Fokkema, 1983).

A survey of the literature reveals similar traits among various other groups of soil animals like lumbricids and oribatid mites (van Straalen, 1985a). A comparison between species of these taxonomic groups shows that surface active species generally tend to have a higher metabolism, than species living deeper in the soil, and this is paralleled by a higher fertility and a higher mobility. It is tempting to interpret this trend by the more unpredictable environment of the soil surface, but this, according to theory (Schaffer

Table 3
Mean metabolic rate (μl_2 O_2 g^{-1} h^{-1}) during a moulting interval of males and females of *Orchesella cincta* and *Tomocerus minor* at 15°C (after Testerink, 1982).

		Mean metabolic rate (±s.e.)
Orchesella cincta	♂	417·0 ± 6·3
	♀	391·2 ± 2·4
Tomocerus minor	♂	345·6 ± 9·3
	♀	356·4 ± 5·4

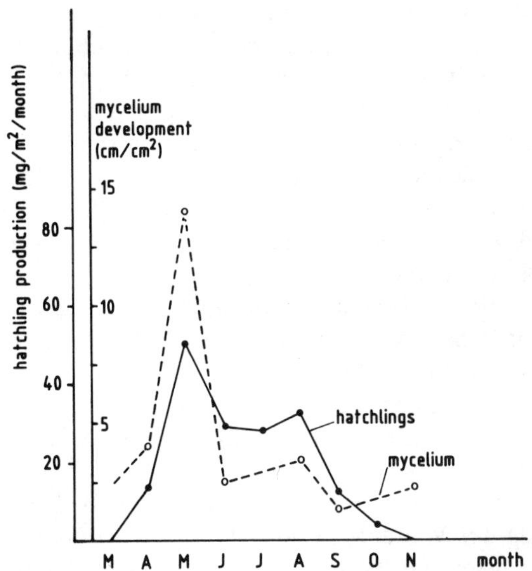

Fig. 16. Seasonal changes in resource utilization by collembolan hatchlings compared with mycelium development in the soil. After Vegter (1985).

1974) should on the contrary select for relatively low fertility. Van Straalen (1983, 1985b) offered an interpretation by assuming that these patterns are caused by the exposure to higher temperatures at the soil surface, which includes a higher energy intensity of these habitats. Kennedy (1928) already stated that the energy intensity of the environment is correlated with the level of metabolism in many insects. This phenomenon could ultimately be seen as a life strategy to optimalize survival and reproduction in an uncertain environment.

Another adaptation to seasonally varying resources has been described for Collembola by Vegter (1985). Various coexisting species appear to be adjusted to the seasonally varying food supply by their phenologies. Among several authors, Witkamp and van der Drift (1961) and Ausmus *et al.* (1976) demonstrated a high production of microfloral resources, food for various species, in early spring, which means a high carrying capacity for species to concentrate recruitment in this period (Fig. 16). Some collembolan species like *Lepidocyrtus lignorum* utilize this microfloral bloom by hatching synchronously early in spring from overwintered diapause eggs (Leinaas and Bleken, 1983). *Tomocerus longicornis* also hatches in spring from dormant eggs produced during the previous summer (Vegter, 1985). This spring hatching may be especially useful as a compensation for nutrient deficiency at other times of the year. Leinaas and Bleken (1983) state with Masaki

(1980) that the diapause egg is a direct adaptive response to exploit this spring bloom. However, the selective forces underlying these timing mechanisms and the seasonal specialization of recruitment is probably not the avoidance of competition as was proposed by Leinaas and Bleken (1983), but adaptation to minimize hatchling mortality (Vegter, 1985), for there are few indications of competitive interactions. In Table 4 seasonal differences between eight species of surface dwelling collembolan species from a mixed woodland on sandy soil in the Netherlands are presented. The niche breadth (for methods see Vegter, 1985) result in a certain overlap with other species' niches. The value indicating the average overlap with other species tells us if the resource utilization by hatchlings of a species is located in a period of utilization by others. A low value means a low common utilization. There appears to be no clear relation between the average overlap and some measure of success like average density or the amount of resources utilized.

In the isopod *Trichoniscus pusillus*, Sutton (1968) observed a cessation of growth during unfavourable seasons when food sources were lost through drying of the litter and the same was found in *Porcellio laevis* in tropical regions (Nair, 1978), where the production of young also fell rapidly. This suggests an influence on population development.

Somewhat unexpected consequences of food shortage for populations of woodlice are described by Brody and Lawlor (1984). It appeared that food shortage, experimentally imposed in the laboratory, increases, rather than decreases the size of hatchlings. When less food was available females reproduced less frequently, but the size of the offspring increased. It is seen as an adaptive, reproductive response to food stress. In the field, spring is generally more favourable for reproductive success, it is more humid, cooler and far more favourable for foraging. During summer foraging is less favourable and summer broods hatch at less favourable times. This brood is smaller but the offspring larger, probably with a higher survival chance, but this has not yet been studied.

D. Mineral Shortage

As White (1978, 1984) stated, species of animals possibly endure not so much food being in short supply as such, but the right type of food, specifically food which contains enough nitrogen for the success of very young animals. Saprophagous soil animals depend for their food supply mainly on litter and microflora. The N content of different plant tissue can range from 0·03 to 7·0% of dry weight. During the season N levels drop sharply (Scriber and Slansky, 1981) and concentrations decline gradually during the course of a growing season until senescence, after which concentrations decline sharply again, owing to minimal protein synthesis, increas-

Table 4

Niche parameters, annual resource utilization by hatchlings and average population density of eight species of Collembola from a mixed woodland on sandy soil in the Netherlands (after Vegter, 1985).

	Average birth date	Niche breadth	Average overlap with other species	Average density of the total population during the sampling season	Annual resource utilization
Lepidocyrtus lignorum	5·75	3·7	0·179	3620	56·9
Orchesella cincta	7·15	5·1	0·164	1008	47·1
Orchesella flavescens	7·60	3·0	0·177	92	3·4
Tomocerus flavescens	7·60	3·4	0·144	323	10·2
Tomocerus vulgaris	6·26	3·8	0·204	357	24·4
Tomocerus longicornis	4·96	1·3	0·236	114	10·7
Entomobrya nivalis	6·53	3·1	0·198	289	8·6
Entomobrya corticalis	7·01	3·4	0·164	351	7·4
Dimensions	Month (January = 1)	Month	$(\text{Month})^{-1}$	nrs/m^2	mg dry weight/m^2/year

ing protein hydrolysis and translocation to perenniating plant parts. When the tissue becomes litter, minimal concentrations occur (0·5–1·5%; Mattson, 1980).

P and K also decline markedly between maturity and senescence of plant material. Data on the movements of nutrients such as Cu, Zn, S, Mn and Fe from leaves differ and are sometimes contradictory (Hill, 1980); some studies indicate that S can move out of the mature leaves in which it was initially deposited.

1. *Behavioural Avoidance Mechanisms*

In general it can be said that organisms on diets with low N consume more food than organisms on diets with high N content as demonstrated by Mattson (1980) on the basis of a great number of data on invertebrate herbivores. *Lumbricus terrestris* gets sufficient N by feeding not on plant litter alone but on associated microbial protein (Satchell, 1983). Cooke and Luxton (1980) suggested that microbial contamination makes a food source attractive to worms. They found with Cooke (1983) that earthworms prefer fungally contaminated cellulose, but some species can also grow on activated sewage sludge, which is composed almost entirely of live and dead bacterial cells (Neuhauser *et al.*, 1980).

Soil fungi have the ability to concentrate biologically important elements such as N, P, Ca, Cu, K, Fe, Mg, Mn, Na and Zn (references in Wallwork, 1983) and it is not surprising that many soil arthropods feed on fungi and so obtain and mobilize the minerals (Reichle *et al.*, 1969). Fungi and algae have relatively high quantities of N (usually 2%). Many fungi can adapt physiologically to a very wide range of available N. The nutritional quality of different species of fungi is therefore likely to be variable (Merrill and Cowling, 1966). It is shown that immature forms of cryptostigmatid mites for instance may be more selective in their feeding habits and not survive if they are prevented from feeding on fungal hyphae soon after hatching (Woodring and Cook, 1962). Similar phenomena appeared from culture experiments with the collembolan *Tomocerus minor* (Testerink, 1982).

It is known from a comparison of mixed deciduous woods and pine woods (Cornaby *et al.*, 1975, Gist and Crossley, 1975) that the rate of utilization of Ca was many times higher in the deciduous woodland where oribatid mites dominated and Crossley (1977) described high body concentrations of Ca in various species of oribatid mites (35 mg g^{-1}), whereas Collembola have less than 3 mg g^{-1}. Three times as much K was utilized in the pine wood, where Collembola dominated over mites, than in deciduous woodlands, indicating a differential need for minerals in the various species. Springtails appear to contain about 20 mg g^{-1} K (Bierenbroodspot, 1978) whereas mites have about 2 mg g^{-1}.

Mineral requirements in arthropods remain, however, probably the most neglected area of research and very little if anything can be said about adaptive traits involved in nutrient imbalance.

2. *Life History Patterns*

Food quality has been found to cause a significant change in the growth rates of populations of *Porcellio olivieri*. Isopods raised on a high quality food had an accelerated rate of growth, which in turn directly affected the timing of first reproduction. Similar results were shown by Merriam (1971) for *Armadillidium vulgare*. Thus reduced availability of a specific nutrient in the field could result in a decrease of the intrinsic rate of natural increase. It was for instance shown by Booth and Anderson (1979), that the N content of the food has a significant effect on fecundity in the collembola *Folsomia candida*.

E. Toxic Compounds

An important factor influencing the food quality of soil animals, is the tannins and lignins; complex phenolic compounds that occur widely in vascular plants. Most tannins are of moderate molecular size and are soluble components of living cells. They show a wide structural divergence. Lignins are polymers with higher molecular weight (Swain, 1979). Tannins and lignins accumulate in decomposing plant material and become incorporated into the humus. The substances are considered to have adverse effects on animals and have been accorded an important role as defensive chemicals that protect plant tissues for herbivore attack. The food may be repellent or unpalatable to animals because of the presence of these compounds. Sclerophylly is characteristic for ecosystems with periodic drought combined with nutrient stress (Specht, 1979) as in dune landscapes. This kind of leaf generally has higher concentrations of organic polymeric substances such as tannins and lignin. Apart from the translocation of nutrients away from leaves in senescing plants, causing a possible nutrient shortage, the litter will have an increased C/N and C/P ratio. In the case of nitrogen this may be in the range of 100 : 1 to 500 : 1.

1. *Behavioural Avoidance Reactions*

Satchell and Lowe (1967) showed that palatability in worms is strongly associated with the tannin content of the food. The concentration of polyphenolics in leaf litter and of other constituents vary between leaves of different ages and species, so a choice of food material is possible for earthworms. They exhibit marked preferences. Palatability of leaf discs to *Lumbricus terrestris* is related to N, soluble carbohydrates and phenolic contents, but also microbial action on the leaves is important (Satchell,

1983). Satchell and Lowe (1967) suggested that microbial action was the main process which removed distasteful substances from dead leaves, and microbial contamination makes the food attractive to worms (Cooke and Luxton, 1980).

2. *Tolerance Mechanisms*

Isopods and diplopods are described as tannin-tolerant species (Neuhauser and Hartenstein, 1978). Leaves with higher concentrations of condensed tannins are generally believed to be less palatable than leaves with lower concentrations. Edwards (1974) suggested that one of the main factors influencing palatability to isopods seems to be the sort and amount of polyphenols in the litter. It appears, however, that the phenolic content is not related to leaf palatability in isopods and diplopods and up to 50% of the phenolic content of the leaves disappeared when consumed by *Oniscus asellus* (Neuhauser and Hartenstein, 1978). Apparently they possess some tolerance mechanism.

Tannins, in most plants belonging to the condensed tannins (Swain, 1979), are known to precipitate proteins (McManus *et al.*, 1983) and it has been suggested that the digestibility of plants is reduced by precipitating the plant proteins or by forming complexes with the digestive enzymes of the animal (Feeny, 1976; Rhoades and Cates, 1976) and bound proteins are inert to hydrolysis. Lignins occur in mature cell walls of all vascular plants and like tannins reduce the availability of carbohydrates and proteins (Swain, 1979).

Martin and Martin (1983) showed that tannic acid and pine oak tannins precipitate proteins at pH values between 6·0 and 8·0, but little or none at pH values above 8·0. It has been suggested that an alkaline gut fluid of animals could reduce the deleterious effects of tannins (Fox and Macauley, 1977) and even an alkaline midgut would reflect adaptation to tannins (Berenbaum, 1980) as shown in lepidopterous larvae.

Very little is known about pH values of the digestive tract of these soil invertebrates. The gut contents seem to have a slightly acid reaction generally in the same range as the digestive enzyme optima, which is 5·0–6·0. There are, however, exceptions for peptidases, which have a pH optimum of about 8·0. Hartenstein (1964) estimated the pH of the digestive tract and hepatopancreas of *Oniscus asellus* as 6·4–6·8 (Fig. 17). By using redox dyes he showed a higher concentration of reducing substances in the papillate midgut. This suggests a somewhat higher pH in this region and distinguishes this part of the gut functionally from the rest of the digestive system.

Isopods feed on decomposing plant materials rich in micro-organisms. The alimentary canal is relatively simple and hindgut fermentation systems bring about inefficient internal assimilation mechanisms. Often the animal's own faeces is ingested. The alimentary canal possesses, however, a subdivided hindgut and a typhlosole (Fig. 17) which allows the animals to

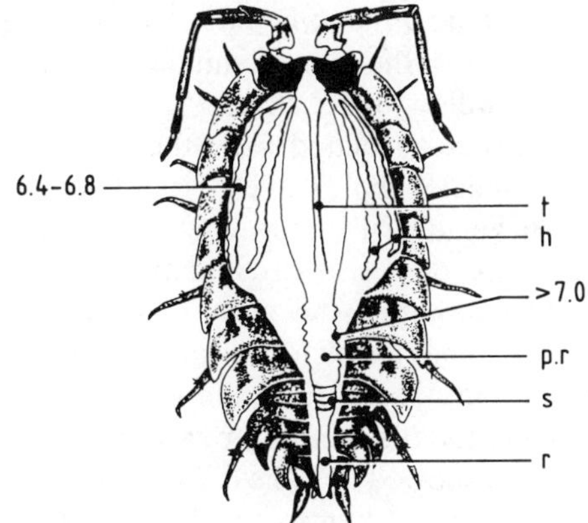

Fig. 17. Diagram of the alimentary canal of an isopod (dorsal view) and pH values in some regions. t = typhlosole; h = hepatopancreas; p.r. = papillate region; s = sphincter; r = rectum. After Hartenstein (1964).

exploit to the maximum the degradative capabilities of the ingested micro-organisms (Hassall and Jennings, 1975). Any available soluble nutrients are immediately passed into the hepatopancreas and absorbed. The solid remains of the meal passes into both parts of the hindgut where the breakdown of the food continues. In these areas absorption occurs. The enzyme acting in the papillate region of the hindgut apparently does not kill the micro-organisms which occur in this region. In ruminant mammals micro-organisms are killed and the animals utilize them as a source of protein. It is conceivable that this special function of the hindgut with the high pH also promotes the tannin resistance. A remarkable coinciding fact is the occurrence of iron in the papillate hindgut, which indicates an uptake of Fe^{2+}, only possible at relatively high pH values.

In Collembola, colorimetric determination of the pH of the digestive tract revealed that the intestinal pH is acidic in the first region of the midgut (5·4–6·4) and alkaline in the second region of the midgut (8·2–8·8). The hindgut is again acidic (Humbert, 1972) (Fig. 18).

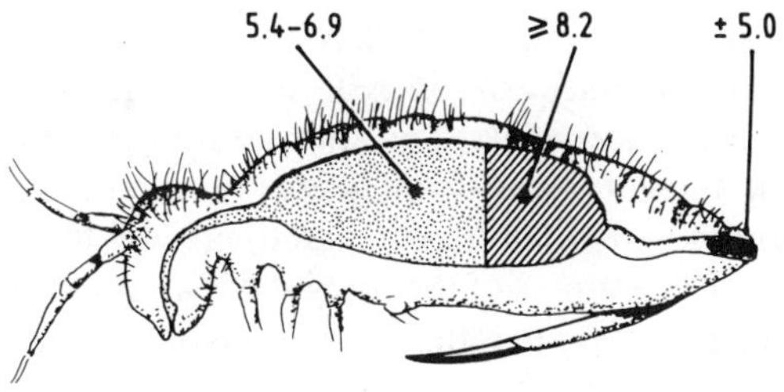

Fig. 18. pH values in the alimentary canal of Collembola. After Humbert (1972).

In the soil-feeding termite *Procubitermes aburiensis* a sharp rise in pH from 6·9–9·7, occurs between the midgut and the first proctodaeal segment as a result of an active inflow of K^+ across the gut wall. Analysis of the populations of actinomycetes isolated from the various parts of the guts suggested that in this species the viability was affected and the alkaline conditions seemed to have adverse effects.

Martin and Martin (1984) demonstrated that the detergent action of insect gut fluid can serve to counteract the protein precipitating capacities of tannins. In every study that has tested for surfactants in insect gut fluids their presence is confirmed. Very low surface tension of the digestive juice of several crustaceans suggested the presence of a sort of bile acids.

Neuhauser *et al.* (1974) showed that the isopod *Oniscus asellus* is capable of cleaving the aromatic ring of benzoic acid, although it was not certain whether this is attributable to gut flora. Neuhauser and Hartenstein (1976a) demonstrated that the species is capable of degrading ringlabelled cinnamic acid (a precursor substance in lignin biosynthesis), and phenol. These substances are regarded as xenobiotics in vertebrates and invertebrates. By this ability the animals may perform a very important function in soil development.

Neuhauser and Hartenstein (1978) and Neuhauser *et al.*, (1978) demonstrated that several species of invertebrates, including earthworms, diplopods, isopods, snails and slugs, were unable to degrade lignin but were able to convert the ring carbon atoms of phenolics such as vanillin, a major aromatic product obtained during the oxidative decomposition of lignins, into CO_2.

The most important system for detoxification of these allochemics is most probably the mixed function oxydases (MFO), bound to the endoplasmic reticulum of cells. They are active in fat body and Malpighian tubules, but especially in the gut to detoxify the substances before they enter the haemolymph (Ahmad, 1983). Neuhauser and Hartenstein (1976b) measured demethylation of methoxyl groups, a major component of lignins in several detritus-feeding invertebrate species. The reaction is catalyzed by O-demethylase, an MFO.

Some species of polyphagous insects (Acridioides) also seem to be quite tolerant to hydrolysable tannins (Bernays, 1981) and even an increased digestibility of food and an increased conversion efficiency was demonstrated in *Schistocerca gregaria*. It appeared to be the result partly from hydrolysis in the gut, partly from being absorbed onto the peritrophic membrane (Bernays and Chamberlain, 1980). It was found that approximately 25% of the free tannins of the foregut of *Schistocerca gregaria* could be taken up. A suggestive point is that in graminivorous insects, which rarely encounter tannin, the peritrophic membrane tends to be thinner (Bernays and Chamberlain, 1980), in spite of the fact that grasses are very hard and silicious. Almost always it is said that peritrophic membranes function in protecting the midgut epithelium from abrasion by food particles, but some species which need it seem to lack a peritrophic membrane and numerous juice feeders possess one (i.e., Lepidoptera) (Richards and Richards, 1977). Peritrophic membranes have been found in Collembola (Humbert, 1979) but not in Isopods (Hartenstein, 1964; Holdich and Mayes, 1975). Humbert states, following Mello *et al.* (1971), that it could play the role of an ultrafilter. There is little to say about the permeability of the peritrophic membrane however. From a few data it appears that particles of up to approximately 9 nm penetrate. For comparison, amino acids measure up to 4 nm (Zhushikov, 1970). Considerable differences may probably be found among insects. Richards and Richards (1977) report no penetration of ferrititoxic metal ions (see below).

Peritrophic membranes produce an ectotrophic space where movement may be slower than in the endotrophic space and uptake of minerals may be enhanced. Degradation of macromolecules only takes place in the endotrophic space, the hydrolysis to dimers and oligomers occurring principally in the ectotrophic space and some terminal digestion occurring intracellularly or mediated by enzymes bound to the apical cell membrane (Terra *et al.*, 1979; Terra and Ferreira, 1981).

3. *Life History Patterns*

Specific phenolic compounds and humic acids and lignins were tested for their effect on the growth of *Eisenia foetida* (Hartenstein, 1982). It appeared that the worm is tolerant to relatively high concentrations of several substances which are quite toxic for rats. It was concluded that the presence of an aldehydic methoxyl, methyl or a hydroxyl group of low molecular weight compounds is potentially toxic to earthworms. High molecular weight materials with these functional groups, such as humic acids and lignin appear not to be toxic.

In some acridic species tannic acid has been shown to have extremely serious effects on growth and survival and caused a delay in maturation (Bernays, 1981; Feeny, 1968; Haukioja *et al.*, 1978). Feeny (1970) argued that all tannins play a key role in determining seasonal distribution of foliage feeding Lepidoptera. Fox and Macauley (1977) correlated the effects of tannin with the availability of nitrogen and found in a lepidopteran that the nitrogen concentration was more important. For soil invertebrates no data are available and many problems remain to be solved regarding the effects and resistance mechanisms to tannins and lignins.

III. POLLUTED ENVIRONMENTS

Widespread concern about the ecological effects of chemicals developed only recently. The sources of pollution are manifold and not only affect

Table 5
Chemical composition of sewage-sludge from a rural and an industrial area (after de Haan, 1975) compared with the Dutch directives for maximal admissible values (1981).

Element (ppm)	Rural	Industrial	Maximal admissible
Zn	1700	5557	2000
Cu	400	1087	600
Pb	414	663	500
Cd	11	135	10
Ni	35	933	100

urban areas but also more remote rural habitats. Pollution from automobile exhausts, incineration of refuse and industrial activities have brought about increased deposition of sulphate, nitrate, ammonium, hydrogen ions and heavy metals. Trees are weakened and as a consequence insect herbivore populations sometimes enhanced (Alstad *et al.*, 1982). The dumping of solid wastes in sanitary landfilling according to different practices has recently lead to many social problems. Composting of solid wastes goes into intensive gardening and into city parks and introduces amounts of heavy metals in the soil environment which are an increasing hazard because of their high persistence (Ernst and Joosse, 1983). The same can be expected from the application of wastewater sludges onto land, which contain inadmissible quantities of heavy metals when it is of urban origin (Table 5). The slurry from intensive cattle and poultry breeding brings an excess of nitrate, phosphate and copper to agricultural soils and natural forest floors. The application of numerous pesticides and herbicides brings toxic organic and mercury-, arsenic- and copper-containing substances into the soil.

This section focuses attention on the effects on and responses of soil invertebrates to air pollution and the most dangerous heavy metals and the consequences of these toxic influences for the functioning of their populations.

A. Air Pollution

The increased emission of air pollutants contains deposits of SO_4^{2-}, NO_3^-, NH_4^+ and H^+ ions and heavy metals. In forest ecosystems the research concerning air pollution is focused first of all on effects of acid deposition on vegetation. Soil biota has been given less attention. A few studies are available on the effects of experimental acidification with diluted sulphuric acid on the abundance of soil fauna. Most of them describe correlative effects. These studies indicate that soil organisms could be sensitive to increased acidity of the rain. Significant effects are described among

Oribatei, Mesostigmata and Collembola by Hågvar and Kjøndal (1981). In the most strongly acidified situations the abundance of Oribatei has increased. Also many collembolan species increased in numbers, but some species like *Isotoma notabilis* and *Lepidocyrtus cyaneus* declined in numbers. It is thus important that in such studies the animals are identified to species level. Although preference experiments in which soil animals were allowed to colonize sterile soil samples support these results (Hågvar and Abrahamsen, 1980), Bååth *et al.* (1980) got different results, he found the fungal feeding collembolan *Tullbergia krausbaueri* in considerable abundance but little change in oribatid mites.

The principal enchytraeid bacterial grazer *Cognettia sphagnetorum* showed marked decreases with low pH and liming, which could be the result of an osmotic shock. Huhta (1984) recognized *Cognettia sphagnetorum* as an acidophilic species, which is stimulated by acidification and negatively influenced by liming with $CaCO_3$. Huhta *et al.* (1983) also found with increasing pH an increase in bacterial feeding nematodes and enchytraeids and a decrease in species typical of acidic conditions. An inhibition of bacterial feeding Protozoa in grassland has been suggested by Coleman (1983). These results may indicate a shift of the microflora in favour of bacteria. It is known that many fertilizers also affect the soil acidity, which results in changes in the soil microflora, the food for fauna. Food availability in turn could thus be a main factor in changes in the abundance of soil fauna (Huhta, 1984; Bååth *et al.* 1980).

It must be concluded that the different results of these authors are as yet not easy to explain. It is not yet clear whether the pH or the hydrogen ion activity in the soil are the causal factors for the observed phenomena. Explanations remain speculative. Effects on invertebrates have only been observed in studies where the applied acid is 10–20 times more concentrated that in a natural acid precipitation (Abrahamsen, 1983). Hågvar (1984) gives as many as 8 hypotheses, related to the direct influence of lime or sulphuric acid, the reduction of vegetation, reduced predation pressure, reduced availability of food (fungal hyphae), fecundity and population growth rate or competitive interactions between species. It seems to be a fact that those species which in general are found in raw humus soil with a natural pH of about 4·0 are the most tolerant. It is thus assumed that the factors initiating the increased abundance of these particular species in the acidification experiments are the same factors which have induced their high abundance in natural acid soils. These species are supposed to be increasingly favoured in the competition process.

Data from laboratory experiments are very scarce. White (1983) assessed effects of SO_2, NO_2, CO and CH_4 on woodlice (*Tracheoniscus rathkei*) in short-term laboratory experiments, which suggested only effects if the gases are combined and in high concentrations.

Most probably the relation between species abundance and pH is thus indirect and could be caused by, among other things, the increasing mobility and biological availability of heavy metals (see below). Adverse effects of man-made inputs of Pb particulates and acid precipitation on the decomposition processes are a growing concern in this context (Johnson and Siccama, 1983). In general an increased leaching of metal cations can be noticed, including those cations like Ca, K, Mg and Mn, required by animal species (Hågvar and Kjøndal, 1981).

B. Heavy Metals

The binding of heavy metals to organic material is strongly influenced by pH and it differs for various metals (Gerritse and van Driel, 1984). The pH dependence of Cu is for instance higher than for Zn, Cu being more strongly fixed to organic matter (Sidle and Kardos, 1977; Sanders 1982) and the solubility of Cu is much reduced by high pH. Also the stability of Pb in the soil is enhanced by the presence of lime and thereby its availability for plants and animals is reduced. Ma (1982b) showed that Pb, and also Cd and Zn, were more strongly accumulated by *Lumbricus rubellus* in soils with a low pH. The uptake of Cu, however, was unaffected, probably by its different binding capacities or speciation. In solution it may exist as the Cu^{2+} or as an organic complex. It will be clear that the chemical form of a metal and thus its availability influences the uptake into biological systems (Coombs and George, 1977; Morgan and Morris, 1982) and is highly dependent on the type of soil. In sandy, acid soils in general animals are more strongly exposed to heavy metals than in organic soils with higher pH.

In places which are extremely contaminated with heavy metals, a variety of organisms can be found, often with considerable amounts of heavy metals in their bodies (Table 6). Metal contents in species appear to be closely correlated with the body weight of the individuals (Williamson, 1980), but usually this relationship is not linear, so that the concentration of the element in the animal will vary with the weight of the animal. Although much of this variation can be explained, individual differences in diet and digestive activity hamper the interpretation of concentration factors.

Recorded concentration factors (CF) for heavy metals, as summarized by Martin and Coughtrey (1982) demonstrate that Cd nearly always has a CF of more than unity. Other metal concentrations, Pb, Fe and Ni, do not exceed those of the litter. In the CF calculations no attention is given to the availability of metals to the animals. In a study of van Straalen (in prep.) the availability of metals in the substrate (food), as measured by extraction with ammonium acetate (NH_4Ac) was compared to the levels of metals in a springtail species *Orchesella cincta*. In Table 7a the data of Pb, Zn, Cu and Cd levels in a disturbed soil of a zinc smelter and of a control soil are compared. The total soil concentrations and the available concentrations

Table 6

Some selected maximal concentrations recorded in soil invertebrates (ppm dw) whole animals (after Martin and Coughtrey 1982, supplemented).

Species	Pb	Zn	Cd	Cu	Fe	Ni	Cr	Hg	Site	Author
Earthworms										
Dendrobaena rubida	7593				2010				Mine sewage sludge	Ireland (1977)
Aporectodea tuberculata					2010				Sewage sludge	Helmke *et al*. (1979)
Lumbricus terrestris								1·29	Chlor/alkali works	Bull *et al*. (1977)
Lumbricus rubellus		3500	202						Smelter	Ma *et al*. (1983)
Lumbricus terrestris		1950							Roadside	Czarnowska and Jopkiewica (1978)
Eisenia foetida				150		46	13		Sewage sludge	Hartenstein *et al*. (1980)
Isopods										
Oniscus asellus			232						Smelter	Coughtrey *et al*. (1980)
Oniscus asellus	1190								Mine	Martin and Coughtrey (1982)
Porcellio scaber		1930							Smelter	Joosse *et al*. (1981)
Porcellio scaber	700								Roadside	van Capelleveen (1986)
Philoscia muscorum					5435				Smelter	Joosse and van Vliet (1984)
Tracheoniscus rathkei				538					Mine	Wieser *et al*. (1976)
Oniscus asellus					4187	206	140		Serpentine	Martin and Coughtrey (1982)
Snails										
Clausilia bidentata			76						Smelter	Martin and Coughtrey (1975)
Cepaea nemoralis	365	714							Mine	Coughtrey (1975)
Helix aspersa				87					Smelter	Coughtrey and Martin (1977)
Springtails										
Orchesella cincta	86								Roadside	Joosse and Buker (1979)
Orchesella cincta					68				Smelter	Joosse (unpubl.)
Orchesella cincta		97		87					Smelter	van Straalen (unpubl.)
Invertebrate predators	856	2189	71						Zinc mill	Dmowski and Karolewski (1979)

Table 7a

Concentration of metals (ppm) in pine litter and springtails (*Orchesella cincta*) and CF in springtails related to the exchangeable metal concentration in the litter (van Straalen, pers. comm.).

	Pb	Zn	Cu	Cd
Undisturbed area:				
Soil total	41	68	3·1	<0·1
Soil exchangeable	1·6	4·8	0·1	–
Springtails	3·9	143	57	–
CF	2·4	30	570	–
Disturbed smelter area:				
Soil total	361	904	30	8·2
Soil exchangeable	51	92	0·5	5·6
Springtails	32	176	80	1·3
CF	0·6	1·9	170	0·2

Table 7b

Concentrations of metals (ppm) in soil and earthworms and CF in worms related to the DTPA-extractable soil metal concentration (after Beyer *et al.*, 1982).

	Pb	Zn	Cu	Cd
Undisturbed soil:				
Soil total	2·2	56	12	0·1
Soil exchangeable	3·0	2·9	1·6	0·07
Worms	17	228	13	4·8
CF	5·7	78·6	8·1	68·6
Sludge amended soil				
Soil total	31	132	39	2·7
Soil exchangeable	4·8	27	13	1·8
Worms	20	452	31	57
CF	4·2	16·7	2·4	31·7

found in the disturbed soil are many times higher than in the undisturbed soil, but the levels of Zn and Cu in the springtails differ very little between the sites. The CF values are lower where the soil concentration is higher. Apparently the uptake of these metals is regulated. The CF values of Pb and Cd were between 0·1 and 1, apparently Pb and Cd are assimilated by Collembola up to concentrations of about the same as the "available" metals.

Metal concentrations in earthworms also appear to be best correlated with extractable metal (Beyer *et al.*, 1982). In Table 7b data of total and exchangeable metal concentration in the soil are compared with worms' body levels. Here again the disturbed soils are loaded several times higher than the undisturbed soils, but the levels of Pb and Cu in worms differ little between the sites. The CF values are again lower where the soil concentration is higher. This could be the result of regulation by the worms. Zn and Cu have been reported to be regulated (Ireland, 1979a) but Pb not (Beyer *et al.*, 1982). Pb can be found in high concentrations in earthworm bodies (Zöttl and Lamparski, 1981). It could be that Pb is less available for the worms in the sludge amended soil. Compared with Collembola, worms appear to concentrate Cd and Zn, but not Pb and Cu. The movement of the metals in springtails and worms is apparently different.

Differences in metal concentrations in invertebrates may only partly be explained by differences in food choice. Beyer *et al.* (1985) detected very low concentrations of Pb in seventeen-year cicadas and in bark beetles, feeding on sap and decaying wood in the surroundings of a smelter. Animals living in soil litter, like millipedes and soil-ingesting earthworms had very high concentrations, and they are confronted with a higher level in their food. But not all differences can be related to feeding habits, as also shown by Carter (1983). The earthworm *Dendrobaena rubida* accumulates Cd, whereas a bark beetle although feeding on the same food beneath bark, did not.

Thirteen species of invertebrates occurring at the same location, a pine forest, where the metal content of the substrate is considered to be equal for each species, have been studied with respect to body levels of Pb, Cd and Zn (van Straalen and van Wensem, 1986). The results, in Table 8, show that the lowest levels are found in springtails and carabid beetles, whereas some

Table 8

Mean concentration of Pb, Cd and Zn in 13 species of forest soil arthropods (n = ±10) (after van Straalen and van Wensum, 1986).

Taxonomic group	Species	Pb (ppm)	Cd (ppm)	Zn (ppm)
Carabidae	*Calathus melanocephalus*	2·5	1·2	71
Carabidae	*Notiophilus biguttatus*	1·1	2·0	82
Carabidae	*Notiophilus rufipes*	2·0	2·4	56
Staphylinidae	*Lathrobium brunnipes*	1·5	5·0	235
Millipedes	*Lithobius forficatus*	6·6	2·2	186
Millipedes	*Schendyla nemorensis*	0·5	16·8	396
Linyphiidae	*Centromerus sylvaticus*	1·0	19·9	286
Pseudoschorpionidae	*Neobisium muscorum*	0	17·4	319
Collembola	*Orchesella cincta*	1·1	1·4	72
Collembola	*Lepidocyrtus cyaneus*	0	2·8	46
Collembola	*Isotoma notabilis*	–	7·4	55
Diplura	*Campodea staphylinus*	0	15·9	205
Oribatidae	*Chamobates cuspidatus*	–	3·1	365
Forest soil		663	11·0	1670

spiders and Diplura contain much higher quantities. It is unfortunate that in general so few data are available about the levels of contamination in the food of the animals, which reflects a lack of knowledge about the feeding habits of soil organisms. But it seems again that species belonging to very different food guilds show comparable levels whereas those belonging to a common guild, like the predating carabid beetles and spiders, have very different metal contents. For *Cepaea hortensis* and woodlice species changes in the availability of metals in senescing plant material (Williamson, 1979a), which can be considerable (Swift *et al.*, 1979; Hill, 1980; Ernst, 1983b), and interaction with Ca concentrations in the environment (Beeby, 1978) or other metals (Coughtrey and Martin, 1979) may be important.

The conclusion suggests itself that the observed differences in metal content between species are mainly related to properties of the species (van Straalen and van Wensem, 1986), i.e., to their capacities to discriminate between food of different qualities and to assimilate it differentially, and/or to varying mechanisms of storage and excretion of the metals by the species.

1. *Behavioural Avoidance Mechanisms*

Capacity to discriminate is evident in decreased consumption of food by woodlice and springtails with high concentrations of Zn (Joosse *et al.*, 1981), Fe (Joosse *et al.*, 1983; Nottrot *et al.*, 1986), Cu (Dallinger, 1977) and Pb (Joosse and Verhoef, 1983) (Fig. 19). This points to an avoidance response. Isopods are even able to select food of an optimum Cu content (Dallinger, 1977) in contrast with springtails, which apparently do not possess an

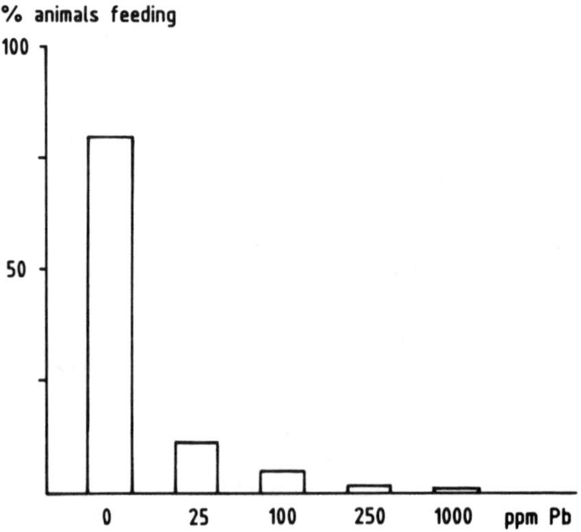

Fig. 19. Avoidance-response of *Orchesella cincta* on Pb-containing food. From Joosse, unpublished.

Table 9

Lead concentration (ppm) in foliage, algae and fungi in the soil along a highway in the Netherlands (after van Straalen, van Capelleveen, van den Berg and van Wachem, 1986).

	Mean concentration (ppm) ± s.d.
Fresh oak leaves	36 ± 34
Old oak leaves	375 ± 62
Fresh American oak leaves	18 ± 6
Old American oak leaves	277 ± 38
Birch, fresh leaves	26 ± 13
Birch, old leaves	240 ± 71
Wood fungi	219 ± 74
Green algae	110 ± 68
Myxomycetes	99 ± 72
Hyphen	80 ± 31
Lichens	70 ± 16
Fruit bodies	40 ± 32

avoidance mechanism, but which have a very efficient excretion mechanism (van Straalen *et al.*, 1985). Earthworms are also able to escape toxic situations (Streit, 1984). At present very little is known about the small-scale horizontal distribution of heavy metals in the soil environment and the chances for the animals to select relatively clean food in a contaminated environment.

In a study by van Straalen *et al.* (1986) it was found that in one site on a small scale the Pb concentration of various plant and fungal materials used as food by soil animals can be very different. In Table 9 a survey is given of Pb concentrations in elements collected along a highway in the Netherlands. Great differences between fresh and old leaves and between various organisms are apparent. The variation (based on four samples) also appears to be great. Besides the avoidance behaviour of an animal the detailed composition of its diet is also needed for information about the animals' exposure to metals.

2. *Physiological Avoidance Mechanisms*

Many factors may influence the assimilation of the metals from the consumed food, including the speciation of the metals in the intestine and properties of the epithelial cells of the intestine for absorption of metal ions. Low assimilation can be expected with an undissolved form of the metals in the intestine. Data on the larvae of the insect *Lucilia cuprina* (Diptera) (Waterhouse and Stay, 1955) suggest that definite places in the intestine absorb Fe ions selectively, related to a very low pH on the spot, which causes the metal compounds to partly dissolve. In the case of Pb, the majority is immediately removed via the faeces and apparently not absorbed in several

Table 10
Some recorded pH values in midguts of insects (after Day and Waterhouse, 1953; Humbert, 1974).

	pH
Collembola	
Tomocerus minor	5·0–8·2
Thysanura	
Ctenolepisma	4·8–7·0
Orthoptera	
Various grasshoppers	5·8–7·5
Odonata	
Anax	6·8–7·2
Hymenoptera	
Apis	6·3

species (Jones and Clement, 1972; Ireland, 1976; Beeby, 1978; van Straalen *et al.*, 1985). Pb is strongly bound to organic macromolecules (Tyler, 1978), which results in relatively low availability. It thus depends on the species of metal and on the animal's properties whether metals can be resorbed. The pH of the insect intestine varies between 5 and 8 (Table 10). Very little is still known about the pH of the intestine of other invertebrate species. Hassall and Jennings (1975) showed that the epithelial cells of the hepatopancreas and the posterior papillate region of the gut of the isopod *Philoscia muscorum* (Fig. 17) contain substantial quantities of Fe, which indicates absorption in these regions. Hartenstein (1964) estimated a high pH in this region of the digestive tract in *Oniscus asellus*. High concentrations of Pb were found in the hindgut and the anterior part of the hepatopancreas by Prosi and Back (1985). Other epithelial cells of the gut were found to be free of Pb. In the region in front of and around the sphincter a higher alkaline phosphatase activity was established, which has a pH optimum of 9 (Kroon *et al.*, 1944). These enzymes are responsible for active transport processes through membranes. Humbert (1972) detected with pH indicators added to the food of Collembola (*Sinella coeca*) an acidic reaction in the first region of the mid-intestine (pH 5·4–6·4) and an alkaline reaction in the second region (pH 8·2–8·8); this could prevent solubility of toxic metals in this region. The relatively high pH of earthworm casts (Reddy, 1983) could indicate a similar function.

A different capacity to assimilate metals may probably also be associated with micro-organisms, used as symbiotic gut flora, which could release the metals (Coughtrey *et al.*, 1980).

A variety of invertebrate species have a peritrophic membrane. Most insects have it in both larval and adult instars. These membranes are though to be present in insects with a diet containing harsh particles, so that the function of the membrane could be the protection of the epithelial cells against sharp substances present in the food. It is also found, however, in insects with soft food as in adult butterflies (Richards and Richards, 1977). The pore diameter of the membrane seems to be different between species and it is as yet unclear whether and when this membrane could act as a filter for metal compounds.

Levels of metals in animal bodies are also related to the balance between uptake and excretion and depends on an eventual compartmentation of the metals in the animals. In several tissues of invertebrates, mainly in hepatopancreas and digestive glands, Malpighian tubules and gut epithelium, metal-containing granules have been described. These metal-containing granules presumably have a metabolic, storage or excretion function.

In species having a hepatopancreas like snails and woodlice, this gland is the site where Cd accumulates (Ireland, 1981, 1982), and Pb and Zn is stored (Coughtrey and Martin, 1976). At least 89% of the total body load of Cd in the woodlouse *Oniscus asellus* is found in the digestive gland (Hopkin and Martin, 1982a). In the snail *Cepaea hortensis* 47% of the body burden of Cd appeared to be present in the midgut gland (Williamson, 1979b). Ireland (1981) found little Cd in the insoluble fractions of a digestive gland homogenate of *Arion rufus*. Fe also is found to accumulate in the woodlouse hepatopancreas in large dense granules (Hryniewiecka-Szyfter, 1972) in cells which can also contain Zn and Pb (Hopkin and Martin, 1982b).

In the hepatopancreas of woodlice two different cell types can be distinguished, which are thought to function in absorption, storage, secretion, excretion, osmoregulation and accumulation of heavy metals (Clifford and Witkus, 1975; Hryniewiecka-Szyfter and Tyczewska, 1978). Prosi *et al.* (1983) demonstrated inclusions containing large amounts of heavy metals in one of these; a small cell type, which shows all signs of an ability to absorb nutrients. The small cells in the hepatopancreas of woodlice also contain granules packed with Cu (cuprosomes), which act as a stock for the haemocyanin in the blood (Brown, 1982). They often contain S and traces of Ca. Woodlice share with Gastropods a special position among the terrestrial invertebrates. Both groups have in common the haemocyanin pigment in the blood to transport oxygen in which Cu is an active element. Isopods are capable of accumulating high concentrations of Cu in the digestive gland up to 1·43% of dry weight (Wieser and Markart, 1961). No mortality occurs even in very high concentrations. The amounts are not related to the quantities needed for the blood pigment (Wieser and Wiest, 1968). The amount of Cu is constant during the intermoult period, but changes very much during the premoult. Then the Cu is withdrawn from the hepatopancreas and glycoproteins are formed. Cu with the glycoproteins becomes soluble and can be transported in a detoxified form (Wieser, 1968) by the haemolymph to other parts of the body to synthesize haemocyanin, and to form a new cuticle. Although the snail *Helix pomatia* also sometimes has very high Cu concentrations (up to 635 ppm), this species is unable to survive higher values (Moser and Wieser, 1979). Gastropods do not possess cuprosomes.

The copper granules of woodlice can also contain Zn, Cd and Pb. These granules are supposed also to perform the function of excretion. They are situated in the anterior end of the glands and thus belong to the oldest cells which might release their content in the gut lumen (Prosi *et al.*, 1983).

Earthworms possess chloragogenous tissue surrounding the major blood vessels and the alimentary canal. The chloragogenous tissue is functionally comparable to the liver of vertebrates and could play a role in excretion of urea and storage of glycogen and lipids. Morgan (1982) attributes two additional functions to this organ. One might be the accumulation of nutrients during diapause, a second role could be the pH-, ion- and osmotic balance. In contaminated soils heavy metals are also found to be stored in the chloragogenous tissue in the electron dense chloragosomes. Morgan and Morris (1982) found metals in different types of granules. Pb, Zn and Ca were found associated with P in the chloragosomes. Ca and P are significantly higher in worms with active Ca glands. A high Ca and P concentration in the tissues of the worms, together with the excretion of $CaCO_3$ by the Ca-glands (Piearce, 1972) enable these species to maintain a stable internal environment in an unstable external environment. These species can tolerate a broad pH range since possibly CO_2, which is not diffused through the body surface, is bound to $CaCO_3$ (Morgan, 1982). Based on the association of Ca and P, Ca in earthworms is assumed to be bound in the form of inorganic Ca orthophosphate or possibly Ca-polyphosphate complexes (Prento, 1979). As the Pb^{2+} ion is very similar to Ca^{2+} in chemical properties, it is probably taken up together with Ca. The availability of Ca determinates the degree of Pb accumulation in earthworms (Anderson, 1979). In Ca rich environments only small amounts of Pb are taken up. Ireland (1978) found that chloragosomes bind Pb more strongly than Ca, but how it is bound is not known. Ireland (1975) concluded that neither high molecular weight proteins nor low molecular weight Pb-metallothionein were involved so one must consider compounds in addition to organic phosphate esters, which are responsible for the Ca-binding properties of the chloragosomes. Gupta (1977) demonstrated that acid mucopolysaccharides have metallic ion binding capacities.

Cd was found associated with S in cadmosomes. A relatively constant ratio of Cd : S indicates that Cd is bound to an S-rich metallothionein-like protein; various characteristics indicate that it is bound as an organic complex (Brown, 1982). From studies made mainly of mammals it is known that Cd is transported to the kidneys with a protein of low metabolic weight: metallothionein. Because of this immobilization the excretion of Cd is low. It is found to be toxic to virtually every system in the vertebrate body. Some insects, like *Chironomus* larvae appear to be highly tolerant to Cd, although they accumulate the metal in high concentrations. Although Cd binding proteins could be induced in these species (Yamamura *et al.*, 1983) the high tolerance was, however, not explicable by induction of these proteins.

The molecular weight of the protein in *Chironomus* is lower, it is rich in cysternyl residue and lacks aromatic amino acids. It contains little Zn and Cu, unlike the metallothioneins in vertebrates (Yamamura *et al.*, 1983). Metallothionein-like compounds have also been detected in other invertebrates (van Capelleveen and Faber, 1987) and even in plants (Lolkema, 1984) and micro-organisms (Otvos *et al.*, 1982). Efforts to characterize the structure of the metal binding sites in metallothioneins sometimes reveal a remarkable resemblance in groups of great evolutionary divergence (Otvos *et al.*, 1982).

At the end of their life earthworm chloragocytes are shed into the coelomic fluid and are concentrated in amorphous nodules or brown bodies (Andersen and Laursen, 1982), which could be a way of eliminating toxic substances. Worm species with inactive calciferous glands contain high concentrations of Pb in the waste nodules (Andersen and Laursen, 1982). When the glands are active, Pb, being competitive with Ca, could be expected to be excreted via this route. Morgan and Morris (1982) however, did not find detectable quantities of Pb, Zn or Cd in extracellular spherites secreted by the calciferous glands. They conclude that the glands are probably not directly involved in heavy metal excretion. Ireland (1975, 1976) found high concentrations of Zn in the front part of the body in the calcium glands and suggests that Zn, just like Ca is accumulated in these glands and together with $CaCO_3$ is excreted in the front part of the gut. In *Arion rufus* ^{65}Zn appeared to be absorbed very quickly in the hepatopancreas (Schoettli and Seiler, 1970) and stored in the Ca cells. The cells fill with spherites, the number of which increases with Zn supply and they become completely covered by Zn. Similar processes have been found in other snail species, in *Helix pomatia* (Meincke and Schaller, 1974), *Helix aspersa* (Coughtrey and Martin, 1976) and *Arion ater* and *Nucella lapillus* (Ireland, 1979b, c).

An interesting fact is that with high Zn or Pb burdens, the glycogen in worms decreases and possibly synthesis is prevented (Ireland and Richards, 1977). Possibly the excretion of Pb requires energy (Richards and Ireland, 1978) and it was suggested by Andersen and Laursen (1982) that the synthesis of metallic-ion binding mucopolysaccharides may use up the glycogen reserves.

In insects concentrations of minerals were found in intestinal cells and in the epithelial cells of the Malpighian tubules (Jeantet *et al.*, 1977). In *Musca*, the housefly, Zn and Ca were found to be deposited in the Malpighian tubules and Cu in the midgut cells (Sohal and Lamb, 1979). Filshie *et al.* (1971) showed that when larvae of *Drosophila* are raised on a Cu-rich diet, the number and the size of the cytolysomes in the cuprophilic cells increased.

In Collembola the midgut is described as a mineral storage organ (Humbert, 1978). The temporary accumulation of concretions in Ca-granules may serve to trap toxic ions. The variable composition of Ca-containing granules in invertebrates and the range of metals they contain could be a specific property of those cells containing granules. Brown (1982) distinguishes between two types of Ca-containing granules: one with pure Ca and others with a combination of Ca and Mg, Mn, P and sometimes Zn, Cd, Pb and Fe and further Fe-containing granules. The majority of granules have been found to contain high concentrations of Fe, Zn, S together with low concentrations of P, K and Na. Such granules with mixed composition operate probably in storage and excretion processes but an important function of these structures may be metal detoxification.

In invertebrates relatively little attention has been focused on the interaction between metals and other elements in the uptake and retention, although single-metal exposition is seldom met. In various mammals a relationship has been demonstrated (Petering, 1978) for instance between Fe and Mn, resulting from a mutual antagonism at the absorptive level (Thomson *et al.*, 1971), which may cause Fe deficiency. In woodlice Fe absorption was also found to be reduced by Mn and vice versa (Table 11). Mn, like Fe binds to N and O containing groups.

In many plants and animal species a competitive interaction between Zn and Cd has been demonstrated. Cd concentrations in worms collected from a clean area may be twice as high as those in worms from plots with Zn and

Table 11

Interaction between Fe and Mn in assimilation: (consumption–faeces/consumption) by *Porcellio scaber* (after van Capelleveen, 1983).

Food	% Fe assimilated $\left(\frac{C-F}{C} \times 100\%\right)$
+ Mn 690 ppm	9·4
+ Mn 1250 ppm	22·8
Food	**% Mn assimilated $\left(\frac{C-F}{C} \times 100\%\right)$**
+ Fe 1600 ppm	7·7
+ Fe 2200 ppm	15·5

apparently the Zn/Cd ratio in the soil determines the Cd uptake by worms (Beyer *et al.*, 1982). A synergistic action between Zn and Cd has also been demonstrated in isopods, with respect to their growth and reproduction (van Capelleveen, 1986a, b).

It could be expected that Cd and Pb absorption is also related to Ca, since Ca blocks the SH groups of proteins which are active binding places for Cd and Pb (Khandelwal *et al.*, 1984). The need for Ca in woodlice and snails is well established (Heely, 1941). Zn, Cd and Pb contamination could thus easily lead to a Ca deficiency. An association between Pb and Ca in the digestive gland of the slug *Arion ater*, could not, however, be found (Ireland, 1979a) and in *Helix aspersa* such a relationship is absent (Beeby and Eaves, 1983). Pb toxicity is often associated with a disturbance in the Ca metabolism, especially in the nervous tissue (Kober and Cooper, 1976). This has been established for amphibians. Although most of the Pb is generally lost through the faeces, the assimilated fraction is affected by the presence of Ca in isopods (Beeby, 1978) and in earthworms (Andersen, 1979), and appears to affect the Ca-levels and vitality of young isopods (Beeby, 1980).

Pb and Cu appear to have an antagonistic interaction in woodlice. The uptake of Pb is suppressed by Cu (van Capelleveen and van Zoest, in prep.) despite an increased consumption following the administration of Pb and Cu at the same time. If the combination is offered, Pb seems to be removed quicker from the body than when Pb alone is given.

3. *Life History Patterns*

The immediate effects of pollutants are on individual organisms, but the ecological significance resides in the indirect impact on the populations of species (Moriarty, 1983). These secondary interactions involve decreasing reproduction, retardation of development and disruption of population processes by which subsequently equilibria with other trophic levels are affected. This means that the conventional and most convenient criterion used in the toxicology, i.e., LD_{50} value (the dose that is lethal for 50% of the population) is not valid. Death is always preceded by other symptoms and those which can be related to the disfunctioning of the population are of primary importance. The ecotoxicologist has further to develop specific skills by which population parameters can be integrated (Kooijman and Metz, 1982). Unfortunately only limited knowledge is as yet available with respect to the causality of effects of metals on populations of invertebrates. Ecophysiological studies can be of help to find measurements which serve as signs of damage, and to analyse the backgrounds of negative effects.

Beeby (1980) could not demonstrate any effect of Pb on number of young *Porcellio scaber*, but he found that the amount of Ca derived from the parent by the young, living in the brood-pouch, declines with a Pb burden, resulting in unsuccessful pregnancies.

The greatest part of Pb is immediately removed via the faeces, but nevertheless Ireland and Fischer (1978) demonstrated symptoms of an effect of Pb on the zinc metalloenzym ALAD (delta-amino laevulinic acid dehydratase), which has a function in the synthesis of haemoglobin, acting as a blood pigment in earthworms. In this pigment Fe^{3+} plays an important role and, as in mammals, the Fe^{3+} uptake is made possible by Fe-binding proteins. The SH groups in these proteins have a high affinity to Pb (Quaterman *et al.*, 1980). The Pb does not persist in the worm body, retention is apparently low. Within some days after transfer of worms to an uncontaminated environment, the Pb appeared to be removed (Ireland, 1975, 1976).

The SH groups are also the binding places for the very dangerous organic mercury compounds, which are applied as fungicides and insecticides, and also result from industrial and fertilizer activities. The total soil in the Netherlands, for instance, receives 25 000 kg Hg per year, originating from atmospheric pollution, fertilizers and harbour sludge (Paul *et al.*, 1981). Methylized mercury, which is formed by bacteria from inorganic Hg (Jensen and Jernelöv, 1969; Beckert *et al.*, 1974) passes through biomembranes very efficiently and is subsequently bound to SH groups (Carty and Malone, 1979). It is only eliminated very slowly from animal tissues. Symptoms in invertebrates, however, are not known, but since Hg compounds appear to be effective enzyme retarders (Vallee and Ulmer, 1972), deleterious effects are very likely.

Abnormally high doses of Pb administered to Collembola did not result in death (Joosse and Buker, 1979). These species have a specific excretion route by regularly removing toxic substances together with regeneration of their intestinal epithelium, as we have seen before. Although the toxicology of Pb is associated with Ca, as we have seen, the toxic effects on invertebrates are not very apparent. Detailed analysis, however, of reproduction and growth in Collembola on a 25 ppm Pb nitrate-containing food showed a decreased growth and metabolism (O_2 consumption) and a slightly lowered egg production (Joosse and Verhoef, 1983). The food (green algae) however, concentrated the Pb strongly, the ultimate available concentration was unknown.

Cd is considered a hazardous metal the occurrence of which in biota is undesirable, but which still can be found in relatively high body concentrations (Table 6). Cd interacts strongly with Zn. Apparently Cd is less toxic for invertebrates than for vertebrates and the wide occurrence of tolerance mechanisms is probable (see below). Nevertheless, sublethal effects of Cd have been noticed in woodlice (van Capelleveen, 1985). When exposed to Cd concentrations of 2 ppm in their food, the growth and the number of offspring decreases. The negative influence on development is more marked when Zn is added together with Cd to the food (Table 12).

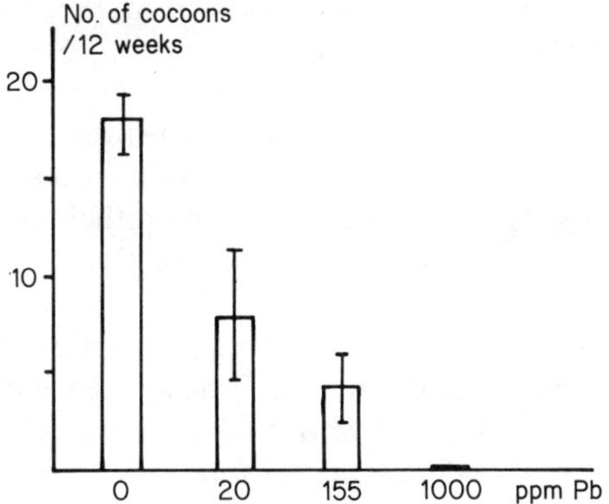

Fig. 20. Effect of Cd on cocoon production in *Lumbricus rubellus*. After Ma (1982a).

Earthworms show a significant decrease in reproductive activity, noticed as a reduction of cocoon production, with 150 ppm Cd (Fig. 20). As the admissible value in the soil is 10 ppm this is a rather high concentration, and earthworms thus appear also to be rather tolerant to Cd. Snails show symptoms of decreased consumption and reproduction when the concentration of Cd exceeds 10 ppm in their food. With high concentrations they start dormancy (Russell *et al.*, 1981).

On the basis of studies on birds and mammals toxic effects of Zn are not to be expected. Zn is an important catalyst in enzyme systems in the cells, which serve a wide range of functions. Nevertheless, Strojan (1978) describes Zn as a suspect metal for invertebrates, and indeed several studies

Table 12

Effects of Zn and Cd* on population parameters of *Porcellio scaber* (summarized after Joosse *et al.*, 1983; van Capelleveen, 1985, 1986b).

	Clean food	2000 ppm Zn	% effect	10 ppm Cd	% effect	2000 ppm Zn 10 ppm Cd	% effect
Consumption (mg dw/mg/90 days)	7·1	6·8	4·2	6·9	1	5·4	23
% weight increase (mg dw/10/67 days)	34	6	82	17	50	−26	177
Growth efficiency (weight increase/C–F)	40	2	95	18	55	−22	155
Clutch-size	26·8	23·1	14	23·2	13	10·6	61
Gestation time	36	38	5	38	5	44	22

* Zn and/or Cd burdening only during gestation.

describe symptoms of toxicity. Joosse *et al.*, (1981) and van Capelleveen (1985) showed that consumption and growth of isopods decreased significantly with Zn in the food. Also the number of progeny declined (Table 12). Sastry *et al.* (1958) found a decrease in growth and enzyme activity in the lepidopteran *Corcyra cephalonica*. In molluscs histopathological phenomena such as hypoplasia and necroses of epithelium and cell structure in intestine, kidney, ovotestis and connective tissue appear (Russell *et al.*, 1981).

Cu is known to be a toxic element to earthworms. Nielson (1951) reported that a spill of $CuSO_4$ in grassland of 260 mg kg^{-1} Cu eliminated nearly all the worms. With 12·7 mg kg^{-1} a negative relation occurs between Cu content of the soil and the abundance of *Allolobophora caliginosa, A. rosea* and *Lumbricus terrestris* (Andersen, 1979; Hartenstein *et al.*, 1980). Ma (1982b) found 60 mg kg^{-1} to be a critical value. With 100–120 ppm Cu in their bodies (dry weight), worms die (Streit, 1984). The type of soil plays an important part in the uptake of metals by worms. Worms can take up more metals in soils with low cation exchange capacity (CEC) (Ma, 1982b). Moreover, the availability of metals for the worms is determined by the organic content of the soil (Ma *et al.*,.1983) (Fig. 21), since metals very easily form complexes

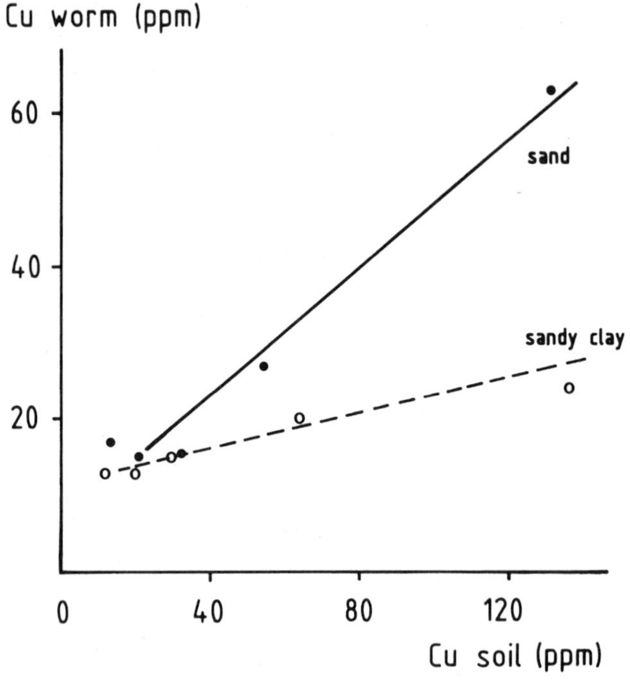

Fig. 21. Cu-uptake in earthworms in sand and sandy clay. After Ma *et al.* (1983).

with organic material and then become unavailable. Ma (1984) demonstrated that a concentration of 370 mg kg^{-1} in sandy soils inhibited growth in *Lumbricus rubellus* and litter breakdown activity at 130 mg kg^{-1} in sandy soils and at 65 mg kg^{-1} in sandy loam. Cocoon production was found to respond even more sensitively.

4. *Tolerance Mechanisms*

Only recently attention has been directed towards genetic differentiation between populations of soil invertebrates in relation to heavy metal resistance. It is now known that besides plant species (Antonovics *et al.*, 1971; Ernst, 1976) invertebrates from metalliferous areas have evolved specific tolerance to heavy metals which are abundant on the site.

Laboratory work indicated naturally occurring differences in population tolerance to Cu and Pb in aquatic isopods (Fraser *et al.*, 1978, Fraser, 1980). In *Asellus meridianus* Pb tolerance seems to be inherited (Brown, 1976). Tolerance may be achieved by excretion of the metals or by rendering them into an inocuous form or, theoretically by changing enzyme structures (Ernst, 1976). Avoidance by storage-detoxification and excretion seems to be the form of heavy metal resistance most commonly evolved.

In the hepatopancreas of Cu-tolerant isopods a marked difference in the storage forms of Cu was observed, compared with non-tolerant animals from natural populations (Brown, 1977). Apart from structures resembling cuprosomes described by Wieser and Klima (1969), granular inclusions bound in spherical vesicles were found in the Cu-tolerant animals.

Pb was absent in the hepatopancreas of Pb-tolerant animals. Brown suggested that cuticle and underlying tissues are equally important for storage of Pb. This was also found for Cd (van Capelleveen and Faber, 1987).

In the species *Asellus meridianus* Cu tolerance seems to confer tolerance to Pb (Brown, 1978). Cu-tolerant isopods appear to be able to detoxify Pb by storing the metal in cuprosomes at the expense of Cu. But Pb-tolerant animals were found to restrict the uptake of both Cu and Pb into the hepatopancreas. Beurskens (1985) found, however, high concentrations of Pb in Pb-tolerant *Porcellio scaber*. The mechanisms operating for metals thus may be different, and perhaps different between species.

Differences in tolerance between populations of terrestrial isopods have also been assessed (Joosse *et al.*, 1983). Tolerant Zn populations regulate the body Zn content at a higher level. Cd tolerance appears to coincide with a higher Cd uptake in the F_1-generation of *Porcellio scaber* (Fig. 22). Fe tolerance was established in the same species and was found to be related to a higher Fe-need and involves a higher consumption in Fe-tolerant populations to meet this need.

Cd-tolerant and non-tolerant populations of *Porcellio scaber* have been found to differ in the occurrence of some proteins of low molecular weight

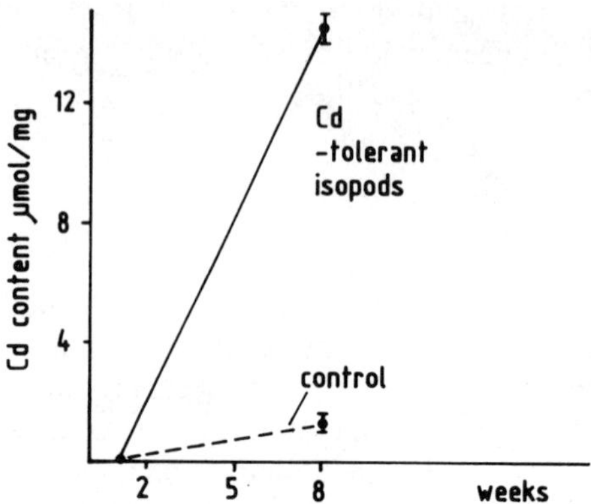

Fig. 22. Cd-content in hatchlings of an F1-generation of *Porcellio scaber* on a 1 ppb Cd diet. From Joosse, unpublished.

(Fig. 23). Two metal-binding proteins could be induced by Cd and Cu (van Capelleveen and Faber, 1987), so the tolerance is imputed to this storage and detoxification mechanism. The production of these proteins is thought to be related to the physiological dependence of Zn- and Fe-tolerant isopod individuals on these metals (Joosse *et al.*, 1983, van Capelleveen, 1985), and laboratory observations as well as field observations on populations of *Porcellio scaber* also indicated a trade-off between tolerance and fitness. Individuals from a Zn/Cd tolerant population were found to be about 20% smaller compared with a non-tolerant population in its natural habitat. Similar costs of tolerance have been described for plant species (Ernst, 1972).

In Collembola a difference in excretion efficiency was demonstrated between Pb-tolerant and control populations but no apparent trade-off between excretion efficiency and life-history traits in springtails could be

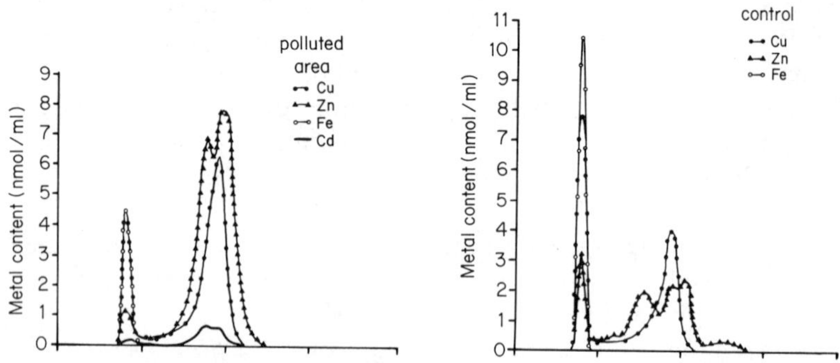

Fig. 23. Differences in metal binding proteins in Cd-tolerant and non-tolerant populations of *Porcellio scaber*. After van Capelleveen and Faber, 1987.

shown up till now (Janssen, 1985). This might be related to the relatively simple excretion system in Collembola, which removes toxic material regularly via renewal of the intestinal epithelium.

The genetic correlation between metal tolerance and other vital characteristics is at present unknown. Some data show that the consequences of such a correlation could be unexpected. Van Capelleveen (1985) found that Zn/Cd tolerant woodlice have a significantly higher transpiration rate than control populations, which essentially increases their sensitivity to drought and thereby affects their fitness.

IV. SUMMARY AND CONCLUSIONS

Attention is directed to ecophysiological research on soil invertebrates to gain insight into the mechanisms of stress resistance in natural environments and in habitats exposed to human disturbances.

To trace the multiple adaptations of organisms in specific habitats, examples are presented of physiological and behavioural avoidance mechanisms, life-history adaptations and tolerance mechanisms in soil inhabiting species of dune and heathlands and of polluted habitats.

The stress factors drought, temperature, food supply, both absence of food and mineral shortage, and the presence of toxic substances are reviewed.

Research on resistance mechanisms of invertebrates to stress factors like drought, cold and heat has been restricted mainly to animals of extreme habitats like deserts and polar regions. Temperate animals, however, meet similar stresses, which can be reflected in the potential of their resistance mechanisms. This review reveals the paucity of such research on temperate soil invertebrates and illustrates the need for similar studies on this group.

Adaptations to food shortage generally appear to be related to drought and temperature. The soil environment is characterized by food (litter) of low nutrient content, with relatively high amounts of naturally occurring phenolic substances. Data about both nutrient shortage and resistance to toxic substances are very scarce and need to be supplemented.

Effects of air pollution on soil biota have received little attention. Adverse effects probably are indirect. Heavy metals can be found in high concentrations in many soil-inhabiting species. Differences between species are mainly ascribed to anatomical and physiological properties of the species.

Behavioural as well as physiological avoidance mechanisms exist in a great number of species. The mechanism of resistance to heavy metals is mainly situated in midgut glands and midgut epithelium. In a number of cases metallothionein-like proteins are found to act as a detoxification mechanism.

Little attention has been given to the interaction of metals and other vita elements: several synergistic, competitive and antagonistic actions arє known or suspected.

Specific skills to integrate effects of metals on life-history parameters havє to be developed. Symptoms suggest a more sensitive response of population: than predicted by LD_{50}-values.

Genetic differentiation between populations of soil invertebrates relatec to heavy metal resistance has only recently been established and appears tc be based on physiological avoidance. The first data indicate the likelihood o unpredictable and negative genetic correlations between metal resistancє and other vital characteristics.

REFERENCES

Abrahamsen, G. (1983). Effects of lime and artificial acid rain on the enchytraeic (Oligochaeta) fauna in a coniferous forest. *Holarctic Ecology* **6**, 247–254.

Ahmad, S. (1983). Mixed function oxidase activity in a generalist herbivore ir relation to its biology, food plants, and feeding history. *Ecology*. **64**, 235–243.

Alahiotis, S. N. and Stephanou, G. (1982). Temperature adaptation of *Drosophilc* populations. The heat shock proteins system. *Comp. Biochem. Physiol.* **73B** 529–533.

Al-Dabbagh K. Y. and Block, W. (1981). Population ecology of a terrestrial isopoc in two breckland grass heaths. *J. Anim. Ecol.* **50**, 61–77.

Alexandrov, V. Ya. (1977). "Cells, Molecules and Temperature; Conformationa Flexibility of Macromolecules and Ecological Adaptation" Springer-Verlag Berlin.

Alstad, D. N. Edmunds Jr, G. F. and Weinstein, L. H. (1982). Effects of ai pollutants on insect populations. *Ann. Rev. Entomol.* **27**, 369–384.

Andersen, C. (1979). Cadmium, lead and calcium content, number and biomass ir earthworms (Lumbricidae) from sewage sludge treated soil. *Pedobiologia* **19** 309–319.

Andersen, C. and Laursen J. (1982). Distribution of heavy metals in *Lumbricu: terrestris, Aporrectodea longa* and *A. rosea* measured by atomic absorption anc X-ray fluorescence spectrometry. *Pedobiologia* **24**, 347–356.

Anderson, J. M. and Healey, I. N. (1972). Seasonal and interspecific variation ir major components of the gut contents of some woodland Collembola. *J. Anim. Ecol.* **41**, 359–368.

Antonovics, J., Bradshaw, A. D. and Turner, R. G. (1971). Heavy metal tolerancє in plants. *In* "Advances in Ecological Research, Vol. 7" (Ed. J. B. Cragg), pp 1–85. Academic Press, London.

Appel, A. G., Reierson, D. A. and Rust, M. K. (1983). Comparative water relation: and temperature sensitivity of cockroaches. *Comp. Biochem. Physiol.* **74A** 357–361.

Arlian, L. G. and Veselica, M. M. (1979). Water balance in insects and mites. *Comp. Biochem. Physiol.* **64A**, 191–200.

Ausmus, B. S., Edwards, N. T. and Witkamp, M. (1976). Microbial immobilizatior of carbon, nitrogen, phosphorus and potassium: implications for forest ecosysterr processes. *In* "The Role of Terrestrial and Aquatic Organisms in Decomposition Processes" (Eds J. M. Anderson and A. Macfadyen), pp. 367–416. Blackwell, Oxford.

Bååth, E., Berg, B., Lohm, U., Lundgren, B., Lundkvist, H., Rosswall, T., Söderström, B. and Wirèn, A. (1980). Effects of experimental acidification and liming on soil organisms and decomposition in a Scots pine forest. *Pedobiologia* **20**, 85–100.

Barker Jørgensen, C. (1983). Ecological physiology: background and perspectives. *Comp. Biochem. Physiol.* **75A**, 5–7.

Bauer, T. (1979). Die Feuchtigkeit als steuernder Faktor für das Kletter-verhalten von Collembolen. *Pedobiologia* **19**, 165–173.

Baust, J. G. and Rojas, R. R. (1985). Review- insect cold hardiness: facts and fancy. *J. Insect Physiol.* **31**, 755–759.

Beckert, W. F., Maghisi, A. A., Au, F. E., Bretthauser, E. W. and McFarlan, M. (1974). Formation of methyl mercury in a terrestrial environment. *Nature* **249**, 674–675.

Beeby, A. (1978). Interaction of lead and calcium uptake by the woodlouse *Porcellio scaber* (Isopoda, Porcellionidae). *Oecologia* (*Berl.*) **32**, 255–262.

Beeby, A. (1980). Lead assimilation and brood-size in the woodlouse *Porcellio scaber* (Crustacea, Isopoda) following oviposition. *Pedobiologia* **20**, 360–365.

Beeby, A. and Eaves, S. L. (1983). Short-term changes in Ca, Pb, Zn and Cd concentrations of the garden snail *Helix aspersa* Müller from a central London car park. *Environ. Pollut.* (**A**) **30**, 233–244.

Berenbaum, M. (1980). Adaptive significance of midgut pH in larval Lepidoptera. *Amer. Nat.* **115**, 138–146.

Bernays, E. A. (1981). Plant tannins and insect herbivores: an appraisal. *Ecol. Entomol.* **6**, 353–360.

Bernays, E. A. and Chamberlain, D. J. (1980). A study of tolerance of ingested tannins in *Schistocerca gregaria*. *J. Insect Physiol.* **26**, 415–420.

Beurskens, K. (1985). De distributie van koper en lood in het lichaam van *Porcellio scaber* onder invloed van hongering en hervoeding. *Int. Rep. V.U. Amsterdam.*

Beyer, W. N., Chaney, R. L. and Mulhern, B. M. (1982). Heavy metal concentrations in earthworms from soil amended with sewage sludge. *J. envir. Qual.* **11**, 381–385.

Beyer, W. N. *et al.* (1985). Metal contamination in wildlife living near two zinc smelters. *Environ. Pollut.* **(A)38**, 63–86.

Bierenbroodspot, A. (1978). Het aandeel van Collembolen in de strooiselafbraak. *Int. Rep. V.U. Amsterdam.*

Bligh, J., Cloudsley-Thompson, J. L. and Macdonald, A. G. (1976). "Environmental Physiology of Animals" Blackwell, Oxford.

Block, W. (1982). Cold hardiness in invertebrate poikilotherms. *Comp. Biochem. Physiol.* **73A**, 581–593.

Bödvarsson, H. (1970). Alimentary studies on seven common soil-inhabiting Collembola of Southern Sweden. *Entomol. scand.* **1**, 74–80.

Booth, R. G. and Anderson, J. M. (1979). The influence of fungal food quality on the growth and fecundity of *Folsomia candida* (Collembola: Isotomidae). *Oecologia* (*Berl.*) **38**, 317–322.

Bowden, J., Haines, I. H. and Mercer, D. (1976). Climbing Collembola. *Pedobiologia* **16**, 298–312.

Brody, M. S. and Lawlor, L. R. (1984). Adaptive variation in offspring size in the terrestrial isopod, *Armadillidium vulgare*. *Oecologia* (*Berl.*) **61**, 55–59.

Brown, B. E. (1976). Observations on the tolerance of the isopod *Asellus meridianus* Rac. to copper and lead. *Water Res.* **10**, 555–559.

Brown, B. E. (1977). Uptake of copper and lead by a metal tolerant isopod *Asellus meridianus* Rac. *Freshwater Biol.* **7**, 235–244.

Brown, B. E. (1978). Lead detoxification by a copper-tolerant isopod. *Nature (London)* **276**, 388–390.

Brown, B. E. (1982). The form and function of metal-containing "granules" in invertebrate tissue. *Biol. Rev.* **57**, 621–667.

Bursell, E. (1970). "An Introduction to Insect Physiology" Academic Press. New York.

Capelleveen, H. E. van (1983). Effects of iron and manganese on isopods. *Proc. Conf. Heavy metals in the environment, Heidelberg*. pp. 666–668.

Capelleveen, H. E. van (1985). The ecotoxicity of zinc and cadmium for terrestrial isopods. *Proc. Int. Conf. Heavy metals in the environment, Athens*. pp. 245–247.

Capelleveen, H. E. van (1986a). The ecotoxicological significance of zinc and cadmium for the woodlouse *Porcellio scaber*. Consumption, growth and accumulation (in prep.).

Capelleveen, H. E. van (1986b). The ecotoxicological significance of zinc and cadmium for the woodlouse *Porcellio scaber*. Reproduction and development (in prep.).

Capelleveen, H. E. van and Faber J. (1987). Woodlice and heavy metal tolerance: the possible role of metal binding proteins (in prep.).

Carter, A. (1983). Cadmium, copper, and zinc in soil animals and their food in a red clover system. *Can. J. Zool.* **61**, 2751–2757.

Carty, A. L. and Malone, S. F. (1979). The chemistry of mercury in biological systems. *In* "The Biochemistry of Mercury in the Environment" (Ed. J. O. Nriaga), pp. 433–481. Elsevier/North Holland, Amsterdam.

Chauvin, G., Vannier, G. and Gueguen, A. (1979). Larval case and water balance in *Tinea pellionella*. *J. Insect Physiol.* **25**, 615–619.

Clifford, B. and Witkus, E. R. (1975). The fine structure of the hepatopancreas of the woodlouse *Oniscus asellus*. *J. Morph.* **135**, 335–350.

Cohen, A. C. and Cohen, J. L. (1981). Microclimate, temperature and water relations of two species of desert cockroaches. *Comp. Biochem. Physiol.* **69A**, 165–167.

Cohen, A. C. and Pinto, J. D. (1977). An evaluation of the xeric adaptiveness of several species of blister beetles (Meloidae). *Ann. ent. Soc. Am.* **70**, 741–749.

Coleman, D. C. (1983). The impacts of acid deposition on soil biota and C-cycling. *Environ. Exp. Bot.* **23**, 225–233.

Cooke, A. (1983). The effects of fungi on food selection by *Lumbricus terrestris* L. *In* "Earthworm Ecology" (Ed. J. E. Satchell), pp. 365–374, Chapman and Hall, London.

Cooke, A. and Luxton, M. (1980). Effect of microbes on food selection by *Lumbricus terrestris*. *Rev. Ecol. Biol. Sol.* **17**, 365–370.

Coombs, T. L. and George, S. G. (1977). Mechanisms of immobilization and detoxication of metals in marine organisms. *In* "Physiology and Behaviour of Marine Organisms" (Eds D. S. Melusky and A. J. Berry), pp. 179–187. Blackwell, Oxford.

Cooper, P. D. (1983). Components of evaporative water loss in the desert tenebrionid beetles *Eleodes armata* and *Cryptoglossa verrucosa*. *Physiol. Zool.* **56**, 47–55.

Cooper, P. D. (1985). Seasonal changes in water budgets in two free-ranging tenebrionid beetles, *Eleodes armata* and *Cryptoglossa verrucosa*. *Physiol. Zool.* **58**, 458–472.

Cornaby, B. W., Gist, C. S. and Crossley, Jr. D. A. (1975). Resource partitioning in leaf-litter faunas from hardwood-converted to pine forests. *In* "Mineral Cycling in Southeastern Ecosystems" (Eds F. G. Howell and H. H. Smith), pp. 588–597. ERDA *Symp. Ser.*

Coughtrey, P. J. and Martin, M. H. (1976). The distribution of Pb, Zn, Cd and Cu within the Pulmonate Mollusc *Helix aspersa* Müller. *Oecologia* (*Berl.*) **23**, 315–322.

Coughtrey, P. J. and Martin, M. H. (1979). Cadmium, lead and zinc interactions and tolerance in two populations of *Holcus lanatus* L. grown in solution culture. *Environ. Exptl. Bot.* **19**, 285–290.

Coughtrey, P. J., Martin, M. H., Chard, J. and Shales, S. W. (1980). Microorganisms and metal retention in the woodlouse *Oniscus asellus. Soil Biol. Biochem.* **12**, 23–27.

Crawford, C. S. (1981). "Biology of Desert Invertebrates" Springer-Verlag, New York.

Crossley, Jr. D. A. (1977). The role of terrestrial saprophagous arthropods in forest soils: current status of concepts. *In* "The Role of Arthropods in Forest Ecosystems" (Ed. W. J. Mattson), pp. 49–56. New York.

Dallinger, R. (1977). The flow of copper through a terrestrial food chain. *Oecologia* (*Berl.*) **30**, 273–276.

Dreisig, H. (1980). Daily activity, thermoregulation and water loss in the tiger beetle, *Cicindela hybrida. Oecologia.* (*Berl.*) **44**, 376–389.

Duman, J. and Horwath, K. (1983). The role of hemolymph proteins in the cold tolerance of insects. *Ann. Rev. Physiol.* **45**, 261–270.

Edney, E. B. (1971). The body temperature of tenebrionid beetles in the Namib desert of Southern Africa. *J. exp. Biol.* **55**, 253–272.

Edney, E. B. (1977). "Water Balance in Land Arthropods" Springer-Verlag, Berlin.

Edwards, C. A. (1974). Macroarthropods. *In* "Biology of Plant and Litter Decomposition" (Eds C. H. Dickinson and G. J. F. Pugh), pp. 533–554. Academic Press, London.

Emlen, J. H. (1973). "Ecology: An Evolutionary Approach" Addison-Wesley, London.

Ernst, W. H. O. (1972). Schwermetallresistenz und Mineralstoffhaushalt. *Forschungsber. Land Nordrhein-Westf.* **2251**, Opladen: 1–38.

Ernst, W. H. O. (1976). Physiological and biochemical aspects of metal tolerance. *In* "Effects of Air Pollutants on Plants" (Ed. T. A. Mansfield), pp. 115–133. Cambridge University Press.

Ernst, W. H. O. (1983a). Ökologische Anpassungsstrategien an Bodenfaktoren. *Ber. Deutsch. Bot. Ges.* **96**, 49–71.

Ernst, W. H. O. (1983b). Element nutrition of two contrasted dune annuals. *J. Ecology.* **71**, 197–209.

Ernst, W. H. O. and Andel, J. van (1985). Autoecologie. *In* "Inleiding tot de Oecologie". (Ed. K. Bakker *et al.*), pp. 69–100. Bohn, Scheltema and Holkema, Utrecht/Antwerpen.

Ernst, W. H. O. and Joosse, E. N. G. (1983) "Umweltbelastung durch Mineralstoffe", G. Fischer Verlag, Jena.

Ernsting, G. and Fokkema, D. (1983). Antennal damage and regeneration in springtails (Collembola) in relation to predation. *Neth. J. Zool.* **33**, 476–484.

Ernsting, G. and Jansen, J. W. (1978). Interspecific and intraspecific selection by the predator *Notiophilus biguttatus* F. (Carabidae) concerning two collembolan prey species. *Oecologia* (*Berl.*) **33**, 173–183.

Ernsting, G. and Joosse, E. N. G. (1974). Predation on two species of surface dwelling Collembola. A study with radio-isotope labelled prey. *Pedobiologia* **14**, 222–231.

Ernsting, G., Marquenie, J. M. and Vries, C. N. de (1977). Aspects of behaviour and predation risk of two springtail species. *Rev. Ecol. Biol. Sol.* **14**, 27–30.

Feeny, P. P. (1968). Effects of oak leaf tannins on larval growth of the winter oak moth *Operophtera brumata. J. Insect Physiol.* **14**, 805–817.

Feeny, P. (1970). Seasonal changes in oak leaf tannins and nutrients as a cause of spring feeding by winter moth caterpillars. *Ecology* **51**, 565–581.

Feeny, P. (1976). Plant apparency and chemical defense. *Rec. adv. in Phytochem.* **10**, 1–40.

Filshie, B. K., Poulson, D. F. and Waterhouse, D. F. (1977). Ultrastructure of the copper-accumulating region of *Drosophila* larval midgut. *Tissue Cell* **3**, 77–102.

Fox, L. R. and Macauley, B. J. (1977). Insect grazing on *Eucalyptus* in response to variation in leaf tannins and nitrogen. *Oecologia.* (*Berl.*) **29**, 145–182.

Fraser, J. (1980). Acclimation to lead in the fresh-water isopod *Asellus aquaticus. Oecologia* (*Berl.*) **45**, 419–420.

Fraser, J., Parkin, D. T. and Verspoor, E. (1978). Tolerance to lead in the freshwater isopod *Asellus aquaticus. Water Res.* **12**, 637–641.

Gehrken, U. (1984). Winter survival of an adult bark beetle *Ips acuminatus* Gyll. *J. Insect Physiol.* **30**, 421–429.

Gerritse, R. G. and Driel, W. van (1984). The relationship between adsorption of trace metals, organic matter, and pH in temperate soils. *J. envir. Qual.* **13**, 197–204.

Giesel, J. T. (1976). Reproductive strategies as adaptations to life in temporally heterogeneous environments. *Ann. Rev. Ecol. Syst.* **7**, 57–79.

Gilby, A. R. (1980). Transpiration, temperature and lipids in insect cuticle. *In* "Advances in Insect Physiology, Vol. 15" (Eds M. J. Berridge, J. E. Treherne and V. B. Wigglesworth), pp. 1–33. Academic Press, London.

Gilmore, S. K. and Raffensberger, E. M. (1970). Foods ingested by *Tomocerus* spp. (Collembola, Entomobryidae), in relation to habitat. *Pedobiologia* **10**, 135–140.

Gist, C. S. and Crossley, Jr. D. A. (1975). Feeding rates of some cryptozoa as determined by isotopic half-life studies. *Environ. Entomol.* **4**, 625–631.

Greenslade, P. (1981). Survival of Collembola in arid environments: observations in South Australia and the Sudan. *J. Arid Environ.* **4**, 219–228.

Grime, J. P. (1977). Evidence for the existence of three primary strategies in plants and its relevance to ecological and evolutionary theory. *Amer. Natur.* **111**, 1169–1194.

Grime, J. P. (1979). "Plant Strategies and Vegetation Processes." Wiley, New York.

Gupta, A. S. (1977). Calcium storage and distribution in the digestive gland of *Bensomia monticola* (Gastropoda, Pulmonata). A histophysiological study. *Biol. Bull.* **153**, 369–376.

Haan, S. de (1975). Land application of liquid municipal wastewater sludges. *J. Water Pollut. Control Fed.* **47**, 2707–2710.

Hadley, N. F. and Quinlan, M. (1982). Simultaneous measurement of water loss and carbon dioxide production in the cricket, *Acheta domesticus. J. exp. Biol.* **101**, 343–346.

Hågvar, S. (1984). "Ecological Studies of Microarthropods in Forest Soils with Emphasis on Relations to Soil Acidity." Norw. For. Res. Inst.

Hågvar, S. and Abrahamsen, G. (1980). Colonisation by Enchytraeidae, Collembola and Acari in sterile soil samples with adjusted pH levels. *Oikos* **34**, 245–258.

Hågvar, S. and Amundsen, T. (1981). Effects of liming and artificial acid rain on the mite (Acari) fauna in coniferous forest. *Oikos* **37**, 7–20.

Hågvar, S. and Kjøndal, B. R. (1981). Effects of artificial acid rain on the microarthropod fauna in decomposing birch leaves. *Pedobiologia* **22**, 409–422.

Hartenstein, R. (1964). Feeding, digestion, glycogen, and the environmental conditions of the digestive system in *Oniscus asellus J. Insect Physiol.* **10**, 611–621.

Hartenstein, R. (1982). Effect of aromatic compounds, humic acids and lignins on growth of the earthworm *Eisenia foetida. Soil Biol. Biochem.* **14**, 595–599.

Hartenstein, R., Neuhauser, E. F. and Collier, J. (1980). Accumulation of heavy metals in the earthworm *Eisenia foetida. J. Envir. Qual.* **9**, 23–26.

Hassall, M. (1983). Population metabolism of the terrestrial isopod *Philoscia muscorum* in a dune grassland ecosystem. *Oikos* **41**, 17–26.

Hassall, M. and Jennings, J. B. (1975). Adaptive features of gut structure and digestive physiology in the terrestrial isopod *Philoscia muscorum* (Scopoli) 1763. *Biol. Bull.* **149**, 348–364.

Haukioja, E., Niemela, P., Iso-livari, L., Ojala, H. and Aro, E. M. (1978). Birch leaves as a resource for herbivores. 1. Variation in the suitability of the leaves. *Rep. Kevo Subarct. Res. Station* **14**, 5–12.

Heely, W. (1941). Observation on the life histories of some terrestrial isopods. *Proc. Zool. Soc. London* **III(B)**, 79–149.

Heinrich, B. (1974). Thermoregulation in insects. *Science* **185**, 747–756.

Hill, J. (1980). The remobilization of nutrients from leaves. *J. Plant Nutr.* **2**, 407–444.

Hoese, B. (1981). Morphologie und Funktion des Wasserleitungsystems der terrestrischen Isopoden (Crustacea, Isopoda, Oniscoidea). *Zoomorphology*. **98**, 135–667.

Holdich, D. M. and Mayes, K. R. (1975). A fine-structural re-examination of the so-called "midgut" of the Isopod *Porcellio. Crustaceana* **29**, 186–192.

Hopkin, S. P. and Martin, M. H. (1982a). The distribution of zinc, cadmium, lead and copper within the woodlouse *Oniscus asellus* (Crustacea, Isopoda). *Oecologia* (*Berl.*) **54**, 227–232.

Hopkin, S. P. and Martin, M. H. (1982b). The distribution of zinc, cadmium, lead and copper within the hepatopancreas of a woodlouse. *Tissue & Cell* **14**, 703–715.

Horwath, K. L. and Duman, J. G. (1984). Yearly variations in the overwintering mechanisms of the cold-hardy beetle *Dendroides canadensis. Physiol. Zool.* **57**, 40–45.

Hryniewiecka-Szyfter, Z. (1971/72). Ultrastructure of hepatopancreas of *Porcellio scaber* Latr. in relation to the function of iron and copper accumulation. *Bull. Soc. Amis Sci. Lettr. Poznan* **D(12/13)**, 135–142.

Hryniewiecka-Szyfter, Z. and Tyczewska, J. (1978). The fine structure of the hepatopancreas of *Mesidotea entomon* L. (Crustacea, Isopoda). *Bull. Soc. Amis Sci. Lettr. Poznan* **D18**, 73–79.

Hubbell, S. P. (1971). Of sowbugs and systems: the ecological bio-energetics of a terrestrial isopod. *In* "System Analysis and Simulation in Ecology" (Ed. B. C. Patten), pp. 269–324. Head Press, New York.

Huhta, V. (1984). Response of *Cognettia sphagnetorum* (Enchytraeidae) to manipulation of pH and nutrient status in coniferous forest soil. *Pedobiologia* **27**, 245–260.

Huhta, V., Hyvönen, R., Koskenniemi, A. and Vilkamaa, P. (1983). Role of pH in the effect of fertilization on Nematoda, Oligochaeta and microarthropods. *In* "New Trends in Biology" (Ed. Ph. Lebrun *et al.*), pp. 61–73. Dieu-Brichart, Ottignies – Louvain-la-Neuve.

Humbert, W. (1972). Etude du pH intestinal d'un Collembole (Insecte, Apterygote). *Rev. Ecol. Biol. Sol.* **11**, 89–97.

Humbert, W. (1978). Cytochemistry and X-ray microprobe analysis of the midgut of *Tomocerus minor* Lubbock (Insecta; Collembola) with special reference to the physiological significance of the mineral concretions. *Cell Tissue Res.* **187**, 397–416.

Humbert, W. (1979). The midgut of *Tomocerus minor* Lubbock (Insecta, Collembola): Ultrastructure, cytochemistry, ageing and renewal during a moulting cycle. *Cell Tissue Res.* **196**, 39–57.

Husain, M. Z. and Alikhan, M. A. (1974). Physiological adaptations in crustacea to the environment: oxygen consumption as a function of body weight and environmental temperature in the terrestrial isopod, *Porcellio laevis* Latreille (Isopoda, Oniscosidea). *Crustaceana* **36**, 277–286.

Hutchinson, V. H. (1961). Critical thermal maxima in salamanders. *Physiol. Zool.* **34**, 92–125.

Hyatt, A. D. and Marshall, A. T. (1985). Water and ion balance in the tissues of the dehydrated cockroach, *Periplaneta americana*. *J. Insect Physiol.* **31**, 27–34.

Ireland, M. P. (1975). Distribution of lead, zinc and calcium in *Dendrobaena rubida* (Oligochaeta) living in soil contaminated by base metal mining in Wales. *Comp. Biochem. Physiol.* **52B**, 551–555.

Ireland, M. P. (1976). Excretion of lead, zinc and calcium by the earthworm *Dendrobaena rubida* living in soil contaminated with zinc and lead. *Soil Biol. Biochem.* **8**, 347–350.

Ireland, M. P. (1978). Heavy metal binding properties of earthworm chloragosomes. *Acta Biol. Acad. sci. hung.* **29**, 385–394.

Ireland, M. P. (1979a). Metal accumulation by the earthworms *Lumbricus rubellus, Dendrobaena veneta*, and *Eiseniella tetraeda* living in heavy metal polluted sites. *Environ. Pollut.* **19**, 201–206.

Ireland, M. P. (1979b). Distribution of essential and toxic metals in the terrestrial gastropod *Arion ater*. *Environ. Pollut.* **13**, 271–278.

Ireland, M. P. (1979c). Distribution of metals in the digestive gland-gonad complex of the marine gastropod *Nucella lapillus*. *J. moll. Stud.* **45**, 322–327.

Ireland, M. P. (1981). Uptake and distribution of cadmium in the terrestrial slug *Arion ater* (L.). *Comp. Biochem. Physiol.* **68A**, 37–41.

Ireland, M. P. (1982). Sites of water, zinc and calcium uptake and distribution of these metals after cadmium administration in *Arion ater* (Gastropoda: Pulmonata). *Comp. Biochem. Physiol.* **73A**, 217–221.

Ireland, M. P. and Fischer, E. (1978). Effect of Pb on Fe tissue concentrations and delta-amino-laevulinic acid dehydratase activity in *Lumbricus terrestris*. *Acta Biol. Acad. sci. hung.* **29**, 395–400.

Ireland, M. P. and Richards, K. S. (1977). The occurrence and localisation of heavy metals and glycogen in the earthworms *Lumbricus rubellus* and *Dendrobaena rubida* from a heavy metal site. *Histochemistry* **51**, 153–166.

Janssen, M. P. M. (1985). Loodresistentie en levensgeschiedenis–kenmerken in bodemdieren. *Int. Rep. V.U. Amsterdam.*

Jeantet, A. Y., Ballan-Dufrançais, C. and Martoja, R. (1977). Insect resistance to mineral pollution. Importance of spherocrystal in ionic regulation. *Rev. Ecol. Biol. Sol.* **14**, 563–582.

Jensen, S. and Jernelöv, A. (1969). Biological methylation of mercury in aquatic organisms. *Nature* **223**, 753–754.

Johnson, A. H. and Siccama, T. G. (1983). Acid deposition and forest decline. *Environ. Sci. Technol.* **17**, 294A-305A.

Jones, L. H. P. and Clement, C. R. (1972). Lead uptake by plants and its significance for animals. *In* "Lead in the Environment" (Ed. P. Hepple), pp. 29–33. *Proc. zool. soc. Lond.*

Joosse, E. N. G. (1971). Ecological aspects of aggregation in Collembola. Rev. Ecol. Biol. Sol. **8**, 91–97.

Joosse, E. N. G. (1981). Ecological strategies and population regulation of Collembola in heterogeneous environments. *Pedobiologia* **21**, 346–356.

Joosse, E. N. G. (1983). New developments in the ecology of Apterygota. *Pedobiologia* **25**, 217–234.

Joosse, E. N. G. and Buker, J. B. (1979). Uptake and excretion of lead by litter-dwelling Collembola. *Environ. Pollut.* **18**, 235–240.

Joosse, E. N. G., Capelleveen, H. E. van, Dalen, L. H. van and Diggelen, J. van (1983). Effects of zinc, iron and manganese on soil arthropods associated with decomposition processes. *Proc. Int. Conf. "Heavy metals in the environment", Heidelberg*. pp. 467–470.

Joosse, E. N. G. and Groen, J. B. (1970). Relationship between saturation deficit and the survival and locomotory activity of surface dwelling Collembola. *Entomol. Exp. and Appl.* **13**, 229–235.

Joosse, E. N. G. and Testerink, G. J. (1977a). The role of food in the population dynamics of *Orchesella cincta* (Linné) (Collembola). *Oecologia* (*Berl.*) **29**, 189–204.

Joosse, E. N. G. and Testerink, G. J. (1977b). Control of number in Collembola. *Ecol. Bull.* (*Stockholm*) **25**, 475–478.

Joosse, E. N. G. and Veltkamp, E. (1970). Some aspects of growth, moulting and reproduction in five species of surface dwelling Collembola. *Neth. J. Zool.* **20**, 315–328.

Joosse, E. N. G. and Verhoef, S. C. (1983). Lead tolerance in Collembola. *Pedobiologia* **25**, 11–18.

Joosse, E. N. G., Wulffraat, K. J. and Glas, H. P. (1981). Tolerance and acclimation to zinc of the isopod *Porcellio scaber*. Latr. *Int. Conf. "Heavy metals in the environment", Amsterdam*. pp. 425–428.

Kennedy, C. H. (1928). Evolutionary level in relation to geographic, seasonal and diurnal distribution of insects. *Ecology* **9**, 367–379.

Khandelwal, S., Ashquin, M. and Tandon, S. K. (1984). Influences of essential elements on manganese intoxication. *Bull. Envir. Contam. Toxicol.* **32**, 10–19.

Kober, T. E. and Cooper, G. R. (1976). Lead competitively inhibits calcium-dependent synoptic transmission in the bullfrog sympathetic ganglion. *Nature* **262**, 704–706.

Kooijman, S. A. L. M. and Metz, J. A. J. (1982). On the dynamics of chemically stressed populations: the deduction of population consequences from effects on individuals. *TNO, The Hague*, 1–38.

Kroon, D. B., Neumann, H. and Krayenhoff Sloot, W. J. (1944). *Enzymologia* **11**, 186.

Leinaas, H. P. (1983). Winter strategy of surface dwelling Collembola. *Pedobiologia* **25**, 235–240.

Leinaas, H. P. and Bleken, E. (1983). Egg diapause and demographic strategy in *Lepidocyrtus lignorum* Fabricius (Collembola; Entomobryidae). *Oecologia* (*Berl.*) **58**, 194–199.

Leinaas, H. P. and Fjellberg, A. (1985). Habitat structure and life history strategies of two partly sympatric and closely related, lichen feeding collembolan species. *Oikos* **44**, 448–458.

Levins, R. and Lewontin, R. (1980). Dialectics and reductionism in ecology. *Synthese* **43**, 47–78.

Levitt, J. (1980). "Responses of Plants to Environmental Stresses. Vol. I. Chilling, Freezing, and High Temperature Stresses" Academic Press, London.

Lindquist, O. V. and Fitzgerald, G. (1976). Osmotic interrelationship between blood and gut fluid in the isopod *Porcellio scaber* Latr. (Crustacea). *Comp. Biochem. Physiol.* **53A**, 57–59.

Lolkema, P. C., Donker, M. H., Schouten, A. J. and Ernst, W. H. O. (1984). The possible role of metallothioneins in copper tolerance of *Silene cucubalus. Planta*, **162**, 174–179.

Ma, W. C. (1982a). Regenwormen als bioindicators van bodemverontreiniging. *RIN/Min. VROM*.

Ma, W. C. (1982b). The influence of soil properties and worm-related factors on the concentration of heavy metals in earthworms. *Pedobiologia* **24**, 109–119.

Ma, W. C. (1984). Sublethal toxic effects of copper on growth, reproduction and litter breakdown activity in the earthworm *Lumbricus rubellus*, with observations on the influence of temperature and soil pH. *Environ. Pollut.* **(A)33**, 207–219.

Ma, W. C., Edelman, Th., Beersum, I. van and Jans, Th. (1983). Uptake of cadmium, zinc, lead, and copper by earthworms near a zinc-smelting complex: influence of soil pH and organic matter. *Bull. Environ. Contam. Toxicol.* **30**, 424–427.

Machin, J. (1981). Water compartmentalisation in Insects. *J. Exp. Biol.* **215**, 327–333.

Martin, J. S. and Martin, M. M. (1983). Tannin assays in ecological studies. *J. Chem. Ecol.* **9**, 285–294.

Martin, M. H. and Coughtrey, P. J. (1982). "Biological monitoring of heavy metal pollution" Applied Science Publ. London and New York.

Martin, M. M. and Martin, J. S. (1984). Surfactants: their role in preventing the precipitation of proteins by tannins in insect guts. *Oecologia* (*Berl.*) **61**, 342–345.

Masaki, S. (1980). Summer Diapause. *In* "Ann. Rev. Entomol. Vol. 25" (Eds T. E. Mittler, F. J. Radovsky and V. H. Resh), pp. 1–25. Annual Reviews Inc. USA.

Mattson, W. J. (1980). Herbivory in relation to plant nitrogen content. *Ann. Rev. Ecol. Syst.* **11**, 119–161.

May, M. L. (1979). Insect thermoregulation. *In* "Ann. Rev. Entomol. Vol. 24" (Eds T. E. Mittler, F. J. Radovsky and V. H. Resh), pp. 313–349. Annual Reviews Inc. USA.

McManus, J. Lilly, T. M. and Haslam, E. (1983). Plant polyphenols and their association with protein. *In* "Plant Resistance to Insects" (Ed. P. A. Hedin), pp. 123–137. Ac. Symp. Series.

McMillan, J. H. (1975). Interspecific and seasonal analysis of the gut contents of three Collembola (Family Onychiuridae). *Rev. Ecol. Biol. Sol.* **12**, 449–457.

McMillan, J. H. and Healey, I. N. (1971). A quantitative technique for the analysis of gut contents of Collembola. *Rev. Ecol. Biol. Sol.* **8**, 295–300.

Meincke, K. F. and Schaller, K. H. (1974). Über die Brauchbarkeit der Weinbergschnecke (*Helix pomatia* L.) in Freiland als Indikator für die Belastung der Umwelt durch die Elemente Eisen, Zink und Blei. *Oecologia* (*Berl.*) **15**, 393–398.

Mello, M. L. S., Vidal, B. C. and Valdrighi, L. (1971). The larval peritrophic membrane of *Melipona quadrifasciata* (Hymenoptera: Apoidea). *Protoplasma* **73**, 349–365.

Merriam, H. G. (1971). Sensitivity of terrestrial isopod populations to food quality differences. *Canad. J. Zool.* **49**, 667–674.

Merrill, W. and Cowling, E. B. (1966). Role of nitrogen in wood deterioration: amount and distribution of nitrogen in fungi. *Phytopathology* **56**, 1083–1090.

Miller, K. (1982). Cold-hardiness strategies of some adult and immature insects overwintering in interior Alaska. *Comp. Biochem. Physiol.* **73A**, 595–604.

Morgan, A. J. (1982). The elemental composition of the chloragosomes of nine species of British earthworms in relation to calciferous gland activity. *Comp. Biochem. Physiol.* **73A**, 207–216.

Morgan, A. J. and Morris, B. (1982). The accumulation and intracellular compartmentation of cadmium, lead, zinc and calcium in two earthworm species (*Dendrobaena rubida* and *Lumbricus rubellus*) living in highly contaminated soil. *Histochemistry* **75**, 269–285.

Moriarty, F. (1983). "Ecotoxicology" Academic Press, London.

Moser, H. and Wieser, W. (1979). Copper and nutrition in *Helix pomatia* L. *Oecologia* (*Berl.*) **42**, 241–251.

Murray, D. R. P. (1968). The importance of water in the normal growth of larvae of *Tenebrio molitor. Entomol. Expl. Appl.* **11**, 149–168.

Nair, G. A. (1978). Some aspects of the population characteristics of the soil Isopod, *Porcellio laevis* (Latreille), in Delhi Region. *Zool. Anz. Jena* **201**, 86–96.

Neuhauser, E. F. and Hartenstein, R. (1976a). Degradation of phenol, cinnamic and quinic acid in the terrestrial crustacean, *Oniscus asellus. Soil Biol. Biochem.* **8**, 95–98.

Neuhauser, E. F. and Hartenstein, R. (1976b). On the presence of O-demethylase activity in invertebrates. *Comp. Biochem. Physiol.* **53C**, 37–39.

Neuhauser, E. F. and Hartenstein, R. (1978). Phenolic content and palatability of leaves and wood to soil isopods and diplopoda. *Pedobiologia* **18**, 99–109.

Neuhauser, E. F., Hartenstein, R. and Connors, W. J. (1978). Soil invertebrates and the degradation of vanillin, cinnamic acid, and lignins. *Soil Biol. Biochem.* **10**, 431–435.

Neuhauser, E. F., Kaplan, D. L., Malecki, M. R. and Hartenstein, R. (1980). Materials supporting weight gain by the earthworm *Eisenia foetida* in waste conversion systems. *Agricultural Wastes* **2**, 43–60.

Neuhauser, E. F., Youmell, C. and Hartenstein, R. (1974). Degradation of benzoic acid in the terrestrial crustacean, *Oniscus asellus. Soil Biol. Biochem.* **6**, 101–107.

Newell, R. C., Wieser, W. and Pye, V. I. (1974). Factors affecting oxygen consumption in the woodlouse *Porcellio scaber* Latr. *Oecologia* (*Berl.*) **16**, 31–51.

Nielson, R. L. (1951). Effect of soil minerals on earthworms. *N.Z.J. Agric.* **83**, 433–435.

Nottrot, F., Joosse, E. N. G. and Straalen, N. M. van (1986). Sublethal effects of pollution with iron and manganese on *Orchesella cincta* (L.) (Collembola). *Pedobiologia* (in press).

Otvos, J. D., Olafson, R. W. and Armitage, I. M. (1982). Structure of an invertebrate metallothionein from *Scylla serrata. J. Biol. Chem.* **257**, 2427–2431.

Parle, J. N. (1963). Micro-organisms in the intestines of earthworms. *J. Gen. Microbiol.* **31**, 1–11.

Paul, P. G., Somers, J. A. and Scholte Ubing, D. W. (1981). Belasting van de bodem in Nederland met zware metalen. *de Ingenieur* **8**, 15–19.

Petering, H. G. (1978). Some observations on the interaction of zinc, copper, and iron metabolism in lead and cadmium toxicity. *Environ. Health Persp.* **25**, 141–145.

Petersen, H. (1971). Collembolernes ernaeringsbiologi og dennes okologiske betydning. *Entom. Medd.* **39**, 97–118.

Petersen, H. (1980). Population dynamic and metabolic characterization of Collembola species in a beach forest ecosystem. *In* "Soil biology as related to land use practices" (Ed. D. E. Dindal), pp. 806–833. Syracusa.

Phillips, J. G. (Ed.) (1975). "Environmental Physiology" Blackwell Scientific, Oxford.

Phillipson, J. and Watson, J. (1965). Respiratory metabolism of the terrestrial isopod *Oniscus asellus* L. *Oikos* **16**, 78–87.

Piearce, T. G. (1972). The calcium relations of selected Lumbricidae. *J. Anim. Ecol.* **41**, 167–188.

Poole, T. B. (1959). Studies on the food of Collembola in a Douglas fir plantation. *Proc. Zool. Soc. London* **132**, 71–82.

Prento, P. (1979). Metals and phosphate in the chloragosomes of *Lumbricus terrestris* and their possible significance. *Cell Tissue Res.* **196**, 123–134.

Price, J. B. and Holdich, D. M. (1980). Changes in osmotic pressure and sodium concentration of the haemolymph of woodlice with progressive desiccation. *Comp. Biochem. Physiol.* **66A**, 297–305.

Prosi, F., Storch, V. and Janssen, H. H. (1983). Small cells in the midgut glands of terrestrial Isopoda: sites of heavy metal accumulation. *Zoomorphology* **102**, 53–64.

Prosi, F. and Back, H. (1985). Indicator cells for heavy metal uptake and distribution in organs from selected invertebrate animals. *Proc. Int. Conf. Heavy metals in the environment, Athens*, pp. 242–244.

Prosser, C. L. (1973). "Comparative Animal Physiology" Vol. I, "Environmental Physiology" W. B. Saunders Company, London.

Quaterman, J., Humphries, W. R., Morrison, J. N., and Morrison, E. (1980). The influence of dietary amino acids on lead absorption. *Environ. Res.* **23**, 54–67.

Rapoport, E. H. and Tschapek, M. (1967). Soil water and soil fauna. *Rev. Ecol. Biol. Sol.* **4**, 1–58.

Reddy, M. V. (1983). Effects of fire on the nutrient content and microflora of casts of *Pheretima alexandri*. *In* "Earthworm Ecology" (Ed. J. E. Satchell), pp. 209–214. Chapman and Hall, London.

Reichle, D. E., Shanks, M. H. and Crossley, Jr. D. A. (1969). Calcium, potassium and sodium content of forest arthropods. *Ann. Entomol. Soc. Am.* **62**, 57–62.

Rhoades, D. F. and Cates, R. G. (1976). Towards a general theory of plant anti-herbivore chemistry. *Rec. adv. in Phytochem.* **10**, 168–213.

Richards, A. G. and Richards, P. A. (1977). The peritrophic membranes of insects. *Ann. Rev. Entomol.* **22**, 219–240.

Richards, K. S. and Ireland, M. P. (1978). Glycogen-lead relationship in the earthworm *Dendrobaena rubida* from a heavy metal site. *Histochemistry* **56**, 55–64.

Riddle, W. A. (1985). Hemolymph osmoregulation in several Myriapods and Arachnids. *Comp. Biochem. Physiol.* **80A**, 313–323.

Russell, L. K., Haven, J. I. de and Botts, R. P. (1981). Toxic effects of cadmium on the garden snail (*Helix aspersa*). *Bull. Envir. Contam. Toxicol.* **26**, 634–640.

Salt, R. W. (1961). Principles of insect cold-hardiness. *In* "Ann. Rev. Entomol. Vol. 6" (Eds T. E. Mittler, F. J. Radovsky and V. H. Resh), pp. 55–74. Annual Reviews Inc. USA.

Sanders, J. R. (1982). The effect of pH upon the copper and cupric ion concentrations in soil solutions. *J. Soil Sci.* **33**, 679–689.

Sastry, K. S., Murty, R. R. and Sarma, P. S. (1958). Studies on zinc toxicity in the larvae of the rice moth *Corcyra cephalonica*. *Biochem. J.* **69**, 425–428.

Satchell, J. E. (1967). Lumbricidae. *In* "Soil Biology" (Eds A. Burgers and F. Raw), pp. 259–322. Academic Press, London.

Satchell, J. E. (1983). Earthworm microbiology. *In* "Earthworm ecology" (Ed. J. E. Satchell), pp. 351–364. Chapman and Hall, London.

Satchell, J. E. and Lowe, D. G. (1967). Selection of leaf litter by *Lumbricus terrestris*. *In* "Progress in Soil Biology" (Eds O. Graff and J. E. Satchell), pp. 102–119. Elsevier, Amsterdam.

Schaffer, W. M. (1974). Optimal reproductive effort in fluctuating environments. *Amer. Nat.* **108**, 783–790.

Schoettli, G. and Seiler, H. G. (1970). Uptake and localization of radioactive zinc in the visceral complex of the land Pulmonate *Arion rufus*. *Experientia* **26**, 1212–1213.

Scriber, J. M. and Slansky, Jr. F. (1981). The nutritional ecology of immature insects. *Ann. Rev. Entomol.* **26**, 183–211.

Sell, D. and Houlihan, D. F. (1985). Water balance and rectal absorption in the grasshopper *Oedipoda*. *Physiol. Ent.* **10**, 89–103.

Sibly, R. M. and Calow, P. (1986). "Physiological Ecology of Animals: an Evolutionary Approach" Blackwell Scientific, Oxford.

Sidle, R. C. and Kardos, L. T. (1977). Adsorption of copper, zinc, and cadmium by a forest soil. *J. Environ. Qual.* **6**, 313–317.

Sohal, R. S. and Lamb, R. E. (1979). Intracellular deposition of metals in the midgut of the adult housefly *Musca domestica*. *J. Insect Physiol.* **23**, 1349–1354.

Sømme, L. (1982). Supercooling and winter survival in terrestrial arthropods. *Comp. Biochem. Physiol.* **73A**, 519–543.

Specht, R. L. (1979). "Heathlands and Related Shrublands" Elsevier, Amsterdam.

Stearns, S. C. (1976). Life-history tactics: a review of the ideas. *Quart. Rev. Biol.* **51**, 3–47.

Stephanou, G., Alahiotis, S. N., Marmaras, V. J. and Christodoulou, C. (1983). Heat shock response in *Ceratitis capitata*. *Comp. Biochem. Physiol.* **74B**, 425–432.

Storch, V. (1982). Der Einflusz der Ernährung auf die Ultrastruktur der grossen Zellen in der Mitteldarmdrüsen terrestrischer Isopoda (*Armadillidium vulgare, Porcellio scaber*). *Zoomorphology* **100**, 131–142.

Storey, K. B., Baust, J. G. and Buescher, P. (1981). Determination of water "bound" by soluble subcellular components during low-temperature acclimation in the gall fly larva, *Eurosta solidaginis*. *Cryobiology* **19**, 180–184.

Straalen, N. M. van (1983). Vergelijkende demografie van springstaarten. PhD thesis, Vrije Universiteit, Amsterdam.

Straalen, N. M. van (1985a). Size-specific mortality patterns in two species of forest floor Collembola. *Oecologia* (*Berl.*) **67**, 220–223.

Straalen, N. M. van (1985b). Comparative demography of forest floor Collembola populations. *Oikos* **45**, 253–265.

Straalen, N. M. van, Burghouts, Th. B. A. and Doornhof, M. J. (1985). Dynamics of heavy metals in populations of Collembola in a contaminated forest soil. *Proc. Int. Conf. Heavy metals in the environment, Athens*, pp. 613–615.

Straalen, N. M. van and Wensem, J. van (1986). Heavy metal content of forest litter arthropods as related to body-size and trophic level. *Environ. Pollut.* Ser. **A42**, 209–221.

Streit, B. (1984). Effects of high copper concentrations on soil invertebrates (earthworms and oribatid mites). *Oecologia* (*Berl.*) **61**, 381–388.

Strojan, C. L. (1978). The impact of zinc smelter emissions on forest litter arthropods. *Oikos* **31**, 41–46.

Sutton, S. L. (1968). The population dynamics of *Trichoniscus pusillus* and *Philoscia muscorum* (Crustacea, Oniscoidea) in limestone grassland. *J. Anim. Ecol.* **37**, 425–444.

Swain, T. (1979). Tannins and lignins. *In* "Herbivores" (Eds G. A. Rosenthal and D. H. Janzen), pp. 657–682. Academic Press, New York.

Swift, M. J. Heal, O. W. and Anderson, J. M. (1979). "Decomposition in Terrestrial Ecosystems" Blackwell Scientific, Oxford.

Terra, W. R., Ferreira, C. and Bianchi, A. G. de (1979). Distribution of digestive enzymes among the endo- and ectoperitrophic spaces and midgut cells of *Rhynchosciara* and its physiological significance. *J. Insect Physiol.* **25**, 487–494.

Terra, W. R., and Ferreira, C. (1981).The physiological role of the peritrophic membrane and trehalase: digestive enzymes in the midgut and excreta of starved larvae of *Rhynchosciata. J. Insect Physiol.* **27**, 325–331.

Testerink, G. J. (1981). Starvation in a field population of litter-inhabiting springtails (Collembola). Methods for determining food reserves in small arthropods. *Pedobiologia* **21**, 427–433.

Testerink, G. J. (1982). Strategies in energy consumption and partitioning in Collembola. *Ecol. Entomol.* **7**, 341–351.

Testerink, G. J. (1983). Metabolic adaptations to seasonal changes in humidity and temperature in litter-inhabiting springtails. *Oikos* **40**, 234–240.

Thibaud, J.-M. (1968). Contribution à l'étude de l'action des facteurs température et humidité sur la durée du développement postembryonnaire et de l'intermue de l'adulte chez les Collemboles, Hypogastruridae. *Rev. Ecol. Biol. Sol.* **2**, 265–281.

Thibaud, J.-M. (1977a). Intermue et températures léthales chez les insectes Collemboles Arthropléones I. – Hypogastruridae et Onychiuridae. *Rev. Ecol. Biol. Sol.* **14**, 45–61.

Thibaud, J.-M. (1977b). Intermue et températures léthales chez les Insectes Collemboles Arthropléones II. – Isotomidae, Entomobryidae et Tomoceridae. *Rev. Ecol. Biol. Sol.* **14**, 267–278.

Thomson, A. B. R., Olatunbosun, D. and Valberg, L. S. (1971). Interrelation of intestinal transport system for manganese and iron. *J. Lab. Clin. Med.* **78**, 642–648.

Townsend, C. R. and Calow, P. (1981). "Physiological Ecology" Blackwell Scientific, Oxford.

Tyler, G. (1978). Leaching rates of heavy metal ions in forest soil. *Water, Air and Soil Pollution* **9**, 137–148.

Vallee, B. L. and Ulmer, D. D. (1972). Biochemical effects of mercury, cadmium and lead. *Ann. Rev. Biochem.* **41**, 91–128.

Vannier, G. (1970). Réactions des microarthropods aux variations de l'état hydrique du sol. Techniques relatives à l'extraction des arthropods du sol. *Rev. Ecol. Biol. Sol.* **7**, 289–309.

Vannier, G. (1983). The importance of ecophysiology for both biotic and abiotic studies of the soil. *Proc. 8th Int. Coll. Soil Zool. Louvain-la-Neuve*, pp. 289–314.

Vannier, G. and Ghabbour, S. I. (1983). Effect of rising ambient temperature on transpiration in the cockroach *Heterogamia syriaca* Sauss. from the Mediterranean coastal desert of Egypt. *Proc. 8th Int. Coll. Soil Zool. Louvain-la-Neuve* pp. 441–453.

Vannier, G. and Verdier, B. (1981). Critères écophysiologiques (transpiration respiration) permettant de séparer une espèce souterraine d'une espèce de surface chez les Insectes Collemboles. *Rev. Ecol. Biol. Sol.* **18**, 531–549.

Vannier, G. and Verhoef, H. A. (1978). Effect of starvation on transpiration and water content in the populations of two co-existing Collembola species. *Comp. Biochem. Physiol.* **60A**, 483–489.

Vegter, J. J. (1983). Food and habitat specialization in coexisting springtails (Collembola, Entomobryidae). *Pedobiologia* **25**, 253–262.

Vegter, J. J. (1985). Coexistence of forest floor Collembola. PhD thesis, Vrije Universiteit, Amsterdam.
Verdier, B. and Vannier, G. (1984). Modifications de la consommation d'oxygène chez les arthropods terrestres à respiration cutanée soumis à différents déficits hygrométriques de l'air. *C. R. Acad. Sc. Paris* **299**, 563–566.
Verhoef, H. A. (1981). Water balance in Collembola and its relation to habitat selection: water content, haemolymph osmotic pressure and transpiration during an instar. *J. Insect Physiol.* **27**, 755–760.
Verhoef, H. A. (1984). Releaser and primer pheromones in Collembola. *J. Insect Physiol.* **30**, 665–670.
Verhoef, H. A. and Li, K. W. (1983). Physiological adaptations to the effects of dry summer periods in Collembola. *Proc. 8th Int. Coll. Soil Zool. Louvain-la-Neuve*, pp. 345–356.
Verhoef, H. A., Nagelkerke, C. J. and Joosse, E. N. G. (1977). Aggregation pheromones in Collembola. *J. Insect Physiol.* **23**, 1009–1013.
Verhoef, H. A. and Nagelkerke, C. J. (1977). Formation and ecological significance of aggregations in Collembola; an experimental study. *Oecologia* (*Berl.*) **31**, 215–226.
Verhoef, H. A. and Selm, A. J. van (1983). Distribution and population dynamics of Collembola in relation to soil moisture. *Holarctic Ecol.* **6**, 387–394.
Verhoef, H. A. and Witteveen, J. (1980). Water balance in Collembola and its relation to habitat selection; cuticular water loss and water uptake. *J. Insect Physiol.* **26**, 201–208.
Verhoef, H. A., Witteveen, J., Woude, H. A. van der and Joosse, E. N. G. (1983). Morphology and function of the ventral groove of Collembola. *Pedobiologia* **25**, 3–9.
Wallwork, J. A. (1983). Oribatids in forest ecosystems. *Ann. Rev. Entomol.* **28**, 109–130.
Waterhouse, D. F. and Stay, B. (1955). Functional differentiation on the midgut epithelium of blow fly larvae as revealed by histochemical tests. *Austr. J. Biol. Sci.* **8**, 253–277.
White, J. J. (1983). Woodlice exposed to pollutant gases. *Bull. Environm. Contam. Toxicol.* **30**, 245–251.
White, T. R. C. (1978). The importance of a relative shortage of food in animal ecology. *Oecologia* (*Berl.*) **33**, 71–86.
White, T. R. C. (1984). The abundance of invertebrate herbivores in relation to the availability of nitrogen in stressed food plants. *Oecologia* (*Berl.*) **63**, 90–105.
White, J. and Strehl, C. E. (1978). Xylem feeding by periodical cidada nymphs on tree roots. *Ecol. Entomol.* **3**, 323–327.
Whitmore, D. H., Gonzalez, R. and Baust, J. G. (1985). Scorpion cold hardiness. *Physiol. Zool.* **58**, 526–537.
Wieser, W. (1968). Aspects of nutrition and the metabolism of copper in Isopods. *Am. Zool.* **8**, 495–506.
Wieser, W. (1972). O/N rates of terrestrial isopods at two temperatures. *Comp. Biochem. Physiol.* **43A**, 859–868.
Wieser, W. and Klima, J. (1969). Compartmentalization of copper in the hepatopancreas of Isopods. *Mikroskopie* **22**, 1–9.
Wieser, W. and Makart, H. (1961). Der Sauerstoffverbrauch und der Gehalt an Ca, Cu und einigen anderen Spurenelementen bei terrestrischen Asseln. *Z.f. Naturforschung* **16B**, 816–819.
Wieser, W. and Wiest, L. (1968). Ökologische Aspekte des Kupferstoffwechsels terrestrischer Isopoden. *Oecologia* (*Berl.*) **1**, 38–48.

Williamson, P. (1979a). Comparison of metal levels in invertebrate detritivores and their natural diets: concentration factors reassessed. *Oecologia* (*Berl.*) **44**, 75–79.
Williamson, P. (1979b). Opposite effects of age and weight on cadmium concentrations of a gastropod mollusc. *Ambio* **8**, 30–31.
Williamson, P. (1980). Variables affecting body burdens of lead, zinc and cadmium, in a roadside population of the snail *Cepaea hortensis* Müller. *Oecologia* (*Berl.*) **44**, 213–220.
Willmer, P. G. (1980). The effects of a fluctuating environment on the water relations of larval Lepidoptera. *Ecol. Entomol.* **5**, 271–292.
Willmer, P. G. (1982). Microclimate and the environmental physiology of insects. *In* "Advances in Insect Physiology, Vol. 16" (Eds M. J. Berridge, J. E. Treherne and V. B. Wigglesworth), pp. 1–57. Academic Press, London.
With, N. D. de and Joosse, E. N. G. (1971). The ecological effects of moulting in Collembola. *Rev. Ecol. Biol. Sol.* **8**, 111–117.
Witkamp, M. and Drift, J. van der (1961). Breakdown of forest litter in relation to environmental factors. *Plant and Soil* **15**, 295–311.
Woodring, J. P. and Blakeney, Jr. E. W. (1980). The role of free amino acids in osmoregulation of cricket blood (*Acheta domesticus*). *J. Insect Physiol.* **26**, 613–618.
Woodring, J. P. and Cook, E. F. (1962). The biology of *Ceratozetes cisalprinus* Berlese, *Scheloribates laevigatus* Kock, and *Oppia neerlandica* (Oudemans) with a description of all stages. *Acarologia* **4**, 101–141.
Woude H. A. van der (1986). Seasonal changes in cold hardiness of temperate Collembola. *Oikos* (in press).
Woude, H. A. van der and Verhoef, H. A. (1986). A comparative study of winter survival in two temperate Collembola *Ecol. Entomol.* (in press).
Yamamura, H., Suzuki, K. T., Hatakeyama, S. and Kubota, K. (1983). Tolerance to cadmium and cadmium-binding proteins induced in the midge larva, *Chironomus yoshimatsui* (Diptera, Chironomidae). *Comp. Biochem. Physiol.* **75C**, 21–24.
Young, S. R. and Block, W. (1980). Experimental studies on the cold tolerance of *Alaskozetes antarcticus*. *J. Insect Physiol.* **26**, 189–200.
Zachariassen, K. E. (1982). Nucleating agents in cold-hardy insects. *Comp. Biochem. Physiol.* **73A**, 557–562.
Zachariassen, K. E. and Hammel, H. T. (1976). Nucleating agents in the haemolymph of insects tolerant to freezing. *Nature* **262**, 285–287.
Zachariassen, K. E. and Husby, J. A. (1982). Antifreeze effect of thermal hysteresis agents protects supercooled insects. *Nature* **298**, 865–867.
Zettel, J. (1984). Cold hardiness strategies and thermal hysteresis in Collembola. *Rev. Ecol. Biol. Sol.* **21**, 189–203.
Zhuzhikov, D. P. (1970). Permeability of the peritrophic membrane in the larvae of *Aedes aegypti*. *J. Insect Physiol.* **16**, 1193–1202.
Zöttl, H. W. and Lamparski, F. (1981). Schwermetalle (Pb, Cd) in der Bodenmakrofauna des Südschwarzwaldes. *Mitt. Dtsch. Bodenkundl. Gesellsch.* **32**, 509–518.

Principles of Predator–Prey Interaction in Theoretical, Experimental, and Natural Population Systems

E. KUNO

I. Introduction 250
II. Mathematical Models for Predator–Prey Interaction 252
A. Framework of Predator–Prey Interaction: Classic Models 252
B. Basic Models 253
C. Extended Models 256
III. Patterns of Predator–Prey Dynamics in Theoretical Population Systems . . . 262
A. Classic Models 262
B. Basic Models 262
C. Extended Models 274
IV. Regulation of Prey Population by Predation: Its Possibility in Theoretical Population Systems 281
A. Prey Population Regulation by Predation – its Definition and Detection . . 282
B. Prey Population Regulation in Classic Models 282
C. Prey Population Regulation in Basic Models 283
D. Prey Population Regulation in Extended Models 283
E. Robustness of Prey Population Regulation in a Varying Environment . . . 288
F. Non-Regulatory Suppression of Prey Density by Predation 296
V. Conflict of Interests between Individual and Population: Analysis of an Evolutionary Paradox in Theoretical Predator–Prey Systems 299
A. Selection Among Prey Individuals 299
B. Selection Among Predator Individuals 301
C. Selection Between "Specialists" and "Generalists" in Either Predator or Prey Species 302
D. Predator–Prey Conflict in their Coevolution 305
VI. Predator–Prey Dynamics in Experimental Population Systems 311
A. Protozoan Predator–Prey Systems in the Laboratory 312
B. Arthropod Predator–Prey Systems in the Laboratory 313
C. Cases of Biological Pest Control in the Field 316
VII. Predator–Prey Dynamics in Natural Population Systems 318
A. Insect Populations – Epidemic Species 319

DVANCES IN ECOLOGICAL RESEARCH Vol. 16
SBN 0-12-013916-2

B. Insect Populations – Endemic Species 320
C. Populations of Birds and Mammals 324
VIII. Discussion and Conclusions 326
Acknowledgements 331
References 331

I. INTRODUCTION

Population interaction between predator and prey, an essential warp of the ecological web of nature, has long been one of the central subjects of study in animal population ecology. Reviewing a vast number of contributions so far accumulated on this subject, one may find three major types of study The first is the theoretical type of study founded by Lotka (1925), Volterra (1926) and Nicholson and Bailey (1935). Here various mathematical models have been constructed in terms of simultaneous differential or difference equations to mimic the behaviour of interacting predator and prey popula tions, and many theories derived concerning ecological principles underlying the interaction. The second is the experimental type of study founded by Gause (1934) in which properties of actual predator–prey population sys tems have been compared with theoretical models, using artificially arranged laboratory populations of convenient animals such as protozoa o arthropods. The third type of study, on the other hand, is represented b investigations in the field in which naturally occurring interactions between predator and prey have been analysed in the light of various principle derived from those theoretical and experimental types of study. Varle (1947) may be the first who intentionally made a field population study alon this line, by analysing life tables of the knapweed gall-fly in relation t interactions with its parasitoids.

It should be noted that many of these studies have been made with common fundamental objective, i.e., to elucidate whether it is possible at a for predators in nature to regulate populations of their prey, and, then, ho and under what conditions regulation has actually occurred. This may b primarily because in the past decades there have been active controversies i the field of animal population ecology between "biotic" and "climatic schools as to whether or not density-dependent regulation is significant i the dynamics of natural animal populations (e.g., Krebs, 1984). Anothe reason may be that the subject has often been studied and discusse especially among entomologists, with special reference to an applied prob lem, namely, biological control of pests, where the regulation of numbers a problem of primary importance (e.g., Thompson, 1939). The reality c population regulation by predators is still disputed (e.g., Dempster, 198

Hassell, 1985), and furthermore, since the pioneer work of Rosenzweig and MacArthur (1963) the same problem has also begun to be actively approached from an evolutionary viewpoint according to which predator–prey dynamics are considered to be a consequence of their coevolution.

Despite there having been such detailed studies, involving very different approaches, the principles of predator–prey interaction do not as yet seem to have been well established nor interpreted, at least so as to be generally accepted. One reason for such difficulty in interpreting the principles may be that information especially from theoretical studies is now too extensive and too diversified for it to be readily interpreted. It may be also noted that some of the theories of predator-prey interaction have advanced so far and so elaborately without necessary feedback to the real world that their appropriate evaluation in the light of facts is no longer feasible.

The intention of this paper is first of all to bridge the huge gap existing between studies of theoretical populations and actual ones. In sections II to V of the paper, existing models and resultant theories of predator–prey interaction are critically reviewed from the biologist's viewpoint; some new simple models are reconstructed based on essential sets of biologically reasonable assumptions; and then fundamental principles concerning various aspects of the interaction are derived systematically from these theoretical models. Sections VI and VII are concerned with interactions in actual population systems. The actual systems may be classified into three categories, i.e., systems in laboratory experiments, those in field experiments of biological pest control, and natural interactive systems in the field. Representative studies in each category are reviewed and the data reanalysed carefully, if necessary, to interpret comparatively the characteristics and actual status of the predator–prey interaction under a range of situations in the light of the theoretical principles derived in the foregoing sections. In the general discussion, the possibility of, the necessary conditions for, and the actual status in nature of the regulation of animal populations by predation are discussed comprehensively on the basis of all these results, together with some related problems such as the conditions for success of biological control programmes for agricultural insect pests.

Since the main purpose of this study is to abstract some fundamental principles of predator–prey interaction out of the enormous amount of information so far accumulated, rather than to make a thorough historical review on the subject, I have not always traced faithfully the development of relevant studies in various directions. Readers who wish to have full details of the course of these studies are advised to refer to some of the excellent review articles so far published on this subject, such as those of Royama (1971), May (1973), Murdoch and Oaten (1975), Hassell (1978), Hassell and Waage (1984) and Taylor (1984).

II. MATHEMATICAL MODELS FOR PREDATOR–PREY INTERACTION

In this section a series of simple mathematical models are presented, to be used as the standard for establishing essential principles of the predator–prey interaction. After a critical review of existing models, new "basic" ones will be derived as either simultaneous differential equations or simultaneous difference ones, restricting constituent assumptions and parameters to only those that are regarded as reasonable biologically and essential for the present purposes. Then, some extensions to these basic models will be made by the addition of subsidiary factors to augment reality or concreteness of the models.

A. Framework of Predator–Prey Interaction: Classic Models

Throughout this study the term "predation" is used in its widest sense. Namely, whenever an animal kills another for food, they may be called "predator" and "prey". As is well known, a framework of predator–prey interaction in this wide sense was established by Lotka (1925) and Volterra (1926) in a set of simultaneous differential equations having the form

$$\begin{cases} \dfrac{dx}{dt} = f(x) - g(x, y) \\ \\ \dfrac{dy}{dt} = u(g(x, y), y) - v(y) \end{cases} \tag{1}$$

where x and y represent densities of prey and predator, respectively. The functions f, g, u and v here represent the rates of prey reproduction, prey death due to predation, predator reproduction, and predator death, respectively, which are given concrete forms, $f = bx - dx$, $g = axy$, $u = caxy$ and $v = ey$, where b and d are prey's birth and death rate in the absence of predator, a is predator's searching efficiency, c is predator's efficiency of prey consumption for reproduction and e is predator's natural mortality rate. The basic assumptions underlying this model are: (1) exponential growth of the prey population in the absence of the predator; (2) exponential decline of the predator population in the absence of prey; (3) random search by predator with unlimited capacity to attack all the prey it can find; and (4) immediate conversion of energy input through predation to birth of more predators.

Another classic model of fundamental importance is a difference equation version of the above Lotka–Volterra model derived by Nicholson and Bailey (1935), the framework of which is

$$\begin{cases} x_{t+1} = F(x_t - G(x_t, y_t)) \\ y_{t+1} = U(G(x_t, y_t), y_t) \end{cases} \tag{2}$$

describing the relation of prey and predator densities at a given generation (x_t, y_t) to those at the next (x_{t+1}, y_{t+1}). The functions F, G and U here correspond to f, g and u in Eqn (1), being specified as $F = R(x_t - G)$, $G = x_t(1 - \exp(-ay_t))$ and $U = CG = Cx_t(1 - \exp(-ay_t))$ where parameters b and d for the prey in Lotka–Volterra model are here unified to the overall reproductive potential per generation, R, and c and e for the predator to the overall efficiency of prey consumption for reproduction, C. The basic assumptions are essentially same as those in the differential Lotka–Volterra model except that the conversion of energy input by predation to the birth of more predators is no longer immediate but has a time delay of one generation. In the original Nicholson–Bailey model, parameter C has been fixed at unity, assuming a specific situation of insect host–parasitoid system. But it is evident that this generalization makes the model much more flexible, so that it can now be treated as a model for predation in a wide sense, i.e., as a difference-equation version of Lotka–Volterra model.

The criticism has been repeatedly made from a biologist's point of view that the above two classic models can now no longer be regarded as a reliable basis for reasoning, because their basic assumptions involve some points that are unreasonable biologically (e.g., Royama, 1971). It is also clear, however, that these two models have provided a basic framework for predator–prey interaction theory that is sound and robust enough to be used for deriving more realistic models as seen below.

B. Basic Models

Now that the above expressions (1) and (2) due to Lotka–Volterra and Nicholson–Bailey models have been accepted as reasonable frameworks for the description of predator–prey interaction, the next step in composing our basic models is to define concrete forms of the individual functions comprising the equations, i.e., $f(x)$ and $F(x)$ describing prey population growth, $g(x, y)$ and $G(x, y)$ describing predator rate, and $u(g(x, y), y)$, $v(y)$ and $U(G(x, y), y)$ characterizing predator's population growth.

Let us start with the differential equation model given in Eqn (1). The original assumption in Lotka–Volterra model that the per capita rate of prey reproduction $(f(x)/x)$ is a constant $(= b - d)$ independent of prey density has been criticized as neglecting the crowding effect or the restriction of population growth due to the limits of habitat's carrying capacity. To meet this argument one can, as Volterra (1926) himself proposed, use the logistic formula as $f(x)$, i.e.

$$f(x) = (b - d)x - hx^2 \tag{3}$$

where h is the parameter defining per capita intensity of crowding effect. Equation (3) has often been used in a different form, $f(x) = r(1 - x/K)x$ where $r = b - d$ and K is the "carrying capacity" (e.g., Rosenzweig, 1971; May, 1973). However, this popular expression is often misleading and considered unsuitable for general use, because K here is, contrary to general belief, not a proper, independent measure of the "capacity" in its true sense, but is a parameter which merely represents the "equilibrium" as the balance between birth and death, and hence which fundamentally can never be independent of reproductive rate, r. The unreasonableness of its use may become clear if one considers the case of $r < 0$ and $K > 0$, where an absurd situation that $f(x)/x$ is positively correlated with x occurs (Fulda, 1981). Thus, it is h in Eqn (3), not K in the usual expression, that should be coupled with r as an independent parameter to describe the degree of mutual interference in relation to carrying capacity of the habitat.

As a general model of population growth, Eqn (3) is still incomplete, since it does not include the effect of undercrowding. In the context of predator–prey models, few studies have so far taken this factor into account, despite its universal importance among bisexual organisms. Gilpin (1974) used a function of the form, $f(x) = ax - bx^2 + cx^3$, in his predator–prey model which can describe both over- and undercrowding effects in the prey population. But since this equation is purely descriptive with little biological basis, I adopt here a different expression to incorporate this factor, replacing birth rate b in Eqn (3) by the function,

$$b(x) = bx/(s_x + x) \quad (4$$

where s_x represents the intensity of undercrowding effect as a dilution factor. Equation (4) gives a curve which rises from the origin and approaches the upper limit, b, asymptotically, as may be expected in normal bisexual populations.

The assumption in the Lotka–Volterra model that $g(x, y) = axy$ has also been open to severe criticism, since it implies that the predator will have an unlimited ability to kill all the prey that it can find, however high the prey density may be. To underline this point, Holling's (1959b) well-known "disc equation" is adopted here, i.e.

$$g(x, y) = axy/(1 + ax/f) \quad (5$$

where a is predator's searching efficiency as before and f, the maximum number of prey a predator can kill within a time unit. While prey density remains rather low, $g(x, y)$ in Eqn (5) may roughly equal axy, but its rate of rise gradually decreases as x increases for fixed y, approaching its upper limit f asymptotically. An alternative equation with similar features, $g(x, y) = f(1 - \exp(-ax/f))$, due to Ivlev (1961) has sometimes been used in place of

Eqn (5) (e.g., Rosenzweig, 1971). But Eqn (5) may be preferable because of its simplicity.

Another point to be considered in our basic model may be predator's undercrowding effect which can be incorporated by rewriting $u(g(x, y), y)$ in the predator equation as

$$u(g(x, y), y) = c \cdot g(x, y) \cdot y/(s_y + y) \tag{6}$$

Since there may be no need for the moment to change the assumption of constant predator death rate, $v(y)$ can be written simply as

$$v(y) = ey \tag{7}$$

Thus, incorporating Eqns (3–7) into the framework of Eqn (3), we have the basic differential equation model of predator–prey interaction,

$$\begin{cases} \dfrac{\mathrm{d}x}{\mathrm{d}t} = \dfrac{bx^2}{s_x + x} - dx - hx^2 - \dfrac{axy}{1 + ax/f} \\ \dfrac{\mathrm{d}y}{\mathrm{d}t} = \dfrac{caxy^2}{(1 + ax/f)(s_y + y)} - ey \end{cases} \tag{8}$$

which now includes as many as 9 fundamental parameters, in contrast to 5 in the original Lotka–Volterra model, each having definite ecological meaning.

The only unrealistic assumption remaining still unmodified in this basic model is that of immediate energy conversion from input (predation) to output (reproduction) on the predator's side. Although inclusion of the time-delay effect into this type of model is not impossible as Hutchinson (1948), Wangersky and Cunningham (1957) and Caswell (1972) showed, it greatly reduces the analytical convenience of the model. In this study, probable effects of time delay, ignored in Eqn (8), will be checked against the results of parallel analysis from its difference-equation version that follows, which structurally involves one-generation time delay.

Translation of Eqn (8) into the difference-equation model framed as Eqn (2) can be made readily. First, assuming that the same process of interference as in Eqn (3) occurs among Rx prey individuals that survived from predation, we can expect the final prey density in the absence of predators to be the solution of $\mathrm{d}x/\mathrm{d}t = -hx$ with the initial condition $x = x_t$, which gives

$$F(x) = Rx/(1 + hRx) \tag{9}$$

This proves to be a different expression of the equation Maynard Smith and Slatkin (1973) derived, i.e., $F(x) = RK/(K + (R - 1)x)$ where K is the equilibrium density. An alternative to Eqn (9) which has been used widely among existing predator–prey models (e.g., Hassell, 1978; Beddington *et al.*, 1978; May *et al.*, 1981) is $F(x) = R^{(1-x/K)}x$, another difference-equation version of the logistic. But Eqn (9) above may be more appropriate than this for use in the basic model, since the latter is virtually a model specifically to describe scramble-type competition (Kuno, 1983) so that its use may bring some superfluous disturbance to the interaction.

To incorporate the undercrowding effect in $F(x)$, it is sufficient to replace R in (9) with the expression

$$R(x) = Rx/(s_x + x) \tag{10}$$

corresponding to Eqn (4). The term of prey reduction due to predation, $G(x_t, y_t)$, can also be derived simply by replacing a in the original Nicholson–Bailey model with the disc equation $ax/(1 + ax/f)$, i.e.,

$$G(x_t, y_t) = x_t(1 - \exp(-ay_t/(1 + ax_t/f))) \tag{11}$$

This is the equation derived originally by Royama (1971) and Rogers (1972) as a model for parasitism, but it will be used here as a phenomenological model for predation in a wide sense.

Finally, the term for predator reproduction, $U(G(x_t, y_t), y_t)$, may be written as follows, considering the undercrowding effect.

$$U(G(x_t, y_t), y_t) = C \cdot G(x_t, y_t) \cdot y_t/(s_y + y_t) \tag{12}$$

Thus, combining all these submodels into the framework Eqn (2), we have

$$\begin{cases} x_{t+1} = \dfrac{RX^2/(s_x + X)}{1 + hRX^2/(s_x + X)} \\[2ex] y_{t+1} = \dfrac{C(x_t - X)y_t}{s_y + y_t} \end{cases} \tag{13}$$

$$(X = x_t \cdot \exp(-ay_t/(1 + ax_t/f)))$$

as the difference-equation version of basic predator–prey model which now includes 7 essential parameters.

C. Extended Models

Beside those essential components of predator–prey interaction that we have considered in the basic models derived above, there also exist a number of subsidiary factors which should be taken into account when considering various concrete problems about interaction in nature (e.g. Hassell, 1978). The effects of these additional factors will be studied here based on several "extended" models which are composed by putting them one by one into the basic models. The factors considered here are: (1) existence of a refuge for prey; (2) mutual interference among predator individuals; (3) accelerating functional response of predator to prey density; (4) constant recruitment of either prey or predators to the system; and (5) existence of alternative prey either with or without predator's habit of "switching".

1. *Existence of a Refuge for Prey*

Existence of a prey refuge in the system corresponds to the situation that some prey individuals are more difficult than others for predators to find, which may occur whenever, as is almost invariably the case with natural population systems, either the environment or the prey population itself is more or less inhomogeneous structurally. Revision of the basic model (8) of differential-equation type to account for this factor is quite easy. It is sufficient just to substitute, following Maynard Smith (1974), $x(1 - p)$ for x in the predation term $g(x, y)$ as Eqn (5) where p is the proportion of the prey being protected from the predator's attack. The same procedure can be used to modify the difference-equation type basic model (13) as to $G(x_t, y_t)$ as Eqn (11). The resultant extended model of differential-equation type is

$$\begin{cases} \dfrac{\mathrm{d}x}{\mathrm{d}t} = \dfrac{bx^2}{s_x + x} - dx - hx^2 - \dfrac{a(1-p)xy}{1 + a(1-p)x/f} \\[2ex] \dfrac{\mathrm{d}y}{\mathrm{d}t} = \dfrac{a(1-p)xy^2}{(1 + a(1-p)x/f)(s_y + y)} - ey \end{cases} \qquad (14)$$

whereas the corresponding model of difference-equation type is obtained by substituting

$$X = x_t(p + (1 - p)\cdot \exp(-ay_t/(1 + a(1 - p)x_t/f))) \qquad (15)$$

for X in the basic model, Eqn (13).

In past studies of the predator–prey system using difference-equation models, several authors have formulated this factor in more elaborate ways (Bailey *et al.*, 1962; Hassell and May, 1974; May, 1978), of which May's approach using the negative binomial distribution may be the neatest. But, for the present purpose of deriving fundamental principles, the much simpler treatment described above may be sufficient.

2. *Mutual Interference Among Predators*

Mutual interference among predator individuals may be another factor of practical importance. It may occur in either of the two processes, searching for prey and consumption of the prey captured. Interference among searching predators has been studied by several workers such as Hassell and Varley (1969), Hassell and May (1973), Hassell and Rogers (1972), and Beddington (1975) with reference to difference-equation models. To put this factor in the basic models, a simple equation due to Beddington (1975) is adopted. Here the modification can be made simply by adding $m_s y$ to the denominator of Eqn. (5) for $g(x, y)$, the resultant model of differential-equation type being

$$\begin{cases} \dfrac{\mathrm{d}x}{\mathrm{d}t} = \dfrac{bx^2}{s_x + x} - dx - hx^2 - \dfrac{axy}{1 + m_s y + ax/f} \\ \dfrac{\mathrm{d}y}{\mathrm{d}t} = \dfrac{caxy^2}{(1 + m_s y + ax/f)(s_y + y)} - ey \end{cases} \tag{16}$$

where m_s is the parameter indicating the intensity of interaction. Similarly, its difference-equation version is obtained by substituting

$$X = x_t \cdot \exp\,(-ay_t/(1 + m_s y_t + ax_t/f)) \tag{17}$$

for X in the basic model (13).

On the other hand, mutual interference in the process of prey consumption has scarcely been considered in past studies, despite its probable importance in nature. The density-dependent decrease of female ratio in insect parasitoids as recently formulated in the context of parasitoid–host interaction by Hassell *et al.* (1983) may be regarded as an example of such interaction. The procedure adopted here to bring this factor into the basic models is to substitute the simple equation,

$$c' = c/(1 + m_c y) \quad \text{or} \quad C' = C/(1 + m_c y_t) \tag{18}$$

for c or C in Eqns (8) or (13), respectively. This modification means the introduction of scramble-type competition among predators the intensity of which is measured by parameter m_c.

It may be worthwhile to refer here to Leslie's (1948) model of predator–prey interaction which expresses the predators' mutual interference in a quite different way. In his model the first (prey) equation has the same predation term as the Lotka–Volterra model, but predator population growth of the second equation has the form of logistic equation with prey density y as the variable carrying capacity, $\mathrm{d}y/\mathrm{d}t = cy - \mathrm{d}y^2/x$. Although this model has sometimes been used in past studies of predator–prey interactions (e.g., Shimazu, 1973; Gilpin, 1974), it is less useful for our model on account of the structural defect that there is no direct relationship between the rate at which a predator feeds and the rate at which it reproduces (Maynard Smith, 1974).

3. *Predator's Accelerating Functional Response to Prey Density*

Coupled with the effect of limited attack capacity f in the basic models, this response usually results in an S-shaped or Holling's (1959a; 1961) type III functional response curve. It has biologically realistic basis and has actually been detected experimentally among various species of predators including both vertebrates and invertebrates (Murdoch and Oaten, 1975; Hassell *et al.*, 1977). The simplest reasonable expression of this response is obtainable after Hassell *et al.* (1977), by replacing searching efficiency a in the original model with $ax/(q + x)$, an increasing function of x where q indicates the degree of decrease of a at low prey densities. Therefore, the modified model of differential type becomes

$$\begin{cases} \dfrac{dx}{dt} = \dfrac{bx^2}{s_x + x} - dx - hx^2 - \dfrac{ax^2y/(q + x)}{1 + ax^2/f/(q + x)} \\ \dfrac{dy}{dt} = \dfrac{cax^2y^2/(q + x)}{(1 + ax^2)/f/(q + x)(s_y + y)} \end{cases} \tag{19}$$

and that of difference-equation type is obtainable by replacing X in the basic model (13) with

$$X = x_t \cdot \exp\,(-ax_ty_t/(q + x_t + ax_t^2/f)) \tag{20}$$

4. *Recruitment of Prey or Predator Individuals to the System*

This extension simply represents the situation in biological pest control that either predators or prey are supplied continuously to maintain the control system successfully. For natural systems it may correspond to the situation that migration of either of them occurs among neighbouring habitat patches where the respective local systems might otherwise behave rather independently of each other. To introduce prey recruitment into the basic model (8), it is sufficient just to add a constant w_x, the rate of recruit, to its prey equation, i.e., to replace it with

$$\frac{dx}{dt} = \frac{bx^2}{s_x + x} - dx - hx^2 - \frac{axy}{1 + ax/f} + w_x \tag{21}$$

and in case of the model (13), simply to substitute

$$x_t' = x_t + w_x \tag{22}$$

for all x_t's in the prey and predator equations.

Models for predator recruitment can similarly be derived readily, either by substituting

$$\frac{dy}{dt} = \frac{caxy^2}{(1 + ax/f)(s_y + y)} - ey + w_y \tag{23}$$

for the predator equation in Eqn (8), or by replacing all the y_t's in the model, Eqn (13) with

$$y_t' = y_t + w_y \tag{24}$$

where w_y is the recruiting rate of predator.

5. *Existence of Alternative Prey*

Since purely monophagous predators are rare in nature, inclusion of this factor also serves to generalize the basic models. Two situations are conceivable here, i.e., the case where "switching" or the shift of predator's preference from rare to dominant prey species invariably occurs and the case where it does not occur. In the case without switching, the modification can be obtained by adding z, the density of alternative prey supplied, to the prey density that is effective for the predator. The resulting extended models of differential and difference-equation type are as follows.

$$\begin{cases} \dfrac{dx}{dt} = \dfrac{bx^2}{s_x + x} - dx - hx^2 - \dfrac{axy}{1 + a(x + z)/f} \\ \dfrac{dy}{dt} = \dfrac{ca(x + z)y^2}{1 + a(x + z)/f)(s_y + y)} - ey \end{cases} \tag{25}$$

and

$$\begin{cases} x_{t+1} = \dfrac{RX^2/(s_x + X)}{1 + hRX^2/(s_x + X)} \\ y_{t+1} = \dfrac{C(x_t - X)y_t}{s_y + y_t} \cdot \dfrac{x_t + z}{x_t} \end{cases} \tag{26}$$

$$(X = x_t \cdot \exp\,(-ay_t/(1 + a(x_t + z)/f)))$$

The models for the case with switching are somewhat more troublesome to formulate. The simplest procedure to introduce this factor may be to multiply the predation term of the model by the factor $x^2/(x^2 + z^2)$ (Murdoch, 1969). The models thus extended from the basic models (8) and (13) are as follows.

$$\begin{cases} \dfrac{dx}{dt} = \dfrac{bx^2}{s_x + x} - dx - hx^2 - \dfrac{a(x + z)y}{1 + a(x + z)/f} \cdot \dfrac{x^2}{x^2 + z^2} \\ \dfrac{dy}{dt} = \dfrac{ca(x + z)y^2}{(1 + a(x + z)/f)(s_y + y)} - ey \end{cases} \tag{27}$$

and

$$\begin{cases} x_{t+1} = \dfrac{RX^2/(s_x + X)}{1 + hRX^2/(s_x + X)} \\ y_{t+1} = \dfrac{Cy_t(x_t - X + z - Z)}{s_y + y_t} \end{cases} \tag{28}$$

$$\begin{pmatrix} X = x_t \cdot \exp\,(-ax_t(x_t + z)y_t/(1 + a(x_t + z)/f)/(x_t^2 + z^2)) \\ Z = z \cdot \exp\,(-az(x_t + z)y_t/(1 + a(x_t + z)/f)/(x_t^2 + z^2)) \end{pmatrix}$$

In Table 1 are listed all the parameters we have involved in the basic and the extended models under study, and are also shown which and how many of them have been incorporated in individual mathematical models of predator–prey interaction so far presented.

Table 1

List of parameters incorporated in the basic and the extended models under study. Their correspondence to the factors included in existing models of predator–prey interaction is also shown regardless of the differences in function form. (DF: differential equation model, DC: difference equation model).

	Basic parameters							Subsidiary parameters						
	b, d					c, e								
Model	R	h	s_x	a	f	C	s_y	p	m_s	m_c	q	w_x	w_y	z
Basic models	*	*	*	*	*	*	*							
Extended models	*	*	*	*	*	*	*	*	*	*	*	*	*	*
								(either one of these is involved)						
Existing models (DF)														
Lotka–Volterra	*			*		*								
Volterra (1926)	*	*		*		*								
Leslie (1948)	*	*		*		*				*				
Rosenzweig (1971)	*	*		*	*	*								
May (1973), Shimazu (1973)	*	*		*	*	*				*				
Maynard Smith (1974)	*			*		*		*						
Gilpin (1974)	*	*	*	*		*	*			*				
Existing models (DC)														
Nicholson-Bailey	*			*		*								
Bailey *et al.* (1962)	*			*		*		*						
Hassell & Varley (1969)	*			*		*			*					
Hassell & May (1974)	*			*		*		*						
Beddington *et al.* (1975)	*	*		*		*								
Beddington *et al.* (1978)	*	*		*		*		*						
May (1978)	*			*		*		*						
Hassell & Comins (1978)	*			*	*	*					*			
Hassell (1980)	*			*	*	*		*	*					
Hassell *et al.* (1983)	*			*		*				*				

b, d: birth and death rates (prey); R: overall reproductive rate (prey); h: coefficient of overcrowding effect (prey); s_x: coefficient of undercrowding effect (prey); a: searching efficiency (predator); f: maximum rate of attack (predator); c, e: reproductive efficiency and death rate (predator); C: overall reproductive efficiency (predator); s_y: coefficient of undercrowding effect (predator); p: proportion of prey protected by refuge; m_s: coefficient of mutual interference in searching process (predator); m_c: coefficient of competition in reproductive process (predator); q: coefficient of accelerating functional response (predator); w_x: rate of prey recruitment; w_y: rate of predator recruitment; z: rate of supply of alternative prey.

III. PATTERNS OF PREDATOR–PREY DYNAMICS IN THEORETICAL POPULATION SYSTEMS

The basic and the extended models we have derived in the preceding section have much more complicated structures than the original classic models, so that it may no longer be possible to elucidate their detailed properties in terms of analytical solutions. In this section, systematic analyses for this purpose will be made by means of more intuitive methods, i.e., Rosenzweig and MacArthur's (1963) method of graphical analysis for differential type models which is based on zero-isoclines for both predator and prey on the y–x plane, and systematic numerical simulations for models of the difference-equation type.

A. Classic Models

Before detailed analysis of the basic and the extended models, it may be worthwhile to give a brief description of the properties of original Lotka–Volterra and Nicholson–Bailey models. They have now been well clarified and may be summarized as follows (e.g., Pielou, 1977). (1) The interaction results in reciprocal oscillations of both predator and prey with some delay in phase by the former, its action being substantially destabilizing. (2) The regular oscillations are maintained forever in Lotka–Volterra model, but this stability is only superficial or "neutral" (May, 1972) without any tendency for "resilience" in Holling's (1973) sense to disturbance from the outside, while in the Nicholson–Bailey model the oscillations are truely unstable with ever-increasing amplitudes. (3) Equilibrium densities for prey and predator in Lotka–Volterra model are e/ac and $(b - d)/a$, while those in Nicholson–Bailey model are $R \log R/((R - 1)aC)$ and $\log R/a$, respectively. This indicates that equilibrium density of either prey or predator largely depends upon its partner's parameters, i.e., a and c in case of the prey or $b - d$ or R in case of the predator, and also that the predator population here is to be confronted with a serious paradox, namely that any improvement of efficiency of its prey searching or consumption results inevitably in a decrease of its equilibrium level. In view of biologically unrealistic assumptions involved in these models, however, it seems indispensable that validity or generality of these conclusions be re-examined carefully in the light of biologically more reasonable nodels.

B. Basic Models

Including as many as 9 (model (8)) or 7 (model (13)) parameters, the basic predator–prey models under study can exhibit various patterns of dynamics depending upon the values of these parameters as well as upon the initial density conditions for interaction. In fact, all the four possible outcomes of the interaction, i.e., (1) extinction of both species; (2) extinction of the predator followed by escape of the prey; (3) coexistence of both species in reciprocal oscillation with stable "limit cycles" (May, 1973); and (4) coexistence of both species in damped reciprocal oscillation converging to respective equilibrium levels, can be readily generated in either the differential- (Eqn (8)) and difference (Eqn (13)) type basic models (Figs. 1 and 2). The problem here is then to analyse the significance and the modes of influence of individual parameters in determining dynamical characteristics of the predator–prey interaction in these models.

Figure 3 shows a typical example of the zero-isoclines for predator and prey on the y–x plane for the differential-type basic model (8), which are given by

$$\begin{cases} \dfrac{\mathrm{d}x}{\mathrm{d}t} = 0: \; y = \dfrac{1}{a}\left(1 + \dfrac{ax}{f}\right)\left(\dfrac{bx}{s_x + x} - hx - d\right) \\ \dfrac{\mathrm{d}y}{\mathrm{d}t} = 0: \; x = \dfrac{e(s_y + y)}{cay - ae(s_y + y)/f} \end{cases} \tag{29}$$

In contrast to linear, orthogonal isoclines in the original Lotka–Volterra model (broken lines in the figure), both predator and prey isoclines here are curvilinear. The prey isocline is a humped curve crossing the abscissa at two points, while the predator one is an inverse J-shaped curve approaching the abscissa asymptotically with increasing x. The coordinates of the intersection point of the two isoclines show the respective equilibrium densities of predator and prey. The slopes of both isoclines at the intersection point, on the other hand, have been proved to indicate local stability of the interactive system about the equilibrium; i.e., either the prey isocline rotated clockwise from the horizontal or the predator one rotated similarly from the perpendicular indicates stabilizing or damped oscillation in both populations, whereas either of the isoclines rotated anticlockwise, indicates destabilizing or diverging oscillation (Rosenzweig and MacArthur, 1963; Maynard Smith, 1974).

In the case of differential equation models, we can therefore analyse the effect of a given parameter on the dynamics of interaction easily by graphical examination of the changes in both the isoclines. It is worth noting, however, that when the isoclines are non-linear as is the case here, local stability indicated by them does not necessarily mean global stability of the system. Namely, the outcome of the interaction may be limit-cycle oscillations or even extinction of either the predator or both predator and prey depending upon the initial conditions for the interaction, even though the isoclines satisfy the conditions for local stability. Unfortunately, there exists no such convenient method of analysis for complex predator–prey models of difference-equation type as Eqn (13), so that computer simulation of the interaction seems to be the only possible way to analyse them.

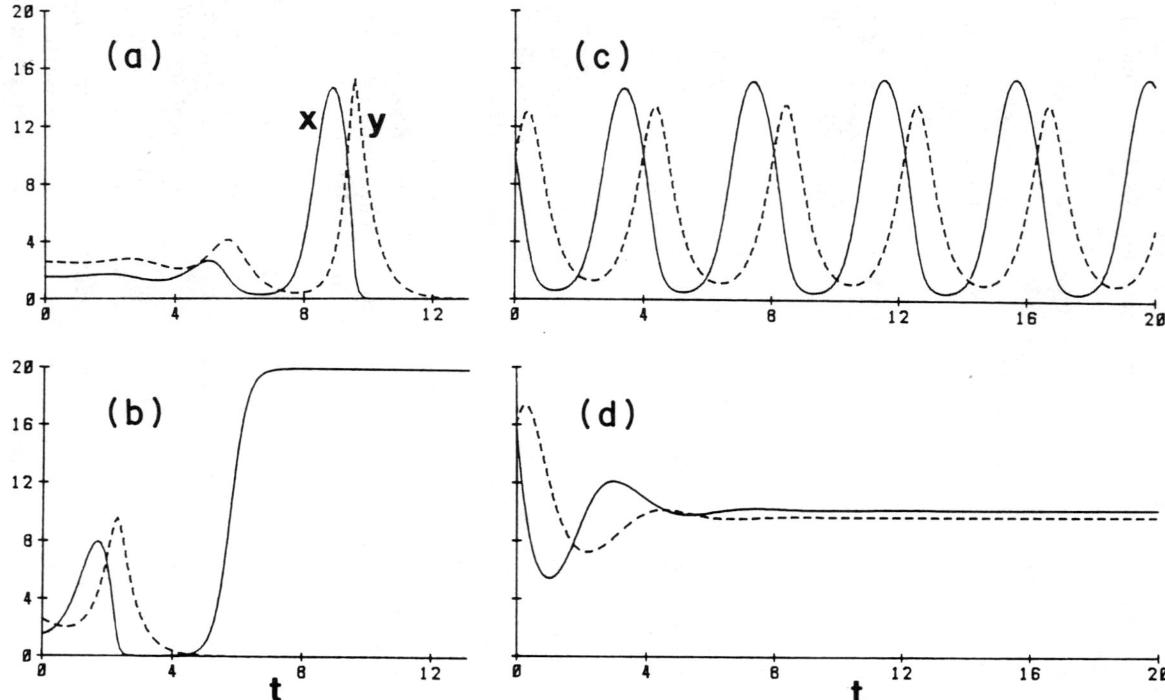

Fig. 1. Four patterns of predator (y)–prey (x) dynamics generated from the basic model of differential-equation type (Eqn (8)). (a) $b = 5$, $d = 1$, $h = 0{\cdot}2$, $a = 2$, $f = 6$, $c = 1$, $s_x = s_y = 0{\cdot}1$, $e = 2$. (b) Same as (a) except that $s_x = 0{\cdot}001$, $s_y = 1$. (c) Same as (a) except that $a = 0{\cdot}714$. (d) Same as (a) except that $a = 0{\cdot}3$.

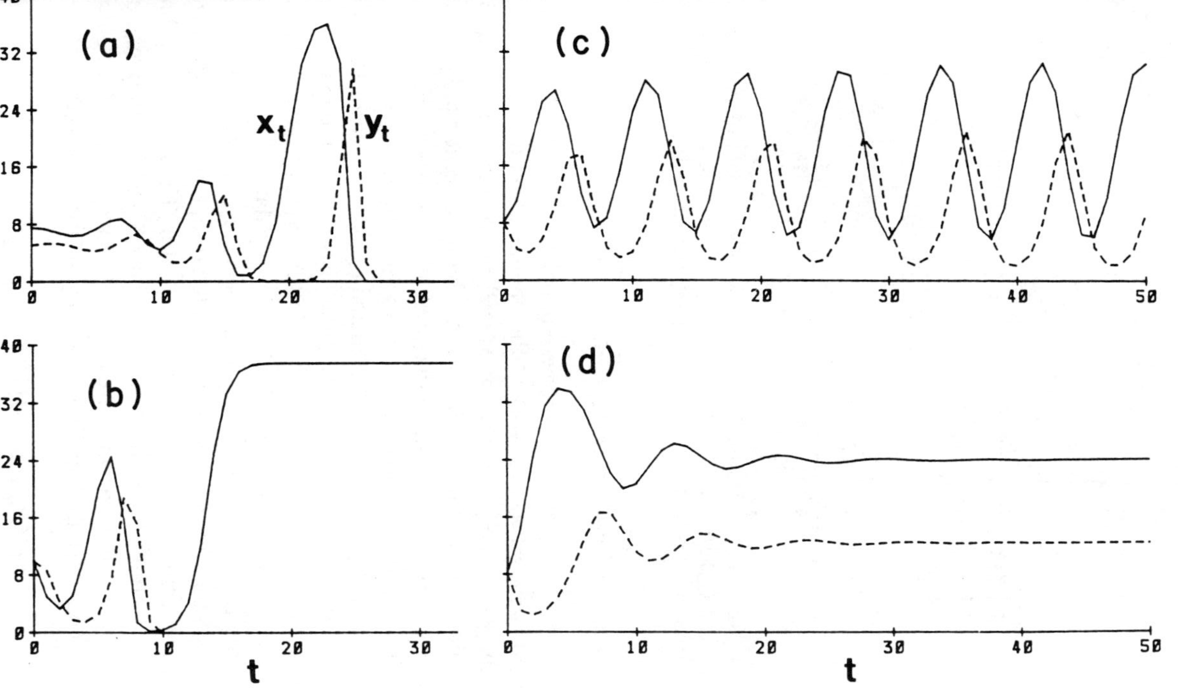

Fig. 2. Four patterns of predator (y_t)–prey (x_t) dynamics generated from the basic model of difference-equation type (Eqn (13)). (a) $R = 4$, $h = 0{\cdot}02$, $a = 0{\cdot}25$, $f = 100$, $C = 1$, $s_x = s_y = 0{\cdot}05$. (b) Same as (a) except that $a = 0{\cdot}2$. (c) Same as (a) except that $a = 0{\cdot}1$. (d) Same as (a) except that $a = 0{\cdot}06$.

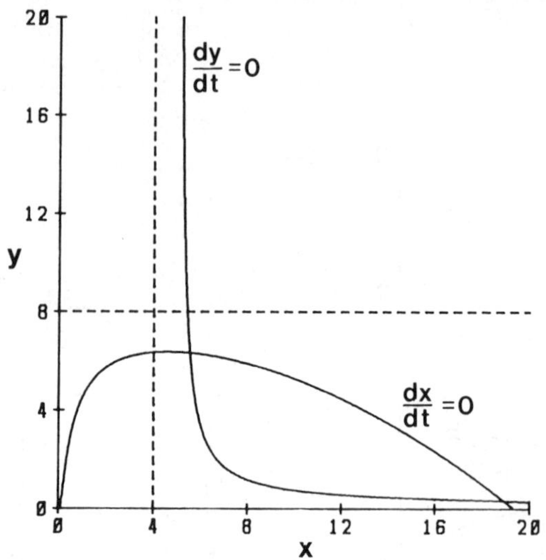

Fig. 3. Zero-isoclines for predator and prey in the basic model of differential-equation type (Eqn (8)) (b = 5, d = 1, h = 0·2, $s_x = s_y$ = 0·5, a = 0·5, f = 10, c = 1, d = 2). Broken lines are those for the Lotka–Volterra model.

1. *Effect of Prey's Reproductive Rate*

The effect of prey's birth rate b on the system behaviour in the differential type model (8) is shown in Fig. 4(a) as the change of predator and prey isoclines on the y–x plane, while that of overall reproductive rate R in the difference type model (13), in Fig. 5(a) as the change of generated population pattern. In the former it is evident that increase of b does not change the predator isocline but steadily shifts the prey isocline upwards, which results in a steady increase of predator equilibrium and remarkable destabilization of the system (as indicated by progressive anticlockwise rotation of the prey isocline at the intersection point). In the latter, however, the result is somewhat different: first, not only the predator population level, but also the prey level are raised with increasing R; and secondly, increase of R does not so much destabilize the system as to lead it to crash, though it appears to have some destabilizing effect for a certain range of R. Thus, it is likely that in the presence of a time-delay effect in the predator's reproduction, higher prey reproductive rate is generally more profitable not only for the predator but also for the prey itself in maintaining their populations steady.

2. *Effect of Carrying Capacity of the Prey*

The effect of the habitat's carrying capacity of the prey on the system dynamics is shown in Figs. 4(b) and 5(b), where parameter h, an inversely proportional index of habitat capacity, is changed progressively. Figure 4(b) shows that gradual decrease of h (i.e., gradual increase of habitat capacity) has a consequence similar to the increase of b just discussed (Fig. 4(a)), resulting in the gradual increase of predator equilibrium coupled with progressive destabilization of the system. The consequence of decreasing h is essentially similar in Fig. 5(b) for the difference equation model (13), too. Thus, too low h, meaning a too large habitat capacity of the prey population, may lead the system, which could otherwise be maintained in a stable interaction, to crash after diverging oscillations (see case E). This is evidently a more strict illustration of the well known "paradox of enrichment" presented by Rosenzweig (1971). For, in his treatment where "K" in the popular expression of the logistic equation was used with r as the index of habitat capacity, not only the effect of habitat capacity but also that of reproductive rate (as in Fig. 4(a)) have been incorporated together in the outcome of the interaction.

3. *Effect of Predator's Searching Efficiency*

Predator's efficiency of prey search, a, is a compound parameter determined by properties of both the species. As seen in Figs. 4(c) and 5(c), its influence on the system dynamics is clear and striking. Increase of a in the differential model (8) shifts the predator isocline to the left and the prey one downward, thereby decreasing remarkably the equilibrium levels of both prey and predator and destabilizing the system progressively (Fig. 4(c)). The situation is essentially the same in the difference equation model, too (Fig. 5(c)). Too low values of a, of course, result in rapid extinction of the predator followed by unimpeded growth of the prey population (case A in both the graphs). The system destabilization due to increasing searching efficiency has already been discussed by Rosenzweig and MacArthur (1963) in reference to differential-equation models, but only the shift of the predator isocline has been considered there. In the present study it has become evident that the prey isocline also is greatly affected by searching efficiency a, in such a way that its overimprovement by the predator may paradoxically drive its own population as well as the prey's to quite unprofitable situations, i.e., oscillations with low and unstable equilibria.

4. *Effect of Predator's Maximum Attack Rate*

The effect of predator's maximum attack rate (f) on the interaction is shown in Figs. 4(d) and 5(d). As seen in Fig. 4(d), its mode of action is somewhat complicated. Gradual increase of f shifts the predator isocline to the left and the prey one downward as in the case of a. But its effect differs from the a's in that the hump-top in the prey isocline moves gradually to the left with increasing f. As a result, the stability of the interactive system decreases initially but increases again thereafter, indicating that changes in parameter

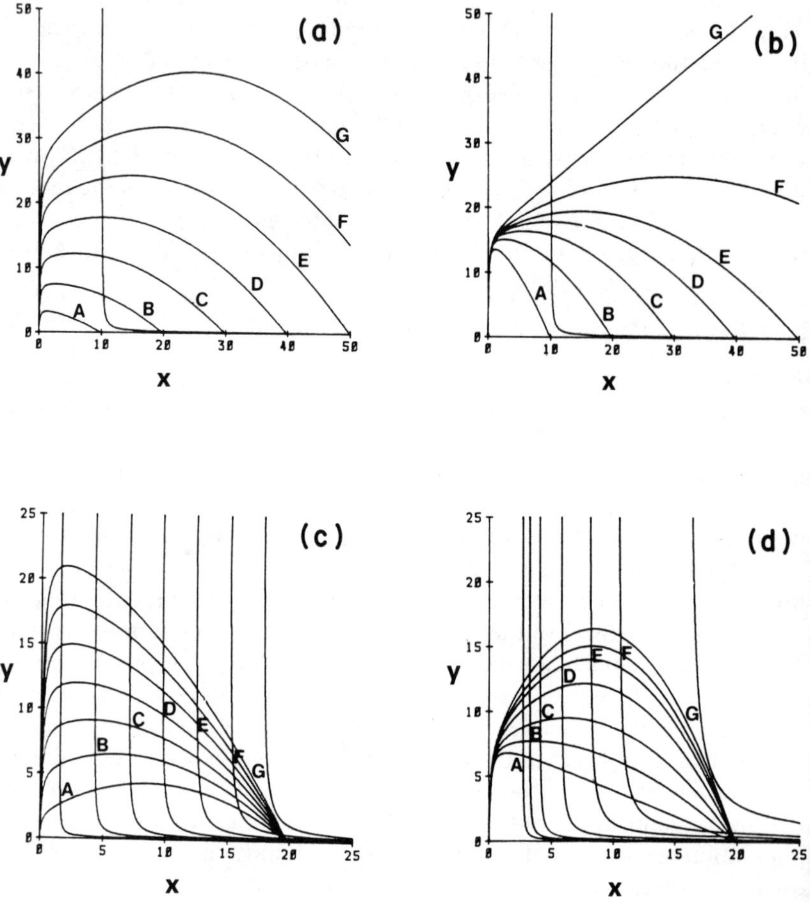

Fig. 4. Effects of individual parameters on the zero isoclines for predator and prey in the basic model of differential-equation type (Eqn (8)). (a) Effect of b (b = 3/5/7/9/11/13/15 (A to G), $d = 1$, $h = 0{\cdot}2$, $s_x = s_y = 0{\cdot}1$, $a = 0{\cdot}5$, $f = 10$, $c = 0{\cdot}6$, $e = 2$). (b) Effect of h (h = 0·8/0·4/0·267/0·2/0·16/0·1/0 (A to G), $b = 9$, the other parameters same as (a)). (c) Effect of a (a = 2/0·714/0·435/0·313/0·244/0·182/0·156 (A to G), $b = 5$, $f = 6$, $c = 1$, the other parameters same as (a)). (d) Effect of f (f = 1000/8/4/2·5/2/1·8/1.6 (A to G), $b = 5$, $c = 1{\cdot}5$, the other parameters same as (a)).

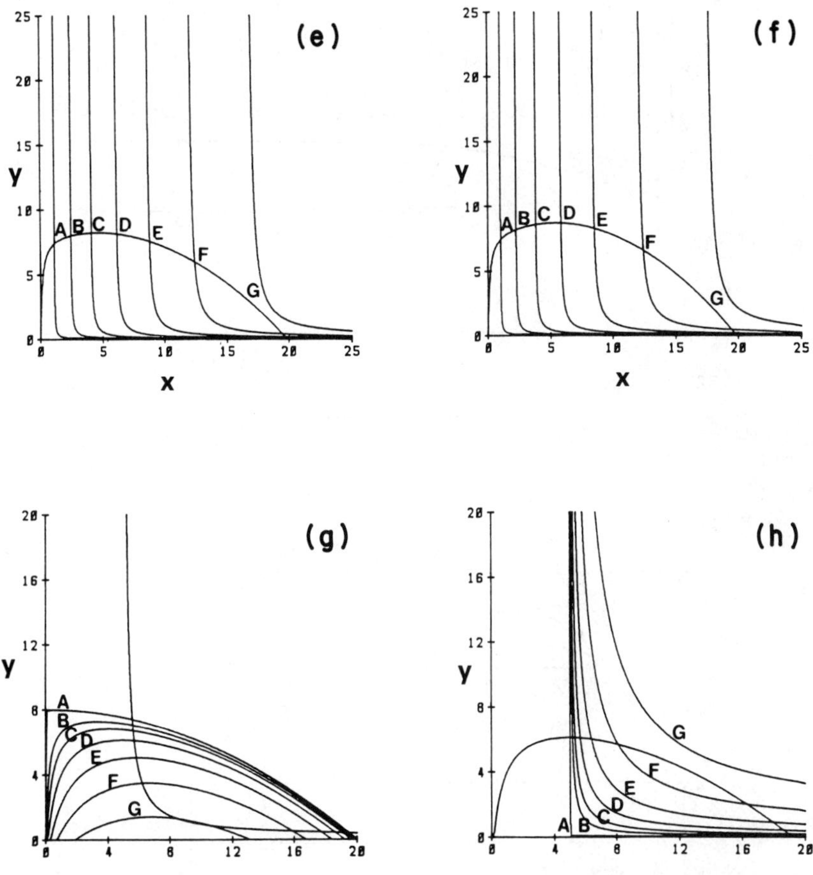

Fig. 4 (*continued*). (e) Effect of c (c = 4/2/1·33/1/0·8/0·667/0·571 (A to G), $b = 5$, $f = 6$, the other parameters same as (a)). (f) Effect of e (e = 1/2/3/4/5/6/7 (A to G), $b = 5, f = 5, c = 2.2$, the other parameters same as (a)). (g) Effect of s_x (s_x = 0·001/0·156/0·313/0·625/1·25/2·5/5 (A to G), $b = 5$, $c = 1$, $s_y = 1{\cdot}6$, the other parameters same as (a)). (h) Effect of s_y (s_y = 0·001/0·156/0·313/0·625/1·25/2·5/5 (A to G), $s_x = 1{\cdot}6$, the other parameters same as (g)).

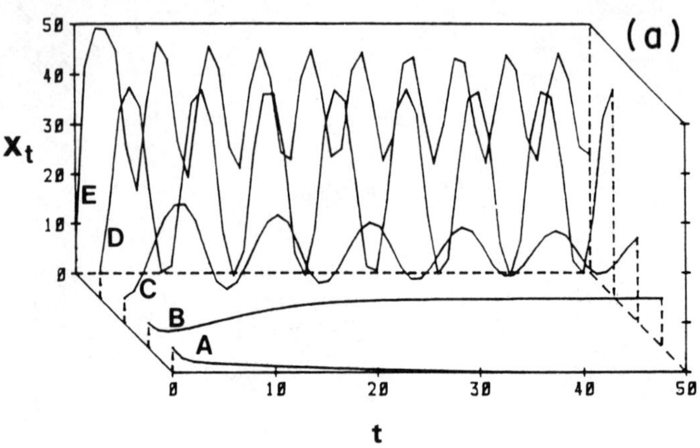
(a)
X_t
50
40
30
20
10
0
E
D
C
B
A
0
10
20
30
40
50
t

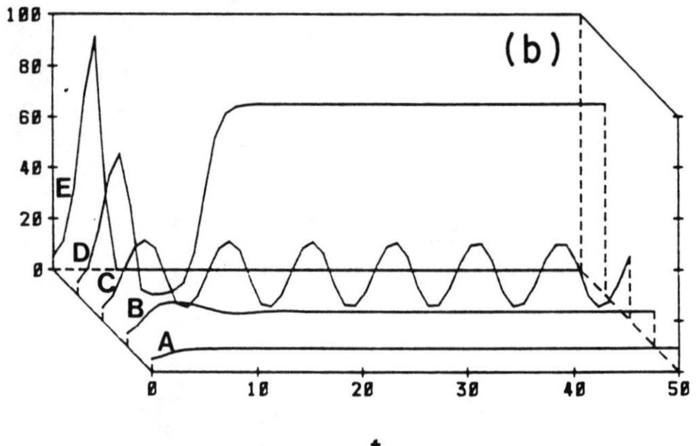
(b)
100
80
60
40
20
0
E
D
C
B
A
0
10
20
30
40
50
t

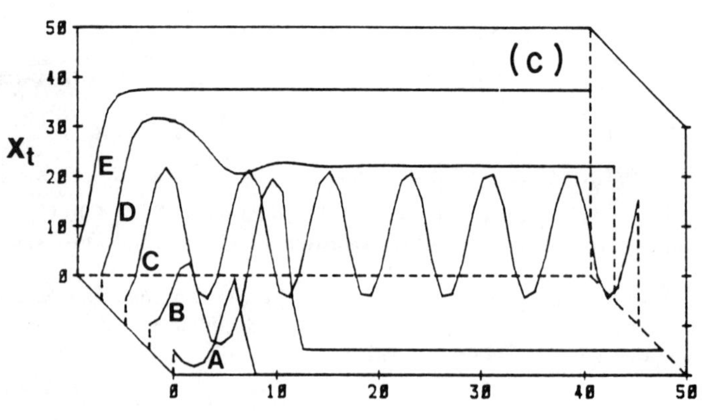
(c)
X_t
50
40
30
20
10
0
E
D
C
B
A
0
10
20
30
40
50
t

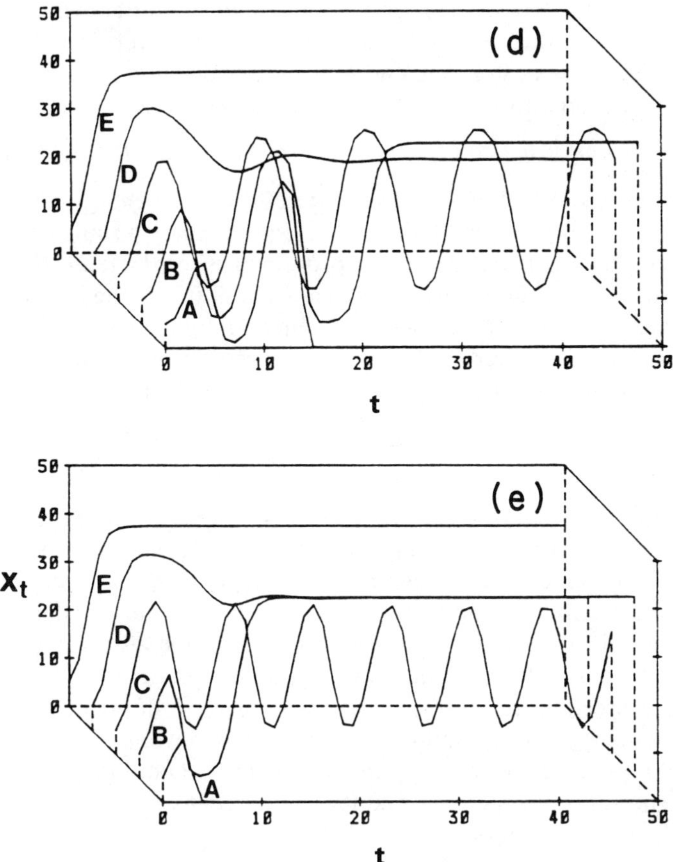

Fig. 5. Effects of individual parameters on predator–prey dynamics in the basic model of difference-equation type (Eqn (13)). (a) Effect of R (R = 1/1·25/2·5/10/80 (A to E), h = 0·02, a = 0·1, f = 100, C = 1, $s_x = s_y$ = 0·05). (b) Effect of h (h = 0·08/0·04/0·02/0·01/0·005 (A to E), R = 4, the other parameters same as (a)). (c) Effect of a (a = 0·4/0·2/0·1/0·05/0·025 (A to E), R = 4, the other parameters same as (a)). (d) Effect of f (f = 16/8/4/2/1 (A to E), R = 4, the other parameters same as (a)). (e) Effect of C (C = 4/2/1/0·5/0·25 (A to E), R = 4, the other parameters same as (a)).

f may have either a stabilizing or a destabilizing effect on the system, depending on the situation concerned. In case of the difference-equation model (13) where inherent stability is much lower, however, it is likely that the stabilizing phase as shown in Fig. 4(d) is concealed usually, increase of f resulting predominantly in system destabilization (see Fig. 5(d)).

5. *Effect of Predator's Reproductive Efficiency and Death Rate*

As shown in Figs. 4(e) and 4(f), predator's efficiency of energy conversion for reproduction (c) and death rate (e) in Eqn (8) share a similar mode of action such that their changes for improvement (i.e., increase for c and decrease for e) do not affect the prey isocline but shift only the predator one progressively to the left, i.e., towards system destabilization. The same can be said also for the difference equation model (13) in which both these parameters are unified to a single parameter C (Fig. 5(e)).

6. *Effect of Vulnerability to Undercrowding in Either Predator or Prey*

Figures 4(g) and 4(h) shows the effects of parameters s_x and s_y, representing the intensity of the undercrowding effect in prey and predator populations, respectively, in the differential model (8). Increase of s_x for prey shifts the prey isocline downwards and its hump to the right, resulting in a decrease of predator equilibrium together with destabilization of the system. Increase of s_y for the predator, on the other hand, shifts the predator isocline in an upper-right direction, resulting in progressive increase of prey equilibrium in addition to system destabilization. Thus, it is obvious that increased vulnerability to undercrowding in the interactive system in either prey or predator species is always unprofitable to the predator, but that this is not necessarily the case with the prey since the destabilization here may readily bring about prey escape preceded by predator extinction. The same conclusion applies to the difference-equation model (13) as well.

The above results on the influences of individual parameters on the outcome of predator–prey interaction are summarized in Table 2. As a whole they impress on us strongly that predator–prey interactions generate a system behaviour which is full of paradox especially for the predator, implying that the struggle for population propagation in each species, such as increase of search efficiency, attack capacity, or reproductive efficiency in the predator, and increase of carrying capacity for the prey or "enrichment of the environment" are apt to be disadvantageous for that very population. It is therefore likely that in interacting predator–prey population systems there inevitably occurs a serious conflict of interests between "population" and "individual" of the same species, especially on the predator's side. The nature of this conflict seems to be of essential importance in determining the characteristics of predator–prey interactions in nature. The problem will be discussed later in more detail in relation to the coevolution of both species.

Table 2

Effects of individual parameters on patterns of predator–prey dynamics in the basic and the extended models under study.

Effect on		Basic parameters							Subsidiary parameters							
		$b-d$					$c, -e$								z^a	
		R	h	s_x	a	f	C	s_y	p	m_s	m_c	q	w_x	w_y	(1)	(2)
Predator isocline or prey equilibrium	DF	0	0	0	−	−	−	+	+	+	+	+	0	−	−	−
	DC	0	0∼−	×	−	−	−	+	+	+	+	+	0∼−	−	−	0
Prey isocline or predator equilibrium	DF	+	−	−	−	−	0	0	+	+	0	+	+	0	+	+
	DC	+	0∼−	−	−	−	0	0	+∼−	+	0	0∼+	+	−	+	+
System stability	DF	−	+	−	−	+∼−	−	−	+	+	+	+	+	+	0	+
	DC	+∼−	+	−	−	−	−	−	+	+	+	+	+	+	+	+

DF: differential equation models; DC: difference equation models.
+: increased with increase of the parameter concerned (shifted upward or to the right in case of the predator or prey isocline).
−: decreased with increase of the parameter concerned.
0: no or little effect observed.
[a] (1): without switching; (2): with switching.

C. Extended Models

As described before, the "extended" models here are those in which either one of the following six factors are added to the basic models (8) or (13) analysed above: prey's refuge, predator's mutual interference, predator's accelerating functional response, constant recruitment of either prey or predator, and existence of alternative prey. In the following analysis we may elucidate how these subsidiary, but biologically important factors affect and change the patterns of predator–prey dynamics observed in the basic models.

1. *Effect of a Refuge for the Prey*

The effect of this factor on the system behaviour in the corresponding differential-equation model (Eqn (14)) is shown in Fig. 6(a). As is evident from the equation, existence of the refuge which protects a certain proportion p of existing prey individuals corresponds exactly to the decrease of search efficiency a for the predator by a factor $(1 - p)$, thus resulting in increases of both prey and predator equilibria associated with stabilization of the interactive system. It is interesting to note here that this stabilizing effect of a refuge never appears when, as Maynard Smith (1974) showed in a modified Lotka–Volterra model, prey population growth is not limited by the habitat's carrying capacity (i.e., $h = 0$), since the prey isocline in this case is not a humped but a monotonically increasing curve. Nevertheless it may be reasonable to conclude that the refuge generally has a stabilizing effect on the predator–prey interaction, since there actually exist no prey species whose populations can be entirely free from the restriction due to carrying capacity of the habitat. The result of analysis on the difference-equation model (Eqns (13) and (15); Fig. 7(a)) is also essentially similar in showing that the refuge brings about system stabilization, but is different in that the stabilizing effect here is much more intense than that in the above differential-type model, permitting under some conditions, the prey population to stabilize at far lower levels than the carrying capacity (compare the figure with Fig. 6(a)). Such a strong tendency for system stabilization produced by the prey's refuge or spatial heterogeneity of the habitat may be regarded as a universal and robust principle underlying predator–prey dynamics, since it has been demonstrated consistently among a variety of other models of difference-equation type (e.g., Bailey *et al.*, 1962; Hassell and May, 1974; May, 1978; Beddington *et al.*, 1978).

2. *Effect of Predator's Mutual Interference*

The effect of interference in the process of prey searching is shown in Figs. 6(b) and 7(b) for the differential (Eqn (16)) and the difference (Eqns (13) and (17)) models, respectively. It is obvious, as has been shown with some different models of difference-equation type by Hassell and Varley (1969),

Rogers and Hassell (1974) and Beddington (1975), that this factor also has a remarkable effect of stabilizing the system. It is also noticeable that in either model increase of interference coefficient m_s changes the system dynamics in such a way that not only its overall stability but also the resultant equilibria of both predator and prey are increased progressively.

In the case that the interference occurs in the process of reproduction or prey consumption (Eqns (8) and (18) or Eqns (13) and (18)), it only affects the predator isocline. Accordingly, although the effect of increasing m_c is also stabilizing, it is generally weaker than that of m_s just described, and it does not increase the predator equilibrium here (Figs. 6(c) and 7(c)). Incidentally it is worth noting that this type of interference corresponds to the scramble-type competition which has been proved to have a marked destabilizing effect in a single species population (e.g., Kuno, 1983), in contrast to such a stabilizing effect as that observed here in predator–prey systems. This discrepancy may be regarded as further evidence illustrating the paradoxical situation to which a predator population is subject in the interactive system with its prey which itself is a reproducing organism.

3. *Effect of Accelerating Functional Response*

In the differential-equation model (Eqn (19)), introduction of this factor not only draws the prey isocline at the top-left end, but also shifts the predator one to the right, resulting in remarkable stabilization of the interaction together with the increase of both predator and prey equilibria (Fig. 6(d)). A similar tendency for system stabilization with increasing parameter q is also observed in the difference-equation model (Eqns (13) and (20)) though its intensity seems to be rather weakened here due to the effect of time delay inherent to the model (Fig. 7(d)). The result obtained here is consistent essentially with those that have been reported before using different models involving this factor (e.g., Murdoch and Oaten, 1975; Hassell and Comins, 1978; Beddington *et al.*, 1978).

4. *Effect of Recruitment of Prey or Predator*

The effect of constant recruitment of prey to the system is shown in Figs. 6(e) and 7(e) for the differential (Eqns (8) and (21)) and the difference (Eqns (13) and (22)) models, respectively. In the former the prey isocline only is changed from a humped curve to a hyperbola-type one, resulting in remarkable system stabilization coupled with the increase of predator equilibrium. The result is similar in the latter, difference-equation model, too. It is noticeable that a continuous supply of prey individuals here does not cause any increase in prey equilibrium, even decreasing it in case of the difference-equation model, and hence it may be regarded as an effective means of keeping the prey population level stable at low levels. This paradoxical principle may be of practical importance in reference to biological pest control programmes.

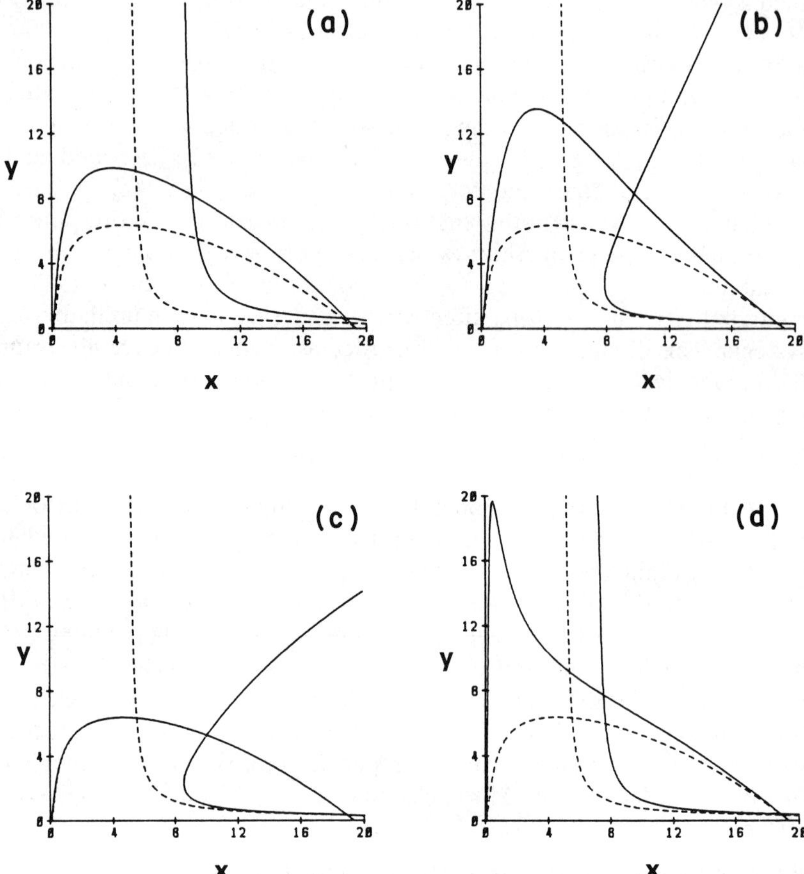

Fig. 6. Effects of additional parameters on the zero-isoclines for predator and prey in the extended models of differential equation type (basic parameters: $b = 5$, $d = 1$, $h = 0{\cdot}2$, $a = 0{\cdot}5$, $f = 10$, $c = 1$, $s_x = s_y = 0{\cdot}5$, $e = 2$). Broken lines are those for the basic model (8) with these parameter values. (a) Effect of p ($p = 0{\cdot}4$). (b) Effect of m_s ($m_s = 0{\cdot}1$). (c) Effect of m_c ($m_c = 0{\cdot}1$). (d) Effect of f ($f = 3$).

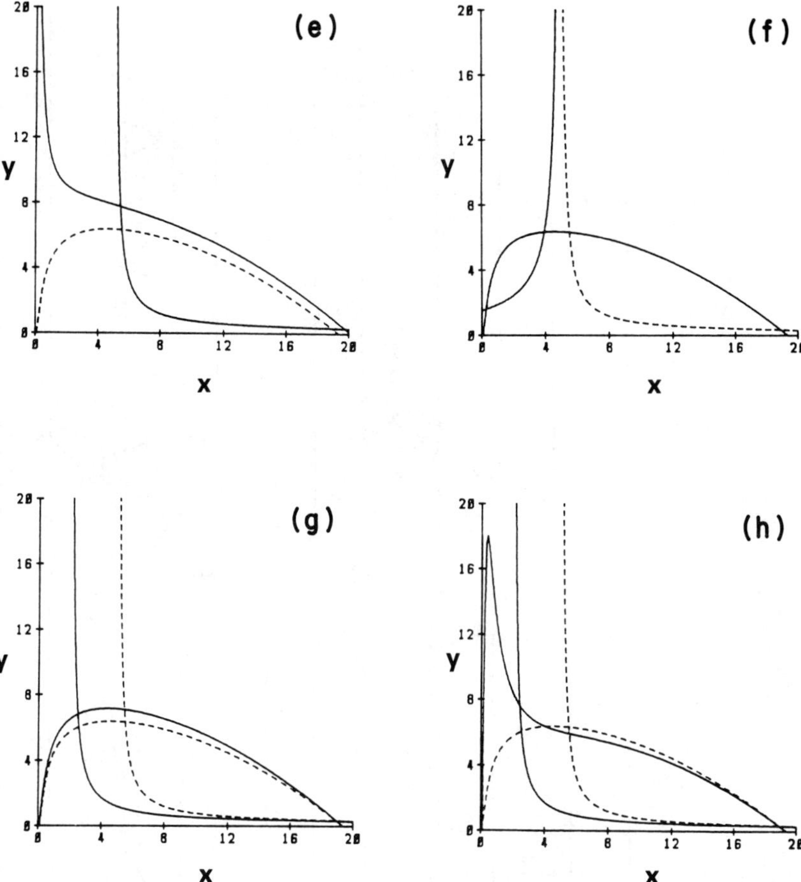

ig. 6 (*continued*). (e) Effect of w_x ($w_x = 3$). (f) Effect of w_y ($w_y = 3$). (g) Effect of ($z = 3$; without switching). (h) Effect of z ($z = 3$; with switching).

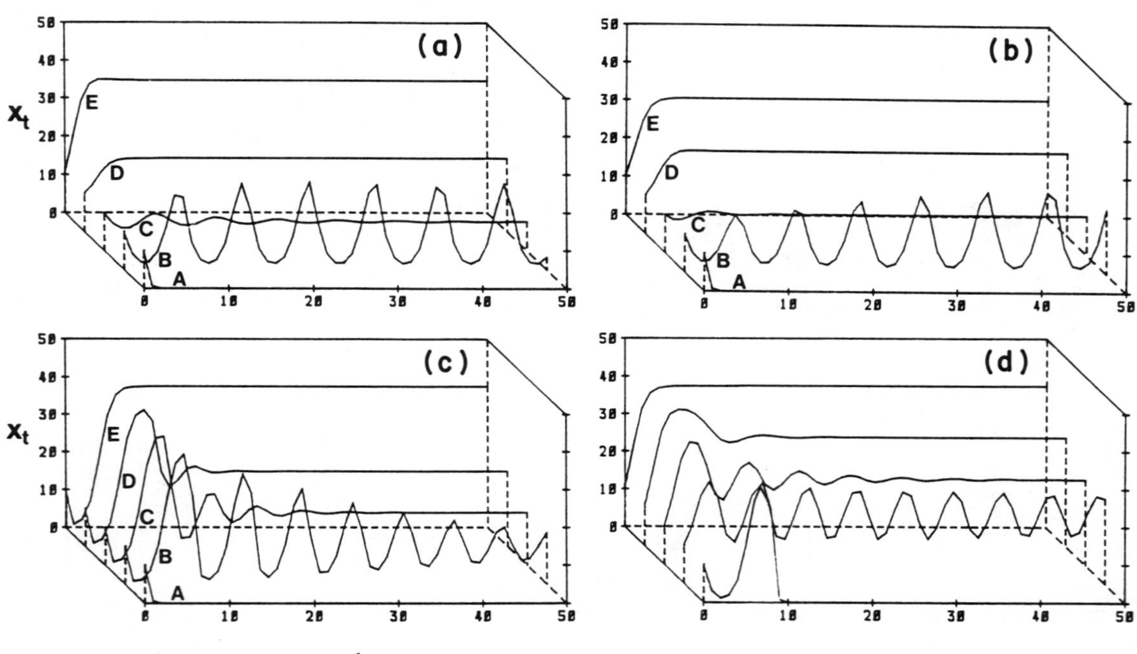
(a)
(b)
(c)
(d)
X_t
t
A
B
C
D
E
0
10
20
30
40
50

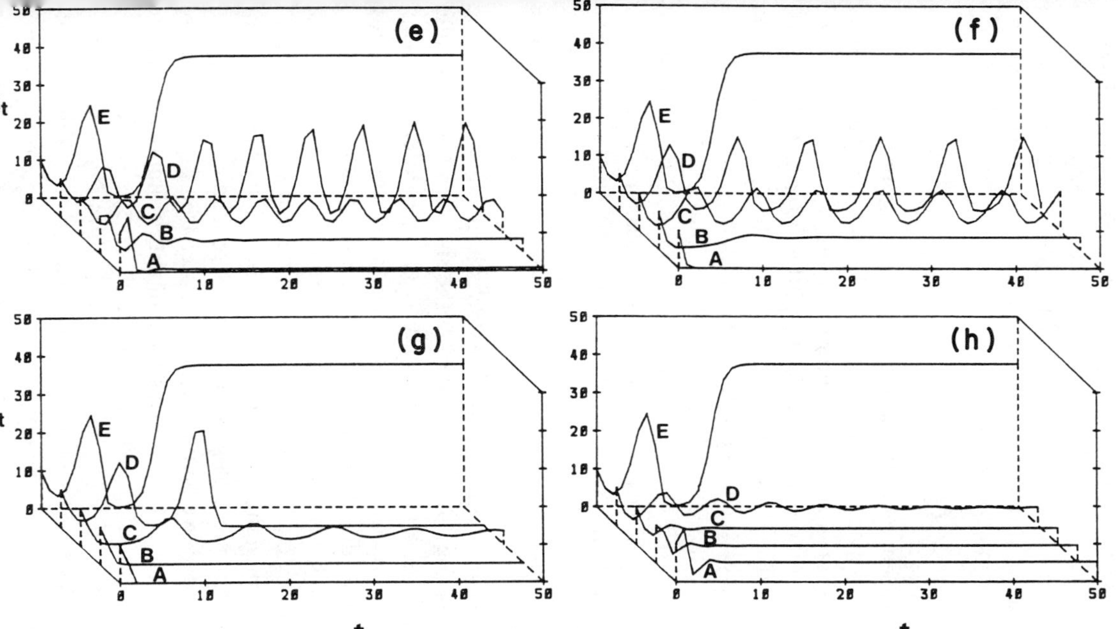

Fig. 7. Effects of additional parameters on predator–prey dynamics in the extended models of difference-equation type (basic parameters: $R = 4$, $h = 0{\cdot}02$, $a = 0{\cdot}4$, $f = 100$, $C = 1$, $s_x = s_y = 0{\cdot}05$). (a) Effect of p ($p = 0/0{\cdot}05/0{\cdot}1/0{\cdot}2/0{\cdot}4/0{\cdot}8$ (A to E)). (b) Effect of m_s ($m_s = 0/0{\cdot}05/0{\cdot}1/0{\cdot}2/0{\cdot}4/0{\cdot}8$ (A to E)). (c) Effect of m_c ($m_c = 0/0{\cdot}4/0{\cdot}8/1{\cdot}6/3{\cdot}2$ (A to E)). (d) Effect of q ($q = 0/32/64/128/256$ (A to E)). (e) Effect of w_x ($w_x = 0/3/6/12/24$ (A to E)). (f) Effect of w_y ($w_y = 0/1/2/4/8$ (A to E)). (g) Effect of z ($z = 0/3/6/12/24$ (A to E); without switching). (h) Effect of z ($z = 0/4/8/16/32$ (A to E); with switching).

Constant recruitment of predators, on the other hand, affects only the predator isocline in the differential-equation model (Eqns (8) and (23); Fig. 6(f)), shifting it to the left and upsetting its slope at the equilibrium from rightwards to leftwards, which results in system stabilization accompanying the decrease of prey equilibrium. A similar result can also be seen in Fig. 7(f) for the difference-equation model (Eqns (13) and (24)). In either case, the recruitment of predators at too high a rate results of course in a crash of the system due to the extinction of the prey.

5. *Effect of Alternative Prey*

The effect of the existence of alternative prey when the predator has no tendency towards switching (i.e., when its attack always occurs at equal probability to both the original and the alternative prey) are shown in Figs. 6(g) and 7(g) for the differential (Eqn (25)) and the difference (Eqn (26)) models, respectively. In the differential model the effect of this factor on system stability is somewhat complicated. The inclusion of this factor moves the predator isocline to the left and the prey one upwards, so that its effect on system behaviour may be either stabilizing or destabilizing depending on the conditions concerned, being not so conspicuous in any case. The situation is similar in the difference model. Although the range of density of alternative prey (parameter z) which can keep the system stable obviously exists here, it is rather narrow, being only from about 5 to 7 in the example of Fig. 7(g).

Once, however, the predator is assumed to have the habit of time-to-time switching from a rare prey to a common one, the situation becomes quite different. Namely, as shown in Figs. 6(h) and 7(h) for the differential (Eqn (27)) and the difference (Eqn (28)) models, respectively, and as has already been clarified in some different models with similar assumptions (e.g., Murdoch and Oaten, 1975), the existence of alternative prey now has a pronounced stabilizing effect on the interactive system, accompanied with the marked suppression of the prey equilibrium. This result again may be of special interest in relation to biological pest control problems.

The results obtained above on these extended predator–prey models are also summarized in Table 2. Overviewing them, it may be important to note that all these modifications contributed generally to the stabilization of predator–prey dynamics, though the degree of stabilization varies widely depending on the factor concerned. Although the results on the differential-equation type models are essentially consistent with those on the corresponding difference-equation type models, some differences have been detected. Particularly it should be noted that the stabilizing effect of the prey's refuge is much more and that of predator's accelerating functional response much less pronounced in the difference-equation models than in the differential-equation ones. Generally speaking, the difference-equation models may be regarded as reflecting natural situations more faithfully. To conclude, it is evident that inclusion of these factors that are regarded as more or less inherent in natural predator–prey systems contributes remarkable stability to the system dynamics which might otherwise be inherently unstable, thereby providing the conditions that makes stable coexistence of both prey and predator much easier.

IV. REGULATION OF PREY POPULATION BY PREDATION: ITS POSSIBILITY IN THEORETICAL POPULATION SYSTEMS

Whether or not predators in nature can regulate populations of their prey has long been a problem under active dispute, forming a part of the well-known controversy about the significance of regulation in animal populations between "biotic" and "climatic" schools. Since the development of classic predator–prey models by Lotka (1925), Volterra (1926) and Nicholson and Bailey (1935), some people such as Nicholson (1954) and Varley (1947) have claimed the possibility of population regulation by predators on the basis of these theoretical models, whereas some others such as Thompson (1939) and Andrewartha and Birch (1954) have denied such a possibility criticizing the simplicity and biological unreality of these models in comparison with the complexity and irregularity underlying natural predator–prey systems. The debate has not yet ended. For example, while Dempster (1983) argues on the one hand that among many studies so far accumulated there is virtually no evidence proving that the regulation by predation occurs in natural insect populations, Hassell (1985), on the other hand, questions this argument, pointing out the difficulties in analytical methods for evaluating the regulation and claiming the necessity for more careful examination of field data before deriving such a conclusion.

It is noticeable, however, that in these discussions the points in dispute have often remained unclarified or confused, especially in the interpretation of field population dynamics. For example, in past controversies the regulation due to predators and the regulation due to intra-specific mechanisms have often been discussed together, but these two should of course be strictly separated when discussing their possibilities. To make these points clear, in this study first the meaning of "regulation by predation" is rigorously defined, and then the possibility and the necessary conditions for regulation by predation so defined to occur in the theoretical predator–prey systems which we have studied in the preceding sections are elucidated. In this connection a close examination will also be made of the statistical methods needed so as to detect the regulation due to predators from population data, to make clear whether or not the conventional techniques of key-factor analysis can be used as effective means of population analysis for the present purpose.

A. Prey Population Regulation by Predation – its Definition and Detection

Among ecologists the term "regulation" is often used loosely in place of the term "control". For example, it has been sometimes concluded that the pest population under study is subject to regulation by its predator when it is suppressed at lower density levels after some predator species was newly introduced to its life system, or, conversely, when its levels are raised considerably by destruction of predator populations due to, say, spraying of some insecticide. Obviously, however, these phenomena do not necessarily prove occurrence of the "regulation" in its strict sense which has been defined as the stabilization of population density by density-dependent, negative feedback processes (Solomon, 1976).

To prove that a given animal population is regulated by its predator, therefore, it is necessary first to show that the population is regulated in the above strict sense, and then to prove that predation actually plays a major role in the process of this regulation. Then, for a given theoretical model how can we examine whether or not the predator is capable of regulating its prey population in this strict sense? For this purpose it may be sufficient to see whether or not the state of "being regulated" in the above definition is still maintained in the prey population even after the other mechanism responsible for the regulation (i.e., intraspecific competition among prey individuals imposed by restricted carrying capacity of the habitat) was eliminated from the model. In the following, we shall examine the possibility of regulation by predation according to this procedure for all the three categories of predator–prey models under study, the classic, the basic, and the extended models.

B. Prey Population Regulation in Classic Models

Neither the Lotka–Volterra model nor the Nicholson–Bailey model assumes any intraspecific mechanisms for the prey population that might cause its self-regulation. Accordingly, any tendency for regulation in the prey population in these classic models, if detected, would be attributable exclusively to predation.

The Lotka–Volterra model has often been regarded as illustrating the possibility of prey population regulation by predation, since in this model the interaction enables both populations, which would otherwise diverge to either plus or minus infinity respectively (in log density), to coexist forever. The apparent stability in this interactive system cannot, however, be regarded as "regulation" in the sense defined above. The cyclic oscillations generated here are readily changed in both amplitude and cycle-length depending on initial conditions as well as on environmental disturbances, and hence the stability observed here is only "neutral" without any negative-feedback mechanisms to resist destabilizing disturbances.

The situation is clearer in the case of Nicholson–Bailey model, where the time-delay effect on predator's reproductive response no longer allows even such neutral stability, the interaction invariably resulting in diverging oscillations of both predator and prey, followed sooner or later by the system crash.

It is therefore evident that in these classic models prey population regulation by predation cannot occur at all, the interaction resulting in mutual disturbance rather than stabilization of individual populations.

C. Prey Population Regulation in Basic Models

Next, let us examine the possibility of prey population regulation in the basic models (8) and (13) where some biologically unrealistic assumptions in the original classic models have been rectified. For this purpose it is necessary and sufficient, as mentioned before, to see whether or not there still exist any conditions stabilizing the prey population oscillations toward a certain finite equilibrium even after the assumption of finite habitat capacity was removed by substituting $h = 0$ in these models.

For the differential-equation model (8), we have seen in Fig. 4(b) that decrease of the interference coefficient h towards zero ultimately removes the hump from prey isocline, and hence the stabilizing force from the interaction. In fact, if $h = 0$ the prey isocline (Eqn (29)) becomes a monotonically increasing curve, the slope of which approaches $(b - d)/f$ for large x. We also see throughout Figs. 4(a)–(f) that in the basic model (8) there can be no condition that rotates the predator isocline in a stabilizing or clockwise manner. It is therefore evident that in this model the predator can not regulate, at least by itself, the population of its prey.

The situation is essentially similar in the difference-equation model (13), too: as seen in Fig. 5(b), reduction of the crowding factor h toward zero always results in diverging oscillation of the system, leading it to crash sooner or later.

We are thus led to a general conclusion that in these basic models, as in the classic models of Lotka–Volterra and Nicholson–Bailey, prey population regulation due to predation cannot be attained substantially, as long as we adopt the strict criterion defined here for judgment.

D. Prey Population Regulation in Extended Models

We have seen before that a variety of additional factors considered in a series of our extended models almost invariably gives stabilizing influences to the system dynamics to a lesser or greater extent. May the regulation due to predation, then, actually become possible in these extended models? The answer is generally "yes".

1. *Models With a Refuge*

In the differential-type model (Eqn (14)), the effect of a refuge has proved to be equivalent to a simple decrease of a in the basic model, indicating that existence of a refuge cannot itself bring about the condition for prey population regulation (see Fig. 6(a)). But the situation is quite different in the corresponding difference-equation model (Eqns (13) and (15)). Namely, as seen in Fig. 8(a) where the response in system behaviour to varying h-values is shown for this model, the predation here is obviously able under appropriate values of refuge parameter p to "regulate" the prey population in the strict sense, i.e., to stabilize it around some fairly low equilibrium levels that are virtually independent of the carrying capacity. (Note that in Fig. 8(a) the carrying capacity ($1/h$) is changed as widely as from 12·5 to 200; the system behaviour when the capacity goes to infinity ($h = 0$) was virtually the same as that when it is 200.) This may be because the time-delay effect in the predator's reproductive response inherent to this type of model has the effect of supplying fairly constant numbers (rather than proportions) of prey individuals to the system at every generation. That the refuge protecting a constant number of prey individuals has a much stronger stabilizing effect on the predator–prey system than that protecting a constant proportion of them has been shown by Maynard Smith (1974) with simple differential-equation models.

2. *Models with Mutual Interference*

We have already seen in Figs. 6(b) and 6(c) that in the models of differential type this factor consistently rotates the predator isocline at the equilibrium point in a clockwise manner, regardless of whether the interference operates in the process of prey searching or in that of prey assimilation. Since this means that the interactive system could be stabilized here even if $h = 0$, i.e., without any help from intraspecific mechanisms, it is evident that prey population regulation by predation has also become possible in these models. In their difference-equation versions (Eqns (13) and (17) or (18)), on the other hand, the possibility of prey population regulation by predation could be proved only for interference among searching predators (Fig. 8(b)), and not for the interference among reproducing predators (Fig. 8(c)), the system for the latter generally showing diverging oscillation towards a crash when $h = 0$, irrespective of the value of the interference coefficient m_c.

3. *Models with an Accelerating Functional Response*

From Fig. 6(c) it is obvious that prey population regulation has also become possible in the differential-type model including this factor (Eqn (18)) where the prey isocline is strongly rotated clockwise. But in its difference-equation version the system again shows diverging oscillations generally when $h = 0$

(Fig. 8(c)), so that our conclusion may be that in the presence of time-delay effect in predator's reproductive response the stabilizing effect of this factor is generally insufficient by itself to allow prey population regulation by predation.

4. *Models with Recruitment of Prey or Predator*

A constant supply of either prey or predators to the system results also in clockwise rotation of the respective isocline in the differential-equation model (Eqns (20) or (22); see Figs. 6(e) and 6(f)), so that predators can regulate the prey population here, too. This is also the case with the difference-equation versions of these models, in either case stable equilibrium of the prey populations being readily attainable even when h approaches 0 (Figs. 8(e) and 8(f)).

5. *Models with Alternative Prey*

This factor also stabilizes the interactive system as shown before, in such a way that prey population regulation due to predation becomes more probable in either the differential- or the difference-type models (Eqns (25–28)), especially if it is accompanied by the predator's habit of switching to the predominant prey species (see Figs 6(g), 6(h), 8(h) and 8(g)). Although the tendency for stabilization is rather weak in the case when no switching in the predator's attack occurs, the example of Fig. 8(g) shows that even here there exist some conditions that enable the predator to regulate prey population.

To conclude from the above results, the regulation of prey population density by predation, which proved to be impossible in either the classic Lotka–Volterra and Nicholson–Bailey models or the more realistic "basic models" derived here, has now become possible in many of these extended versions of the basic models to which a number of subsidiary but biologically important assumptions are added separately (see Table 3 for summary).

Of these conditions that proved to make prey population regulation possible, the most likely to occur among natural predator–prey systems may be the existence of a refuge for the prey, since for any predator in nature the environment in which it interacts with its prey is almost invariably heterogeneous in structure, assuring a considerable number of refuges for the prey at any one time.

Next to this, the existence of alternative prey coupled with a predator's habit of switching to dominant prey species, would be also worthy of consideration, since many predators in nature are polyphagous and the habit of such switching which inherently has a strong stabilizing effect is also common among these predators (e.g., Murdoch, 1969).

In contrast, conditions such as the predator's mutual interference in the process of either searching or consuming prey or accelerating functional response seem to be of rather minor importance when considering the

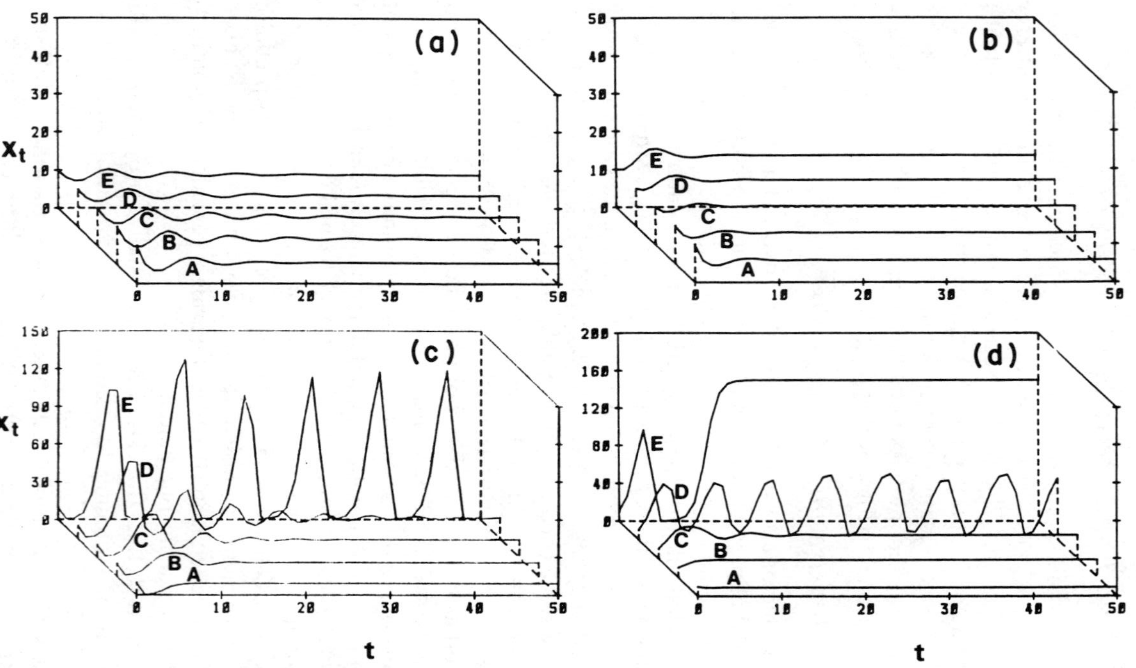

(a)
(b)
(c)
(d)
X_t
X_t
t
t
A
B
C
D
E

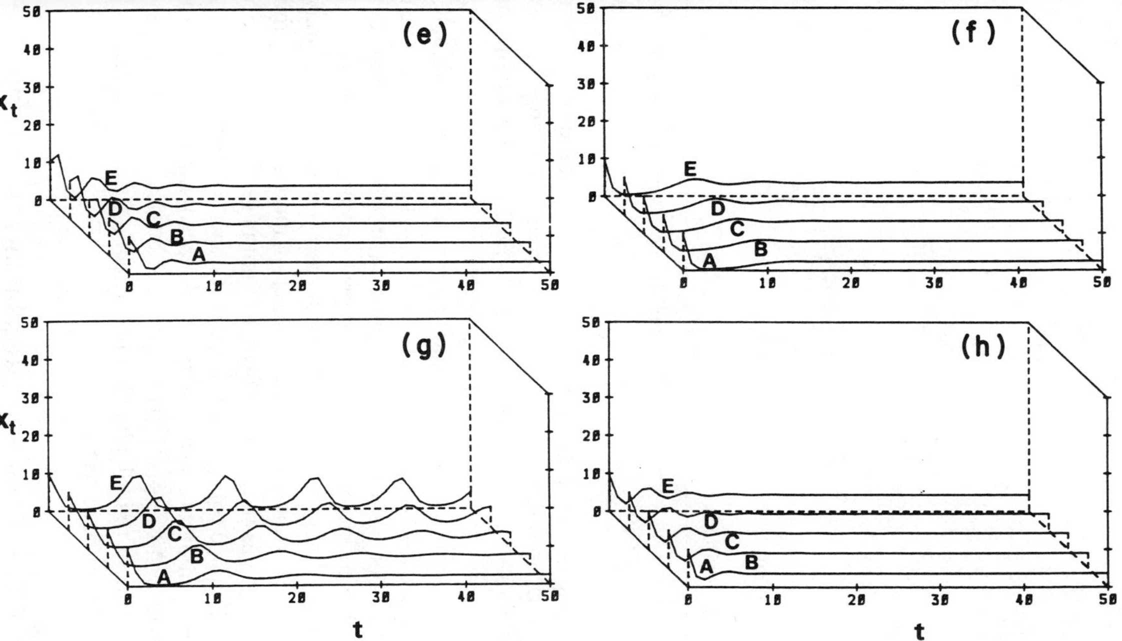

Fig. 8. Detection of prey population regulation in the extended models of difference-equation type by varying h-values (h = 0·08/0·04/0·02/0·01/0·005 (A to E); basic parameters: $R = 4$, $a = 0{\cdot}4$ (1–3) or 0·2 (4–8), $f = 100$, $C = 1$, $s_x = s_y = 0{\cdot}05$). (a) Refuge model ($p = 0{\cdot}2$). (b) Model with mutual interference in prey searching ($m_s = 0{\cdot}2$). (c) Model with mutual interference in prey consumption ($m_c = 0{\cdot}8$). (d) Model with accelerating functional response ($q = 64$). (e) Model with prey recruitment ($w_x = 12$). (f) Model with predator recruitment ($w_y = 4$). (g) Model with alternative prey; no switching ($z = 6$). (h) Model with alternative prey; switching ($z = 6$).

Table 3
Possibility of prey population regulation by predation in the models under study.

Models	Differential-equation type	Difference-equation type
Classic models[a]	No	No
Basic models	No	No
Extended models with		
prey refuge	No	Yes
predator interference (searching process)	Yes	Yes
predator interference (reproductive process)	Yes	No
accelerating functional response	Yes	No
prey recruitment	Yes	Yes
predator recruitment	Yes	Yes
alternative prey (without switching)	No	Yes
alternative prey (with switching)	Yes	Yes

[a] Represent Lotka–Volterra and Nicholson–Bailey models for the differential and the difference-equation types, respectively.

problem of prey population regulation. For, although these are rather common phenomena among natural predators, their regulating force is assessed from the above analysis to be too weak to function in nature as an effective mechanism for prey population regulation.

The time-to-time recruitment of additional prey or predator individuals, on the other hand, proved to have a rather strong regulating effect against the prey population, but this would be of interest mainly from the practical viewpoint in relation to biological control of pest populations by natural enemies. For natural interactive systems these recruitment models may correspond, though in a greatly simplified manner, to models describing the effect of migration such as that which Maynard Smith (1974) considered to explain the maintenance of predator–prey systems in heterogeneous habitats. High and moderate rates of prey and predator dispersal, for example, may correspond respectively to high and moderate rates of prey and predator recruitment to individual habitat patches, resulting in marked stabilization of the interactive system.

E. Robustness of Prey Population Regulation in a Varying Environment

The next problem is to examine the robustness of the possibility of prey population regulation by predation which has been proved theoretically in these extended models, against environmental disturbances which are likely to prevail in natural predator–prey systems. For this purpose we analyse here the behaviour of predator–prey systems generated by a stochastic version of the difference-equation model with a refuge for the prey, which represents a most likely situation for the regulation to occur in nature, i.e.

$$\begin{cases} x_{t+1} = \dfrac{R_t X^2/(s_x + X)}{1 + hR_t X^2/(s_x + X)} \\ y_{t+1} = \dfrac{C(x_t - X)y_t}{s_y + y_t} \end{cases} \tag{30}$$

$$(X = x_t(p + (1 - p)\cdot\exp(-ay_t/(1 + a(1 - p)x_t/f))))$$

where prey reproductive rate R_t is no longer a constant but a random variable the logarithm of which fluctuates from generation to generation following the normal distribution with mean R and standard deviation σ_R measuring the degree of environmental disturbance.

For quantitative analysis of the system behaviour simulated from this model, an empirical approach was adopted comprising the following steps of examination: (1) whether prey population is being kept stable at levels much lower than the habitat's carrying capacity; (2) whether predation is causing a high mortality on the average to the prey population; (3) whether predation is functioning as a key factor for prey population fluctuation; (4) whether the prey population is regulated as a whole by some density-dependent processes; and (5) whether the predation is actually playing a major role in those density-dependent regulatory processes, if ever, for the prey population.

Analysis of step (3) above will be made by means of Podoler and Rogers' (1975) method using the b-value in the regression of log S_p on log I where $I = x_{t+1}/x_t$ and S_p is the proportion of the prey which survived predation, while analyses of steps (4) and (5), based on the regression of log I and log S_p on log x_t together with Kuno's (1973) method of variance analysis.

1. *Effect of Environmental Fluctuation*

Results of the simulation based on Eqn (30) for varying values of σ_R (0, 0·15, 0·3 and 0·6) are given in Fig. 9 and Table 4(a). The σ_R-values of 0·15, 0·3 and 0·6 correspond approximately to the fluctuation ranges (as the high end/low end ratio for $P = 0{\cdot}95$) of 4-, 16- and 250-times, respectively. The parameter conditions for this simulation are such that they bring about complete regulation of prey population, i.e., that they keep the prey population stable at low levels even if $h = 0$ (as case A in Fig. 7(a)) when there is no environmental disturbance on R. In fact, under the condition that $\sigma_R = 0$ the analyses consistently showed that the prey population in this system is completely regulated by the predator. Namely, it can be seen that: (1) the

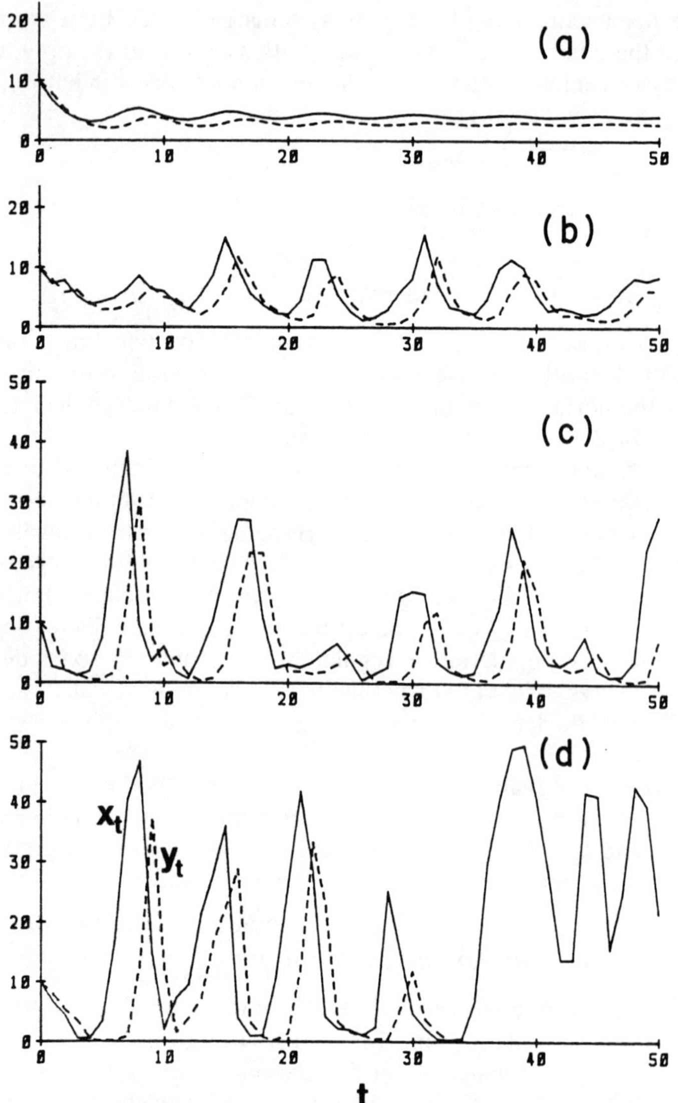

Fig. 9. Predator–prey dynamics generated from the stochastic refuge model (30) with varying σ_R-values ($R = 4$, $h = 0{\cdot}02$, $a = 0{\cdot}8$, $f = 100$, $C = 1$, $s_x = s_y = 0{\cdot}05$, $p = 0{\cdot}2$). (a) $\sigma_R = 0$. (b) $\sigma_R = 0{\cdot}15$. (c) $\sigma_R = 0{\cdot}3$. (d) $\sigma_R = 0{\cdot}6$.

prey population is kept stable at a level much lower than the carrying capacity ($1/h = 50$) (Fig. 9(a)); (2) the percentage of the prey killed by the predator is fairly high; (3) predation is a key factor causing prey population fluctuation, judging from the high b-value in the log S_p-on-log I regression; (4) the prey population is regulated as a whole as indicated by the negative b-value in the log I-on-log x_t regression; and (5) this regulation is caused mainly by predation, the b-value in the log S_p-on-log x_t regression being much higher than that in the log R_r-on-log $(x_t S_p)$ regression and variance of the prey density after predation being smaller than that before predation (see Table 4(a) for results (2–5)).

All these necessary conditions for detecting regulation still hold, as is seen in Fig. 9(b) and Table 4(a), when σ_R is increased to 0·15, so that at this level of environmental fluctuation prey population regulation still remains possible, though the population here shows a consistent tendency for oscillation accompanied to some extent by random fluctuations. If, however, σ_R is increased further to 0·3, then the situation changes considerably. The prey population now shows large oscillations often reaching levels near the carrying capacity (Fig. 9(c)), the importance of predation as a factor causing prey population fluctuation has become less than that of the residual factors and its contribution to overall regulation of the prey population has also become smaller than that of the residual regulatory factor, intraspecific competition (see Table 4(a)). It is therefore evident that when the environmental fluctuation is increased to this level the predator here can no longer regulate the prey population, and yet the populations of both prey and predator can now coexist stably in the presence of continuous interactions.

Further increase of σ_R to 0·6, however, results in population fluctuations that are too violent for the interactive system to persist, followed usually by extinction of the predator and survival of the prey (Fig. 9(d)). We therefore may conclude that for the prey population the state of regulation by a predator is not so robust against the disturbance due to stochastic fluctuation in environmental conditions, the escape from it being rather probable in a varying environment.

2. *Effect of Changes of Parameters* p, R *and* a

We shall examine the robustness of prey population regulation against changes of some of the basic parameters involved, using the same stochastic model (30). In Fig. 10 and Table 4(b) are shown the results of a simulation in which parameter p, the proportion of prey individuals protected by a refuge, was changed in three levels, 0·1, 0·2 and 0·4. The conditions for $p = 0{\cdot}2$ here are the same as those for Fig. 9(b), and we see that the prey population is kept stable at low levels under nearly complete regulation by the predator

Table 4

Statistical analysis of results of simulations based on the stochastic refuge model to examine the possibility of prey population regulation by predation.

(a) Experiment on varying σ_R, variability of prey reproductive rate log R (Eqn (30); Fig. 9).

Condition for σ_R	Detection of key factor		Detection of regulation					Mean rate of predation
	b_{pI}	b_{rI}	b_{Ix}	b_{px}	b_{rx}	V_B	V_A	$1 - \overline{S}_p$
$\sigma_R = 0$	0·985	0·015	−0·459	−0·423	−0·056	0·00589	0·00333	0·721
$\sigma_R = 0.15$	0·569	0·431	−0.327	−0·288	−0·059	0·0680	0·0543	0·714
$\sigma_R = 0.30$	0·394	0·606	−0·368	−0·117	−0·175	0·210	0·210	0·653

b_{pI} and b_{rI} represent b-values in the regressions of log S_p (S_p; percentage survived from predation) and log R_r ($R_r = I/S_p$) on log I ($I = x_{t+1}/x_t$), while b_{Ix}, b_{px} and b_{rx} are those in the regressions of log I, log S_p and log R_r on log x_t, respectively. V_B and V_A are variances of prey densities before (log x_t) and after (log (x_{t+1}/R_r)) predation. Mean rate of predation is $1 - \overline{S}_p$ where $\overline{S}_p$ is the geometric mean of S_p over generations. Parameter conditions: $R = 4$, $h = 0{\cdot}02$, $s_x = s_y = 0{\cdot}05$, $a = 0{\cdot}8$, $f = 100$, $C = 1$, $p = 0{\cdot}2$.

(b) Experiment on varying p, proportion of prey protected by refuge. (Eqn (30); Fig. 10).

Condition for p	b_{pI}	b_{rI}	b_{Ix}	b_{px}	b_{rx}	V_B	V_A	$1 - \overline{S}p$
$p = 0{\cdot}1$	0·858	0·142	−0·300	−0·150	−0·156	0·205	0·258	0·736
$p = 0{\cdot}2$	0·534	0·466	−0·356	−0·278	−0·130	0·0589	0·0494	0·717
$p = 0{\cdot}4$	0·003	0·996	−0·396	−0·008	−0·392	0·0139	0·0137	0·600

Parameter conditions are same as Experiment (a) except for p and σ_R (= 0·15).

(c) Experiment on varying a, searching efficiency of predator (Eqn (30); Fig. 11).

Condition for a	b_{pI}	b_{rI}	b_{Ix}	b_{px}	b_{rx}	V_B	V_A	$1 - \bar{S}p$
$a = 0{\cdot}8$	0·604	0·396	−0·324	−0·299	−0·051	0·0158	0·0448	0·721
$a = 0{\cdot}2$	0·616	0·384	−0·416	−0·150	−0·217	0·0506	0·0611	0·617
$a = 0{\cdot}05$	0·135	0·865	−0·392	0·013	−0·422	0·0145	0·0172	0·227

Parameter conditions are same as Experiment (a) except for a and σ_R (= 0·15).

(d) Experiment on varying R, reproductive rate of prey (Eqn (30); Fig. 12).

Condition for R	b_{pI}	b_{rI}	b_{Ix}	b_{px}	b_{rx}	V_B	V_A	$1 - \bar{S}p$
$R = 4$	0·607	0·393	−0·381	−0·247	−0·125	0·0563	0·0524	0·701
$R = 6$	0·200	0·800	−0·257	−0·152	−0·113	0·0370	0·289	0·783
$R = 8$	0·020	0·980	−0·276	−0·018	−0·258	0·0218	0·0211	0·799

Parameter conditions are same as Experiment (a) except for R and σ_R (= 0·15).

(e) Experiment illustrating non-regulatory suppression of prey density by predation (Eqn (31); Fig. 13).

Condition for h	b_{pI}	b_{rI}	b_{Ix}	b_{px}	b_{rx}	V_B	V_A	$1 - \bar{S}p$
$h = 0{\cdot}02$	0·000	1·000	−0·539	−0·001	−0·538	0·0122	0·0157	0·800
$h = 0{\cdot}002$	0·001	0·999	−0·323	−0·002	−0·320	0·0182	0·0247	0·800

Parameter conditions are same as Experiment (a) except for h and R (= 8).

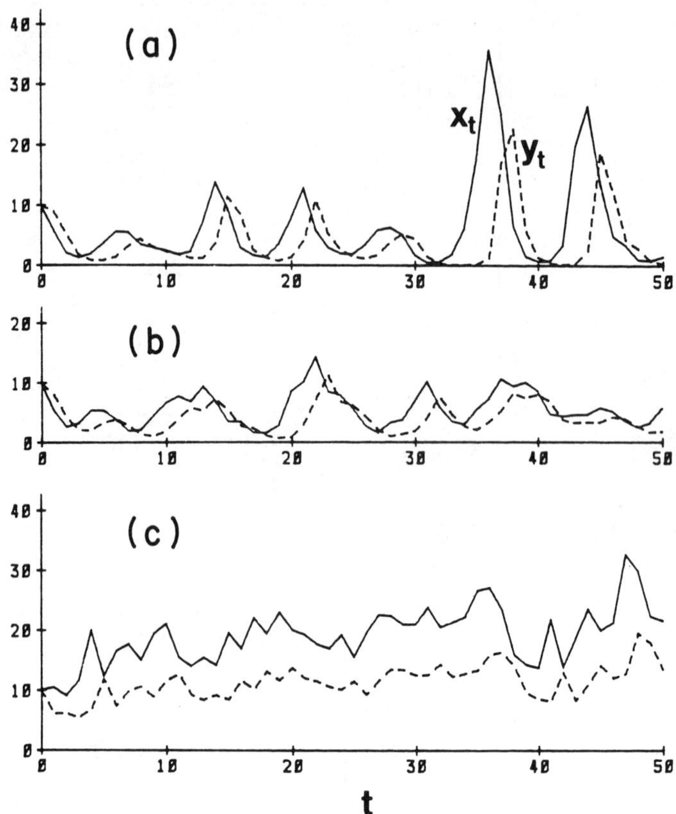

Fig. 10. Predator–prey dynamics generated from the stochastic refuge model (30) with varying p-values ($R = 4$, $\sigma_R = 0{\cdot}15$, $h = 0{\cdot}02$, $a = 0{\cdot}8$, $f = 100$, $C = 1$, $s_x = s_y = 0{\cdot}05$). (a) $p = 0{\cdot}1$. (b) $p = 0{\cdot}2$. (c) $p = 0{\cdot}4$.

(Fig. 10(b), Table 4(b)). But once the p-value is increased to 0·4, the predator obviously fails to regulate the prey population, there being virtually no sign that the predation here is functioning as either a key- or a regulatory factor in the prey population dynamics (Fig. 10(c), Table 4(b)). Conversely, it is also shown that the value of p as low as 0·1 is again not optimal for attaining prey population regulation: although the predation here is undoubtedly an important mortality- and key factor for the prey population, it no longer has the ability to stabilize the prey equilibrium at low levels, working a a destabilizing, rather than a regulating factor (Fig. 10(a) and Table 4(b)).

The restriction in the predator's ability to regulate the prey population is

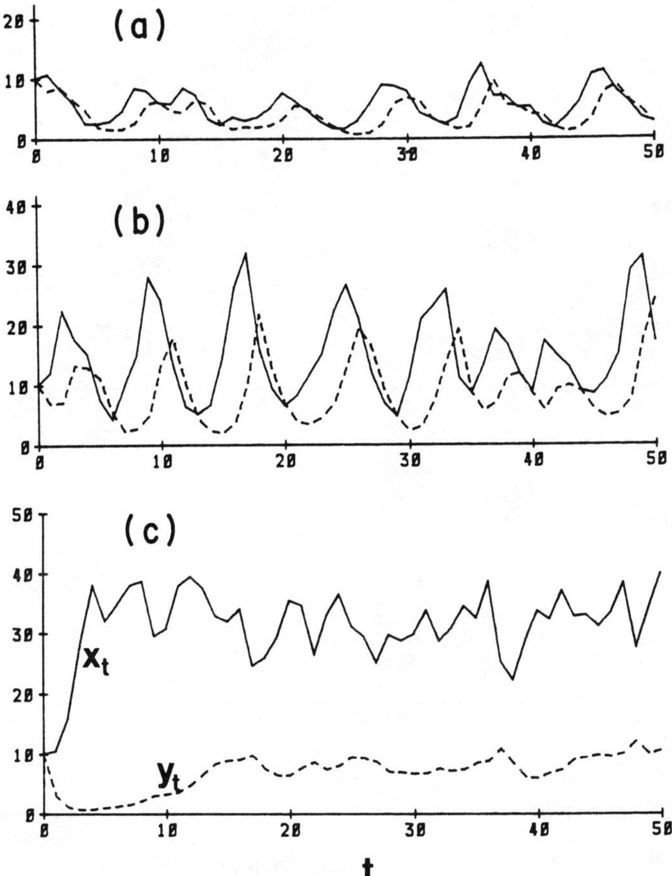

Fig. 11. Predator–prey dynamics generated from the stochastic refuge model (30) with varying a-values ($R = 4, \sigma_R = 0{\cdot}15, h = 0{\cdot}02, f = 100, C = 1, s_x = s_y = 0{\cdot}05, p = 0{\cdot}2$). (a) $a = 0{\cdot}8$. (b) $a = 0{\cdot}2$. (c) $a = 0{\cdot}05$.

similarly striking in the simulations in which parameters a (predator's searching efficiency) and R (prey's reproductive rate) are decreased and increased, respectively, from the control levels, 0·8 and 4, at which the regulation is nearly complete. That is to say, in contrast to the results under the control conditions (Figs. 11(a) and 12(a)), the prey population becomes relatively free from regulatory suppression by the predator when either a or R is changed to 0·2 or 6 (Figs. 11(b) and 12(b)), and nearly completely when a is decreased to as low as 0·05 or R is increased to as high as 8 (Figs. 11(c) and 12(c), see Table 4(c) and (d) for numerical analysis).

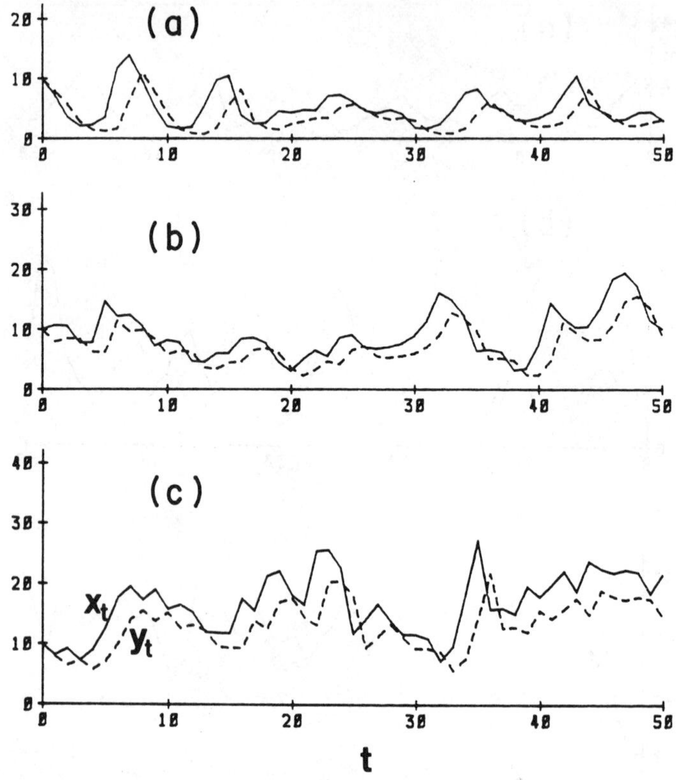

Fig. 12. Predator–prey dynamics generated from the stochastic refuge model (30) with varying R-values ($\sigma_R = 0{\cdot}15$, $h = 0{\cdot}02$, $a = 0{\cdot}8$, $f = 100$, $C = 1$, $s_x = s_y = 0{\cdot}05, p = 0{\cdot}2$). (a) $R = 4$. (b) $R = 6$. (c) $R = 8$.

F. Non-Regulatory Suppression of the Prey Population by Predation

A general conclusion we have derived from the above analyses is that in theory predators can indeed regulate the populations of their prey in the strict sense, but the conditions that allow this situation are very restricted. Namely, for the regulation to be attained, the predator's searching efficiency must first of all be very high, the prey's reproductive rate must be fairly low, environmental fluctuation must not be large, and yet habitat heterogeneity or protecting capacity of a refuge for the prey (in case of the refuge model) should be moderate; if either one of these conditions is violated, the state of regulation may be broken very easily.

At the same time, however, it is noticeable from these analyses that in many cases the populations of both predator and prey can persist in a state of stable coexistance, regardless of whether the prey population is strictly "regulated" by the predator or not. Since in these cases the prey density is generally kept more or less lower than the level to be attained in its single population, it is evident that the prey population level being kept stable and low, as well as the predation rate itself remaining high, is not a sufficient, but only one of the necessary conditions for verifying regulation. Such non-regulatory suppression of the prey population by predation may appear more strikingly under a particular condition, i.e., when the prey density following a predator's attack, instead of that before the attack as in the above-mentioned models, is used to represent the prey population, x_t. For natural predator–prey systems this may correspond, for example, to a commonly occurring situation when an intraspecific regulatory mechanism for the prey population operates during the adult stage, taking the form of, say, density-dependent dispersal. The refuge model (30) used above can be readily modified to cover this point, i.e.

$$\begin{cases} x_{t+1} = X\left(p + (1-p)\cdot\exp\left(\dfrac{-ay_t}{1 + a(1-p)X/f}\right)\right) \\ y_{t+1} = \dfrac{C(X - x_{t+1})y_t}{s_y + y_t} \end{cases} \qquad (31)$$

$$X = \frac{R_t x_t^2/(s_x + x_t)}{1 + hR_t x_t^2/(s_x + x_t)}$$

Notice that Eqn (31) is structurally equivalent to the original Eqn (30), the difference arising simply from the stage of prey taken to represent the density x_t for each generation.

To show a typical example of non-regulatory suppression of a prey population, a simulation was made using Eqn. (31) after the parameter conditions were set to be same as Fig. 12(c), i.e., the conditions such that the prey population becomes nearly entirely free from the predator's regulation due to its high reproductive rate ($R = 8$). The result (Fig. 13(a)) clearly illustrates that the prey population here is suppressed to a level far lower (less than one-tenth) than the habitat's carrying capacity (50) or the level to be attained in the absence of a predator (indicated by $x_t(s)$ in the graph), and yet the numerical analysis (see Table 4(e)) of course shows that the predator here is actually not regulating the prey population at all. That no regulation by the predator occurs here may be realized more clearly if one compares the result with that of the simulation in which parameter h was decreased from 0·02 to 0·002 (Fig. 13(b)), where the prey population level also showed a remarkable increase nearly proportional to this 10-times increase of the

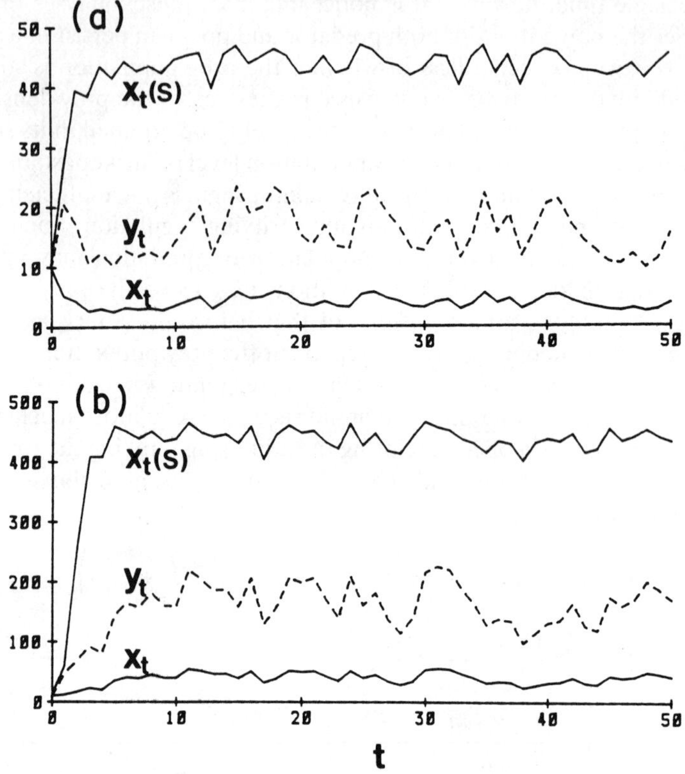

Fig. 13. Examples of non-regulatory suppression of prey density (x_t) by predation generated from the stochastic refuge model (31) ($R = 8$, $\sigma_R = 0{\cdot}15$, $a = 0{\cdot}8$, $f = 100$, $C = 1$, $s_x = s_y = 0{\cdot}05$, $p = 0{\cdot}2$). $x_t(S)$: prey density in the predator-free system under the same conditions. (a) $h = 0{\cdot}02$. (b) $h = 0{\cdot}002$.

carrying capacity $1/h$, instead of keeping the same low equilibrium independently of the capacity.

We notice in these examples that the prey population here is actually preyed upon by the predator at very high rates of nearly 80% (indicating that almost all prey individuals outside the refuge are killed) at every generation, and yet it has succeeded in escaping from regulation by predator, its overall regulation being almost exclusively attributable to intraspecific competition, not to predation (Table 4(e)). It must be stressed again that high predation rates or the resulting low, stable prey population levels do not, contrary to general belief, necessarily mean that the predator is regulating the prey population.

V. CONFLICT OF INTERESTS BETWEEN INDIVIDUAL AND POPULATION: ANALYSIS OF AN EVOLUTIONARY PARADOX IN THEORETICAL PREDATOR–PREY SYSTEMS

In the foregoing analyses we have seen that the population interaction between predator and its prey may inevitably produce a conflict of interests between individuals and a population of the same species, which may affect the predator more seriously than the prey species. For example, it has been shown in the basic models that any tendency of predator individuals towards "prosperity", such as raising search efficiency (a), attack capacity (f) or consumption efficiency (c or C), inevitably produces results that are quite unprofitable for its own population, e.g., considerable system destabilization coupled with a lowering of equilibrium population levels, which may in some cases even lead ultimately to their extinction (Table 2). Similarly, a prey's tendency to increase propagation by raising reproductive rate ($b - d$ or R) or by reducing a crowding effect (h) may also have an undesirable effect of reducing system stability, at least under certain situations, which Rosenzweig (1971) named the "paradox of enrichment".

How, then, and towards which direction may the coevolution of predator and prey, if any, occur in these paradoxical systems? Is it really probable in theory that some predator–prey systems are led as a natural outcome of the individual selection in either species to their self-destruction ultimately against the "interests" of their populations? How high is the probability that coevolution brings about such situations that regulation of a prey population by a predator occurs readily in their interactions?

Since Rosenzweig and MacArthur (1963), some attempts have been made to answer the problems of this kind (see Slatkin and Maynard Smith (1979) for a review). In these studies, however, the trends of very few relevant parameters – usually only predator's searching efficiency – have been considered in the frame of simple differential-equation type models, so that the conclusions obtained often remain too abstract to evaluate their ecological reliability. In the following analysis these problems are approached from a different standpoint, using more concrete models of predator–prey interaction, i.e., the basic models (8) and (13) derived before.

A. Selection Among Prey Individuals

For simulating the selection which occurs among prey individuals in its simplest form, we assume here that: (1) the prey population is composed of several strains which are more or less different from each other in either one of the basic population parameters; (2) individuals of all the strains are subjected to common effects of over- and undercrowding, and (3) they produce progeny each of which inherits all the characteristics of the parental strain faithfully. Then, the corresponding model of differential-equation type based on Eqn (8) will be

$$\begin{cases} \dfrac{dx_i}{dt} = \dfrac{b_i x_i x}{s_x + x} - d_i X_i - h_i x_i x - \dfrac{a_i x_i y}{1 + \Sigma_{j=1}^{n} a_j x_j / f_j} \\ \dfrac{dy}{dt} = \dfrac{\Sigma_{i=1}^{n} c_i a_i x_i y}{(1 + \Sigma_{j=1}^{n} a_j x_j / f_j)(s_y + y)} - ey \end{cases} \tag{32}$$

where the first equation represents a series of prey equations for the respective densities x_i of n strains each being characterized with parameters with suffix i ($i = 1, 2, \ldots, n$) and x is the total density $\Sigma_{i=1}^{n} x_i$ of the prey population. Similarly, the difference-equation version of Eqn (32) is obtained from (13) as

$$\begin{cases} x_{i,t+1} = \dfrac{R_i X_i'}{1 + h_i \Sigma_{j=1}^{n} R_j X_j'} \\ y_{t+1} = \dfrac{\Sigma_{j=1}^{n} C_j (x_{j,t} - X_j) y_t}{s_y + y_t} \end{cases} \tag{33}$$

$$\left[\begin{aligned} X_i &= x_{i,t} \cdot \exp\left(-a_i y_t \Big/ \left(1 + \sum_{j=1}^{n} a_j x_{j,t} / f_j\right)\right) \\ X_i' &= s_x X_i \sum_{j=1}^{n} X_j \Big/ \left(s_x + \sum_{j=1}^{n} X_j\right) \end{aligned} \right]$$

In these models it may be intuitively evident that among the different strains existing, those resulting in higher ultimate population growth rate (dx_i/dt in (32) or x_{t+1}/x_t in (33)) or fitness than the others become dominant in the inter-strain competition. Namely, the selection here will proceed towards such a direction that the prey's reproductive rate (R) is increased and both crowding effect (h) (to say nothing of undercrowding effect, s_x) and predator's searching efficiency (a) are reduced. To prove this numerically, simulations were made using the difference-equation model (33) where 5 prey strains (i.e., $n = 5$) that are different in either one of the basic parameters, R, h, a, f and C were assumed to compete with each other. The parameter conditions for each simulation were set up in such a way that they correspond as a rule to those in each of the foregoing simulations (Fig. 5) to analyse the modes of effects of these parameters on the system behaviour (see Table 5 for details).

The result was quite clear, as is to be expected. It can be seen in Table 5(a) that the strain with highest R, that with lowest h, and that with lowest a win very rapidly in the respective competition experiments, excluding all other strains nearly completely within as few as some 10 generations. The direction of individual selection in the prey as to both R and a in this case are regarded as obviously adaptive for its population also, since the levels, as well as the stability of its population, are expected to be raised by these changes (see Fig. 5). But the situation is different in case of h, the coefficient of interference: decrease of h towards 0 results necessarily in system destabilization as well as an increase of prey population level (Fig. 5(b), so that selection here proceeds in a way unfavourable for the population. It may even lead the population to crash as seen in the present example, unless some stabilizing mechanisms have come into operation.

The simulation experiments also showed that parameters f and c are entirely neutral against the selection among prey individuals, despite the fact that they are joint parameters determined not only by the predator's but also by the prey's ecological characteristics (Table 5(a)). Theoretically this is readily recognizable since parameter C, the predator's reproductive efficiency, appears only in the predator equation while f, the upper limit of the predator's attack rate, is always effective against total prey density rather than the densities of the individual strains.

Biologically, however, this seems to be the result worthy of special note, since it means that such prey strains that are less valuable to the predator in either handling or assimilating efficiency, even if they happened to appear, cannot propagate themselves by selection on the prey's side. The individual–population conflict of interests is again appearing here. Thus, in the selection experiment among strains with different C-values in Table 5(a), we see that the prey population "evolved" in such a way as to crash despite the existence of "safer" strains with lower C-values which would enable their population to persist in stable interaction with the predator.

Thus, if a given prey strain with lower C- or f-values is to become dominant through evolution, it might be only when the predator has evolved the ability to discriminate between prey strains with different favourableness as to C or f, or from the prey's side, only when that prey strain has succeeded in advertising its unfavourableness among others to the predator. The development of such a strategy on the prey's side may not be unrealistic, as is evident from the acquisition of warning colour by many "unfavourable" prey animals.

B. Selection Among Predator Individuals

The models to simulate selection in a predator population may be given general expressions,

$$\left\{\begin{aligned} \frac{dx}{dt} &= \frac{bx^2}{s_x + x} - dX - hx^2 - \frac{\Sigma_{j=1}^n a_j x y_j}{1 + \Sigma_{j=1}^n a_j x / f_j} \\ \frac{dy_i}{dt} &= \frac{c_i a_i x y_i y}{(1 + \Sigma_{j=1}^n a_j x / f_j)(s_y + y)} - e_i y_i \end{aligned}\right. \tag{34}$$

and

$$\begin{cases} x_{t+1} = \dfrac{RX^2/(s_x + X)}{1 + hRX^2/(s_x + X)} \\ y_{i,t+1} = \dfrac{Y_i}{\Sigma_{j=1}^n Y_i} \cdot \dfrac{C_i(x_t - X)y_t}{s_y + y_t} \end{cases} \tag{35}$$

$$\begin{bmatrix} Y_i = a_i y_{i,t}/(1 + a_i x_t/f_i) \\ X = x_t \cdot \exp\left(-\sum_{j=1}^{n} Y_j\right) \end{bmatrix}$$

for the differential- and the difference-equation type, respectively, where the total predator population (y) now consists of n strains having respective densities and parameters indicated with suffix i ($i = 1, 2, \ldots, n$).

In these equations it is evident that the parameters b, d, R and h characterizing prey population growth are all neutral in terms of selection, since they are involved only in the prey equation. But the other parameters, a, f, C, c and e are manipulable for the predator population and so can be changed by selection on the predator's side. We can also readily infer that the direction in this selection will be such that it increases predator's efficiency of searching, handling and consuming its prey together with its survival rate, i.e., such that it increases a, f and c or C and decreases e. Numerical simulations based on the difference-equation model (35) clearly confirmed this conclusion as seen in Table 5(b). Here five predator strains were assumed to exist (i.e., $n = 5$) for each of the parameters a, f and C, and as in the case of prey selection, the conditions for parameters including these were made comparable for each simulation to the corresponding ones in the experiments of Fig. 5 (see Table 5(b) for details). It can be seen that the predator strain, given the highest value of either a, f or C among others, soon excluded all the other strains, and as the result, the predator's equilibrium density as well as prey's was lowered and destabilized until its own population, which might otherwise have been able to persist stably, became extinct sooner or later. Here we can see striking examples of the serious conflict of interests occurring between individual and population, in which the priority is always given to the former, regardless of whether the population would be led to ruin in the long run.

C. Selection Between "Specialists" and "Generalists" in Either Predator or Prey Species

We have been concerned so far only with selection in one predator–one prey system. Next, further analyses will be done concerning some other aspects of selection in both predator and prey which are somewhat more complicated but may also be important ecologically. These are the selection between the predator strains specialized to consume one prey species ("specialist predator") and those that can prey on two or more prey species ("generalist predator") and the selection between the prey strains attacked only by one predator species (called "specialist prey" here) and those that can be preyed upon by two or more predator species (called "generalist prey").

For a specialist predator to become a generalist, it may be inevitable that it should sacrifice considerably its specialization towards the original prey in terms of high efficiency of searching, handling or consumption. Conversely, a prey species being attacked by a single specialized predator species at high rates may be able to relieve the predator's attack by diversifying its mode of life in both time and space. But such diversification would also be accompanied necessarily with the demerit of becoming vulnerable to the attack of new predators and thereby enriching its predator fauna, undergoing the change from a "specialist prey" to a "generalist prey" in the above definition. In either case, then, toward which direction is selection more likely to proceed, to favour generalists rather than specialists or vice versa?

Suppose now a simplest situation in which the system is composed of two prey- and one predator species and the latter includes two strains, one being a generalist capable of attacking both species equally and the other, a specialist capable of attacking only one of the two prey species. Then the difference-equation model to simulate the inter-strain selection in the predator population will be given from Eqn (13) by

$$\left\{\begin{aligned}
x_{1,t+1} &= \frac{R_1X_1}{1 + h_1R_1X_1}\\
x_{2,t+1} &= \frac{R_2X_2}{1 + h_2R_2X_2}\\
y_{1,t+1} &= \left(\frac{Y_{11}}{Y_{11} + Y_{12}}\cdot C(x_{1,t} - X_1') + \frac{Y_{21}}{Y_{21} + Y_{22}}\cdot C(x_{2,t} - X_2')\right)\cdot\frac{y_t}{s_y + y_t}\\
y_{2,t+1} &= \left(\frac{Y_{12}}{Y_{11} + Y_{12}}\cdot C(x_{1,t} - X_1') + \frac{Y_{22}}{Y_{21} + Y_{22}}\cdot C(x_{2,t} - X_2')\right)\cdot\frac{y_t}{s_y + y_t}
\end{aligned}\right. \tag{36}$$

$$\left[\begin{aligned}
Y_{11} &= a_{11}y_{1,t}/(1 + (a_{11}x_{1,t} + a_{21}x_{2,t})/f)\\
Y_{21} &= a_{21}y_{1,t}/(1 + (a_{11}x_{1,t} + a_{21}x_{2,t})/f)\\
Y_{12} &= a_{12}y_{2,t}/(1 + (a_{12}x_{1,t} + a_{22}x_{2,t})/f)\\
Y_{22} &= a_{22}y_{2,t}/(1 + (a_{12}x_{1,t} + a_{22}x_{2,t})/f)\\
X_1' &= x_{1,t}\cdot\exp(-Y_{11} - Y_{12});\; X_2' = x_{2,t}\cdot\exp(-Y_{21} - Y_{22})\\
X_1 &= X_1'^2/(s_{x,1} + x_{1,t});\; X_2 = X_2'^2/(s_{x,2} + x_{2,t})
\end{aligned}\right]$$

where suffix (1 or 2) in the prey equations indicates different species while that in the predator ones, different strains, and parameter a_{12}, for example, indicate the searching efficiency for the predator of strain 2 against the prey of species 1.

Similarly, the corresponding model to describe the selection between prey strains when there are two predator species can be described as

$$\left\{\begin{aligned} x_{1,t+1} &= \frac{RX_1}{1 + h(RX_1 + RX_2)} \\ x_{2,t+1} &= \frac{RX_2}{1 + h(RX_1 + RX_2)} \\ y_{1,t+1} &= \left(\frac{Y_{11}}{Y_{11} + Y_{12}} \cdot C_1(x_{1,t} - X_1') + \frac{Y_{21}}{Y_{21} + Y_{22}} \cdot C_1(x_{2,t} - X_2')\right) \\ &\quad \times \frac{y_{1,t}}{s_{y,1} + y_{1,t}} + w \\ y_{2,t+1} &= \left(\frac{Y_{12}}{Y_{11} + Y_{12}} \cdot C_2(x_{1,t} - X_1') + \frac{Y_{22}}{Y_{21} + Y_{22}} \cdot C_2(x_{2,t} - X_2')\right) \\ &\quad \times \frac{y_{2,t}}{s_{y,2} + y_{2,t}} + w \end{aligned}\right. \tag{37}$$

$$\left[\begin{aligned} Y_{11} &= a_{11}y_{1,t}/(1 + (a_{11}x_{1,t} + a_{21}x_{2,t})/f_1) \\ Y_{21} &= a_{21}y_{1,t}/(1 + (a_{11}x_{1,t} + a_{21}x_{2,t})/f_1) \\ Y_{12} &= a_{12}y_{2,t}/(1 + (a_{12}x_{1,t} + a_{22}x_{2,t})/f_2) \\ Y_{22} &= a_{22}y_{2,t}/(1 + (a_{12}x_{1,t} + a_{22}x_{2,t})/f_2) \\ X_1' &= x_{1,t} \cdot \exp(-Y_{11} - Y_{12}) \\ X_2' &= x_{2,t} \cdot \exp(-Y_{21} - Y_{22}) \\ X_1 &= X_1'(X_1' + X_2')/(s_x + X_1' + X_2') \\ X_2 &= X_2'(X_1' + X_2')/(s_x + X_1' + X_2') \end{aligned}\right]$$

where the suffix for the prey now indicates different strain while that for the predator, different species. In this model, notice that constant recruitment (w) to the system of both species of predators has been assumed. This modification is simply to avoid the extinction of either predator which otherwise would occur readily in this system.

The outcome of simulations based on these models would of course be critically affected by the setting of the difference in parameter a between specialist and generalist strains in either prey or predator species. In the present experiment, the a for the specialist strain in either prey or predator was always set as 0·1, whereas that for generalist strain as 0·05, as low as half the value for specialist. The other parameters were all made equal between the competing strains. The results of these simulations are given also in

Table 5(c) and (d) together with the detailed conditions for each of the relevant parameters.

Results obtained were rather clear-cut in both of these simulations. Notwithstanding the remarkable reduction of a to half its original value, the selection in either prey or predator population strikingly favoured the "generalist" strain, the entire replacement in the population having occurred within as few as 20 generations in either case.

Of course we may expect in some cases a "specialist" prey or predator rather than a "generalist" one to win in this evolutionary competition, specifically when the reduction of a by becoming generalist is much greater, or when the primary prey's R in the case of selection in the predator (R_1 in model (36)) is much higher than the secondary prey's (R_2). But generally speaking, it may be safe to conclude that, other things being equal, selection in either prey or predator species has a tendency to occur in favour of generalists, i.e., towards the direction of enriching both prey and predator fauna comprising the interactive system. The underlying principle to account for this tendency may be that of "spreading of risk" which has been often used in order to explain the persistence of animal populations in varying environments (den Boer, 1968).

D. Predator–Prey Conflict in their Coevolution

As is summarized in Table 6, selection in the prey population proved to occur consistently in such a direction as to increase R and decrease both h and a, while selection on the predator's side increases a, f and C, the other parameters (f and C in prey selection and R and h in predator selection) being neutral as regards selection in either case. In terms of the zero-isocline on the y–x plane, this indicates that selection in the prey has a consistent tendency to shift the prey isocline upward (R, h, a) and the predator one to the right (a), whereas quite the reverse is the case with selection in the predator, shifting the prey and the predator isoclines downward (a, f) and to the left (a, C, f), respectively.

Thus, selection in the prey may proceed generally in the direction favourable for the populations of both predator and prey, raising their levels and assuring their stable persistence. But selection in the predator, by contrast, may occur in such a way that is quite dangerous for not only the prey's but also the predator's own population, reducing their levels and destabilizing their dynamics remarkably. Serious conflict may therefore occur generally between a predator and its prey in their coevolution, as has been postulated by some people using simpler models of interaction (e.g., Rosenzweig, 1973; Schaffer and Rosenzweig, 1978; Dawkins and Krebs, 1979).

Table 5

Results of simulations for competition among prey or predator strains differing in either one of the basic parameters.
(a) Competition among prey strains (model (33)).

Experiment (a)	t	Total prey density (x_t)	Predator density (y_t)	Proportion of prey strain with R-value of 1·00	1·25	2·50	5·00	10·00
$h = 0{\cdot}02$	0	10·00	10·00	0·200	0·200	0·200	0·200	0·200
$a = 0{\cdot}1$	5	25·51	31·26	0·000	0·000	0·001	0·030	0·969
$f = 100$	10	40·22	14·91	0·000	0·000	0·000	0·002	0·998
$C = 1$	15	30·73	3·08	0·000	0·000	0·000	0·000	1·000
$s_x = s_y = 0{\cdot}05$	50	41·79	13·15	0·000	0·000	0·000	0·000	1·000

Experiment (b)	t	x_t	y_t	Proportion of prey strain with h-value of 0·08	0·04	0·02	0·01	0·005
$R = 4$	0	10·00	10·00	0·200	0·200	0·200	0·200	0·200
$a = 0{\cdot}1$	5	33·42	23·69	0·002	0·018	0·096	0·295	0·589
$f = 100$	10	25·95	0·26	0·000	0·005	0·054	0·263	0·678
$C = 1$	15	0·002	48·89	0·000	0·000	0·003	0·086	0·911
$s_x = s_y = 0{\cdot}05$	20	0·00	0·00	—	—	—	—	—

Experiment (c)	t	x_t	y_t	Proportion of prey strain with a-value of 0·40	0·20	0·10	0·05	0·025
$R = 4$	0	10·00	10·00	0·200	0·200	0·200	0·200	0·200
$h = 0·02$	5	34·98	4·25	0·000	0·007	0·088	0·314	0·591
$f = 100$	10	34·55	6.47	0·000	0·000	0·014	0·204	0·783
$C = 1$	15	35·61	4·37	0·000	0·000	0·002	0·114	0·884
$s_x = s_y = 0·05$	50	37·48	0·00	0·000	0·000	0·000	0·039	0·961

Experiment (d)	t	x_t	y_t	Proportion of prey strain with f-value of 16	8	4	2	1
$R = 4$	0	10·00	10·00	0·200	0·200	0·200	0·200	0·200
$h = 0·02$	5	27·89	9·34	0·200	0·200	0·200	0·200	0·200
$a = 0·2$	10	13·20	11·86	0·200	0·200	0·200	0·200	0·200
$C = 1$	15	24·64	9·47	0·200	0·200	0·200	0·200	0·200
$s_x = s_y = 0·05$	50	16·90	11·92	0·200	0·200	0·200	0·200	0·200

Experiment (e)	t	x_t	y_t	Proportion of prey strain with C-value of 4	2	1	0.5	0·25
$R = 4$	0	10·00	10·00	0·200	0·200	0·200	0·200	0·200
$h = 0·02$	5	8·66	13·00	0·200	0·200	0·200	0·200	0·200
$a = 0·1$	10	15·44	16·70	0·200	0·200	0·200	0·200	0·200
$f = 100$	15	17·46	3·78	0·200	0·200	0·200	0·200	0·200
$s_x = s_y = 0·05$	50	37·48	0·00	0·200	0·200	0·200	0·200	0·200

continued

(b) Competition among predator strains (model (35)).

Experiment (a)	t	x_t	y_t	Proportion of predator strain with a-value of				
				0·4	0·2	0·1	0·05	0·025
$R = 4$	0	10.00	10·00	0·200	0·200	0·200	0·200	0·200
$h = 0·02$	5	3·96	0·41	0·067	0·032	0·001	0·000	0·000
$f = 100$	10	0·00	2·74	0·999	0·001	0·000	0·000	0·000
$C = 1$	15	0·00	0·00	—	—	—	—	—
$s_x = s_y = 0·05$								

Experiment (b)	t	x_t	y_t	Proportion of predator strain with f-value of				
				16	8	4	2	1
$R = 4$	0	10·00	10·00	0·200	0·200	0·200	0·200	0·200
$h = 0·02$	5	14·51	9·75	0·551	0·308	0·114	0·025	0·003
$a = 0·2$	10	12·47	1·96	0·698	0·254	0·045	0·003	0·000
$C = 1$	15	0·49	4·29	0·847	0·145	0·008	0·000	0·000
$s_x = s_y = 0·05$	20	29·86	0·00	—	—	—	—	—

Experiment (c)	t	x_t	y_t	Proportion of predator strain with C-value of				
				4	2	1	0·5	0·25
$R = 4$	0	10·00	10·00	0·200	0·200	0·200	0·200	0·200
$h = 0·02$	5	0·087	1·74	0·969	0·030	0·001	0·000	0·000
$a = 0·1$	10	16·72	0·00	—	—	—	—	—
$f = 100$								
$s_x = s_y = 0·05$								

(c) Competition between predator strains of "specialists" and "generalists" when there are two species in the prey (model (36)).

Parameter conditions	t	Prey density of species 1 ($x_{1,t}$)	Prey density of species 2 ($x_{2,t}$)	Total predator density (y_t)	Proportion of predator strain: generalists ($a_{11} = a_{21} = 0.05$)	Proportion of predator strain: specialists ($a_{12} = 0.1$, $a_{22} = 0$)
$R_1 = R_2 = 4$	0	5·00	5·00	10·00	0·500	0·500
$h_1 = h_2 = 0{\cdot}02$	5	27·06	32·71	25·53	0·715	0·285
$f = 100$	10	8·05	23·10	5·42	0·991	0·009
$C = 1$	15	8·71	9·42	31·18	0·998	0·002
$s_{x,1} = s_{x,2} = 0{\cdot}05$	50	28·78	28·88	8·35	0·998	0·002
$s_y = 0{\cdot}05$						

(d) Competition between prey strains of "specialists" and "generalists" when there are two species in the predator (model (37)).

Parameter conditions	t	Total prey density (x_t)	Predator density of species 1 ($y_{1,t}$)	Predator density of species 2 ($y_{2,t}$)	Proportion of prey strain: generalists ($a_{11} = a_{12} = 0.05$)	Proportion of prey strain: specialists ($a_{21} = 0.1$, $a_{22} = 0$)
$R = 4$	0	10·00	5·00	5·00	0·500	0·500
$h = 0{\cdot}02$	5	23·93	15·71	2·29	0·741	0·259
$f_1 = f_2 = 100$	10	25·68	9·22	3·35	0·970	0·030
$C_1 = C_2 = 1$	15	23·80	9·22	5·25	0·992	0·008
$s_x = 0{\cdot}05$	50	24·49	7·26	7·09	0·999	0·001
$s_{y,1} = s_{y,2} = 0{\cdot}05$						
$w = 1$						

Table 6

Summary of directions of selection pressure in either prey or predator species to be imposed on the individual parameters of the basic model of predator–prey interaction.

	Parameters						
Selection in	R	h	s_x	a	f	C	s_y
Prey	+	−	−	−	0	0	0
Predator	0	0	0	+	+	+	−

+: increase; −: decrease; 0: neutral.

How, then, and at what levels of compromise would the conflict generally reach a state of balance? If the prey has priority in this evolutionary game relative to its predator, then the system as a whole would be predicted to persist stably and not only the prey's but also the predator's population may be able to maintain high levels. But if, conversely, the predator has priority, then such stable persistence of the system would become difficult to attain for both populations against their common interests.

From a theoretical viewpoint, it seems certain that in the coevolution of predator and prey, the prey takes in most cases the initiative relative to the predator, i.e., that the interactive system has a general tendency to evolve towards stable coexistence of both species. A very simple argument about this is as follows.

In the process of such conflict between a predator and its prey in determining the values of joint parameters such as searching efficiency, a, characterizing the system behaviour, a simple principle of "blind man's buff", so to speak, will usually operate, thus assuring the prey's potential priority in relation to the predator in their evolutionary competition, provided that any mutation for the coevolution occurs in random directions.

Suppose, for example, that the prey has n potential escape strategies to reduce the a-value from the original level a_1 to a_2, each having an equal probability of being acquired by mutation, p, per generation, and that there also exist the corresponding counter strategies which the predator is capable of developing with the same probability p and which, once developed, would restore the reduced a to its original level a_1 again. It is then obvious that the probability of the predator developing the specific counter strategy necessary to restore the reduced a-value is as small as only $1/n$ as compared with that for the prey to develop any one of the n escape strategies to reduce the original a-value to a_2. This is to say that the expected total length of the period during which the a-value is kept at the reduced level a_2 throughout this evolutionary seesaw game would be n times as long as that during which it remains at the original level a_1. In this "blind man's buff" game, therefore, the inherent priority of the escaping prey over the pursuing predator seems to be apparent, although the prey's complete escape would not occur here since it may always wait, once escaped, for the pursuing predator until it is again caught up with, owing to the resultant relaxation of selection pressure on the prey's side.

Beside these mechanisms for single prey–single predator systems, the selection in either predator or prey for increasing the number of prey or predator species comprising the whole interactive system as discussed before may also be regarded as important and as another probable driving force working for system stabilization against the selective pressure on the predator to raise its searching efficiency.

The above reasoning may give an answer to commonly asked questions such that: "Why does a predator not evolve to be so efficient that it drives its prey and ultimately itself to extinction?"; or "Why does the prey not evolve an escape strategy that drives the predator to extinction?" (Slatkin and Maynard Smith, 1979). There seems to be little need to invoke here such explanations by "group selection" as did Gilpin (1975) whose arguments are based on rather limited assumptions.

In the foregoing section we have seen that for regulation of prey population density by predation to occur requires some very restricted conditions to hold, such as high a, and rather low and stable R, even in some of the extended models which have the background for regulation to occur. In this connection, the above results lead us to the conclusion that predator–prey coevolution may rarely, if ever, give rise to systems which would enable predators to regulate the populations of their prey. Namely, the most likely situation to be realized in the coevolution of a predator and its prey may be their stable coexistence without regulatory control of the latter by the former. Validity of such predictions will be discussed in the following sections with reference to the dynamics in natural interactive systems.

VI. PREDATOR–PREY DYNAMICS IN EXPERIMENTAL POPULATION SYSTEMS

The development of mathematical theories of predator–prey interaction since Lotka (1925), Volterra (1926) and Nicholson and Bailey (1935) has been followed by a number of attempts to test their reality in experimental population systems in the laboratory, such as those of Gause (1934) using protozoan systems and those of Utida (1953, 1957a) and Huffaker (1958) using arthropod systems. Apart from these intentional experimental studies, there also have been many attempts among entomologists to release predators or parasites for biological control of insect pests, which may be interpreted also as artificial realization of predator–prey interactions under semi-natural conditions.

In this section we shall overview these experimental attempts, focusing attention to the reality and the actual conditions for (1) reciprocal population oscillations of predator and prey; (2) stable persistence of their interactive system and (3) regulation of prey population by predation, in reference to the theoretical principles deduced in preceding sections.

A. Protozoan Predator–Prey Systems in the Laboratory

As is well known, Gause (1934) made an important pioneer work in this field using *Paramecium caudatum* as prey and *Didinium nasutum* as predator. In this study it is particularly notable that a reciprocal population oscillation of predator and prey could be actually generated in such a way as the Lotka–Volterra model had predicted on a theoretical basis, but that the persistence of such interaction for more than one cycle proved to be virtually impossible in his simple, uniform experimental universe, unless regular recruitment of prey and predator individuals was made to the system.

Further development of this classic study has been made by Luckinbill's (1973) elaborate experiments using *Paramecium aurelia* as prey and the same *D. nasutum* as predator. In this study he illustrated that suppression of the mobility of both predator and prey by addition of methyl cellulose to the culture medium remarkably stabilizes the interaction, allowing the system which would otherwise have been led to crash to persist in limit-cycle oscillations for a long period over several cycles. He also could promote stability of the interactive system by controlling the supply of bacteria as prey's food, as well as by increasing the size of experimental universe (Luckinbill, 1974).

These protozoan systems in which both predator and prey reproduce themselves by binary fission may correspond to the differential-equation models such as Eqn (14) in Section II, and the main results summarized above seem to be explicable in terms of the basic principles derived from these models. The observed system stabilization due to their introduced refuge or increased size of the experimental universe may correspond to the stabilizing effect of increasing the p-value in the model (14), whereas the stabilization due to methyl cellulose, to the stabilizing effect of decreasing the a-value (see Figs. 4(c) and 6(a)). Also, the stabilization due to limiting bacterial food for the prey may be interpreted as representing the stabilizing effect of increasing the h-value, i.e., as realization of the "paradox of enrichment".

As for the prey population regulation due to predation, it is unlikely that in any of these experiments the predators have succeeded in "regulating" the prey density in the strict sense defined before, since the prey populations there were generally kept at fairly high levels often exceeding half the control level at peaks of the oscillation. This may be because the experimental universe was inevitably so small and so homogeneous that there was little room for appropriate refuges to exist consistently and thus attain the regulation of prey density at low levels.

It may therefore be necessary if these predators are to regulate the density of their prey that the universe be more heterogeneous in structure and yet the searching efficiency still remain fairly high. One example suggesting such a situation can be found in van den Ende's (1973) experiment in which stable persistence with decreasing amplitude of oscillation of the protozoan predator, *Tetrahymena pyritorensis*, and its bacterial prey, *Krebsiella aelogones*, was observed over a period as long as 1100 hours (about 40 predator generations). It seems probable that in this system the prey population was actually subjected to regulation by the predator, since its density had been consistently kept at very low levels – as low as about 1/10 000 of the control level – throughout the study. Van den Ende (1973) suggested the reason for this stable persistence of the interactive system to be that the wall of the experimental vial had functioned as effective refuge for the prey.

Another interesting suggestion that van den Ende (1973) made in this study is that some genetical change might have occurred in the prey's population in such a way that the tendency for prey individuals to be attracted to the wall was increased, and thereby the oscillations were damped.

B. Arthropod Predator–Prey Systems in the Laboratory

For arthropod predator–prey systems the first attempt to test the reality of theories from mathematical models was made by DeBach and Smith (1941), who succeeded in fitting their experimental data on the house fly, *Musca domestica* and its chalcid parasitoid, *Nasonia vitripennis* to the Nicholson–Bailey model for the period of seven generations. Later, Watanabe (1950) and Burnett (1958), using the same approach, failed in one case and succeeded in another in fitting the Nicholson–Bailey model, using different combinations of insect host–parasitoid pairs. However, these experiments may not be regarded as any realization of the predator–prey interaction with which we are concerned here, but simply as hypothetical simulations to assess its probable outcomes, because the density of the prey population was artificially set every generation by multiplying the density of survivors in the previous generation by a predetermined constant R, and hence it is only the constancy of search efficiency a against varying prey densities and nothing more that they could test in this procedure of "fitting" to the model.

Utida (1953, 1957a, 1957b), on the other hand, made elaborate experiments using the bean weevil (*Callosobruchus chinensis*)–braconid wasp

(*Heterospilus prosopidis*) system, in which the true interactions between them could be realized and maintained for as long as more than 110 generations with fairly regular, reciprocal oscillations of limit-cycle type. There can be no doubt that these definite oscillations were those that had been generated entirely by predator–prey interactions, having provided a clear illustration of the prediction from mathematical models. It should be noted, however, that even in his bean weevil–parasitoid systems realization and persistence of such regular interactions was not so easy. For example, distinct cycles could not be generated in the combination of the same host with a different parasitoid, *Neocatolaccus mamezofagus* (Utida, 1957b), and even in replication experiments of the same *Callosobruchus–Heterospilus* combination, some systems failed to persist for more than a few cycles (Utida, 1953).

In Utida's *Callosobruchus–Heterospilus* system, the percentage of parasitism was around 60% on average, and the host population oscillated around a level which was fairly high relative to the food resource quantity, reaching several hundreds per 20 g of beans at oscillation peaks. It is therefore difficult to be certain that the weevil population there was under the parasitoid's regulation. Even here the main mechanism for its regulation is assumed to have been the intraspecific competition for a limited food resource.

After Utida a number of similar attempts have been made by using experimental predator–prey systems of different grain insects (e.g., Takahashi, 1959, 1969; Hassell and Huffaker, 1969; Podoler, 1974; Benson, 1974), in most of which long-term persistence of the predator–prey system could be attained under appropriate conditions. Of these Takahashi's (1959) study may be worthy of special note in reference to the theories discussed in Section III, since he showed clearly that the oscillation due to predator–prey (ichneumonid parasitoid, *Nemeritus* (*Exidechtis*) *canescens* and flour moth, *Ephestia* (*Cadra*) *cautella*) interaction can be stabilized and the system persistence ensured, simply by increasing the depth of culture medium (rice bran), and thereby providing a refuge for moth larvae to which the wasp's ovipositor can no longer reach. It was also shown not only in Takahashi's (1969) but also in other workers' studies that the prey density levels in these interactive systems are generally too high relative to the carrying capacity to assume that they are being effectively regulated by the predator's activity.

Beside these studies on insects, there also have been a number of important contributions to this subject since Huffaker (1958) in which phytophagous mites and their mite predators were used as experimental materials. As is well known, Huffaker (1958) succeeded by elaborate experiments in maintaining regular reciprocal oscillations for a period of three cycles in the interactive system of the predatory mite, *Typhlodromus occidentalis*, and its prey, *Eotetranychus sexmaculatus*, on a number of orange fruits. Actually, however, it was by no means easy here, too, to enable the system to persist for a long period; he was obliged to use a special treatment to limit the predator's migration among different patches (oranges) with Vaseline barrier and to promote the prey's migration with the aid of an electric fan.

The conditions for persistence in mite predator–prey systems were elucidated more clearly by Takafuji (1977) with reference to the between-patch migration of both predator and prey within the habitat. Using *Phytoseiulus persimilis* as predator and *Tetranychus kanzawai* as prey in the experimental universe composed of a number of kidney bean plants, he proved that the length of the period of persistence of this interactive system is negatively correlated to how large (and hence how few) are the individual patches of host plants comprising the universe and how closely the individual plants within each patch are connected to each other. From these studies, it has now become evident that reduction of the local extinction rate of the prey population by promoting its migration as well as by limiting the predator's movement is one of the essential conditions for stable coexistence of interacting predator and prey populations (Takafuji *et al.*, 1983).

In the context of theoretical models described in Section II, such effects of migration as illustrated in these experimental systems may be interpreted in terms of parameters a and p. Namely, limiting predator migration may correspond to reduction of search efficiency a, while promoting prey migration, to increase of the population of prey protected by a refuge, p, both having stabilizing effects on the system behaviour. As we have seen in Section IV, the changes in parameters a and/or p in these directions produce situations where it is generally difficult for predators to regulate the population of their prey consistently at low levels. In fact, it can be seen here again that prey population levels in these mite predator–prey systems, having persisted in a simple experimental universe, were too high in general to allow one to interpret them as the outcome of effective regulation by their predators.

Finally, a series of studies Pimentel and his collaborators have made on their housefly–parasitoid (*Nasonia vitripennis*) system may be worth referring to. Their studies differ from those we have discussed above in two points. First, the systems set up are not truly "interactive", since the prey pupae were supplied at a constant rate to the predator regardless of the number so far attacked. Secondly, their main interest lies in detecting some evolutionary changes in predators, rather than describing patterns of actual population interaction, which may result from the interaction through "genetic feedback mechanisms" in Pimentel's (1961)'s sense. In these studies, it has been shown that by artificial selection over 20 generations for the individuals with lower ability for host searching, the parasitoid population was changed so as to show dynamic patterns that are remarkably different from those of the non-selected control one, resulting in considerably lower average levels due to lower reproductive ability (Olson and Pimentel, 1974; Pimentel *et al.*, 1978). These findings are of special interest in reference to our models in Section V, since they have proved the possibility that such changes along the line of reciprocal evolution as postulated in those models could actually occur within such a short period of time.

C. Cases of Biological Pest Control in the Field

In the field of applied entomology there have been many attempts to achieve the biological control of pests by introducing new effective natural enemies to their natural (or semi-natural) population systems, many having attained successes, whereas many others having ended in failure (e.g. DeBach, 1964; Huffaker and Messenger, 1976).

These attempts can all be regarded as field experiments, regardless of whether or not the results were successful, on predator–prey interactions in the real world. In fact several suggestions may be obtained by reviewing the records of these trials that may be helpful for interpretation of the principles of predator-prey interaction in nature. First of all it may be noted that persistent interactions with such clear patterns of reciprocal cyclic oscillations of predator and prey as mathematical models predict have never been found in these field studies. This may be because there are so many disturbing factors in the field, which are enough to mask any tendencies for actual systems to show regular fluctuations inherent in the predator–prey interaction.

In the meantime, however, it is also an evident fact that in not a few cases introduced predators have succeeded in maintaining very close interactions with their prey, not only suppressing their populations at levels far lower than the original ones temporarily, but also maintaining such low prey levels for a long period thereafter. For example, Beddington *et al.* (1978) estimated the reduced population levels of pests to be as low as some 0·3 to 3% of their original capacities in several cases of successful biological control, such as the winter moth (*Operophthera brumata*)–parasitic fly (*Cyzenus albicans*) system in Canada (Embree, 1965, 1966), the larch sawfly (*Pristophora erichsoni*)–parasitic wasp (*Olesicampe benefactor*) system in Canada (Ives, 1976), and several systems of coccids and their parasitoids in USA (e.g., DeBach *et al.*, 1971). The situation common to all these cases is, of course, that the predator plays a very important role as a mortality factor in the prey population dynamics, which means that the predator's search efficiency in terms of parameter a has been kept extraordinarily high throughout. There can be no doubt that a high value of a is among the most important conditions that enables predators to suppress their prey's populations to low levels through intimate interaction. The historical fact that successes in biological control have been more readily realized in sedentary pests such as coccids than in other free-living insects may also indicate this.

In theoretical models we have seen before that an increase of a-value surely results in more effective suppression of population levels of the prey, but this can be attained only in exchange for a marked reduction in system stability. Why, then, did the above-mentioned pest–enemy interactive systems succeed in persisting for fairly long periods with such low pest densities? Two explanations have been given to this problem, one assuming habitat heterogeneity which provides refuges for prey (e.g., Beddington *et al.*, 1978) and the other postulating local aggregation of predators to habitat patches with high prey densities (e.g., Hassell, 1980). Notice, however, that from the mathematical viewpoint, there is no essential difference in the mechanism leading to system stability between these two probable causes that are ecologically different, in either case generating the situation that there always exist some prey individuals that are less likely to be attacked by predators than others.

The former explanation has been supported so far by many authors and it really seems to be reasonable, in view of the aforementioned theoretical result showing intense stabilizing effects of the habitat heterogeneity, which in fact is a condition prevailing generally among field populations of insects. As for the latter explanation, on the other hand, there still remain some questions about its reality. For example, Murdoch *et al.*, (1984) and Reeve and Murdoch (1985) expressed doubt about this hypothesis, due originally to Hassell and May (1974), since in their analyses of field population systems of coccid pests (*Parlatoria oleae* and *Aonidiella aurantii*) being well suppressed by their introduced wasp parasites (*Aphytis paramaculicornis* and *Aphytis melinus*, respectively), they could in neither case detect any marked tendency for the parasitoids to aggregate in parts with high prey densities within the habitat.

Another point of interest in these successfully controlled field population systems is whether the pest populations are being truly "regulated" by their predators or they are being merely suppressed by the non-regulatory killing power of the predators as seen in simulations of Fig. 13. There are two opposing viewpoints about this problem, too. Authors such as Beddington *et al.* (1978) and Hassell (1980) apparently tend to accept regulation by predators. Some others such as Murdoch *et al.* (1985), in contrast, seem to be opposed to this assumption, expressing doubt about the premise itself that these systems are really in a state of stable equilibrium.

At present we have no reliable data to provide any definite answer to this problem for any one of these species, since even in such well studied cases as the winter moth or the larch sawfly in Canada no consistent follow-up surveys have been made after the success of biological control, unfortunately. But in view of these case studies, it seems to be not unnatural to assume, at least in some of them, that predators are really regulating populations of their prey at low levels. This is because the situations characteristic of these systems, i.e., able predators attacking their prey in more or less heterogeneous habitats, appear to be such that predators' regulation of prey density is rather likely to occur, as the theories in Section IV have shown.

Beddington *et al.* (1978) made an interesting comparison of the prey population levels attained through the interaction, between successfully established laboratory systems and these field systems of insect host–parasitoid interaction, and found that the ratio of the prey population levels to those to be attained in the absence of predators were much higher generally in the former than in the latter. The implication of their finding seems to be clear. In the laboratory systems such as Utida's (loc. cit.), the universe appears to have been so small and homogeneous that only the systems with parasitoids which had low a-values, and hence low ability to regulate prey populations, could attain their long-term persistence with high prey densities, whereas quite the reverse might have been the case in the field systems under successful biological control as discussed above. It thus may be concluded that in these field systems where environmental heterogeneity is an invariable attribute of the habitat, high searching efficiency of the introduced predators is the most important condition among others that enables them to control (or possibly to regulate in some cases) the pest populations successfully at low levels.

VII. PREDATOR–PREY DYNAMICS IN NATURAL POPULATION SYSTEMS

Up to the present a large number of intensive field studies have been accumulated on the dynamics of natural animal populations, though there is an apparent bias in the object species chosen to some specific taxonomic groups, such as insects, birds and mammals. The final step of this study is to explain and interpret the behaviour of natural population systems as revealed in these studies in the light of the predator–prey theories we have derived in the preceding Sections. The points to make clear here are: (1) whether population oscillations due to the predator–prey interaction can really be observed in nature as mathematical models predict; (2) whether the predation as a whole is actually important as a mortality- and/or a key factor in the dynamics of individual animal populations; and (3) whether some animal populations in nature are actually "regulated" by their predators as has been often believed so far.

A. Insect Populations – Epidemic Species

By "epidemic species" we mean here those species whose populations periodically reach, through progressive outbreaks or "gradation", levels which are so high that they may exhaust much of their food resource in the habitats. Although such species are rather few in number, they include a number of famous forest and agricultural insect pests on which intensive field population studies have been made. In discussing population dynamics of these species, an intimate relationship with their natural enemies may be postulated, since they often show cyclic population fluctuations of large amplitude, at the declining phase of which the mortality due to natural enemies generally shows a marked increase.

As typical examples of those epidemic pests, we refer here to the three well-known species, the larch bud moth (*Zeiraphera diniana*) in Switzerland, the spruce budworm (*Choristoneura fumiferana*) in Canada, and the black-headed budworm (*Acleris variana*) in Canada.

The larch bud moth is famous for its beautiful population oscillations with a regular cycle of some 9-years and extremely large amplitude of more than 10,000 times (Baltensweiler, 1984). Auer's (1968) analysis proved that the natural enemies such as parasitic wasps comprise its important mortality factors whose action has a general tendency to be strengthened with increasing density of the bud moth, larval parasitism, e.g., amounting sometimes to 80% or so around outbreak peaks. He thus postulated the special importance of these natural enemies in determining such a characteristic pattern of the bud moth population dynamics.

The key-factor analysis made later by Varley and Gradwell (1971), however, has clearly proved that it is not the natural enemies such as parasitoids or diseases but some "residual" factors that mainly determine the dynamic oscillations characterizing this epidemic bud moth population. The actual contents of this "residual" component has recently been proved to be physiological degeneration of the host tree caused from the defoliation after the moth outbreaks, which in turn suppresses markedly the moth population thereafter and requires a period of several years to recover entirely (Benz, 1974; Fischlin and Baltensweiler, 1979). It is therefore evident now that the natural enemies in the population system of this species, despite their general importance as mortality agents, neither function as the key factor governing its dynamic oscillations nor as the regulatory factor to keep it stable at low levels. The bud moth population dynamics are thus considered to be governed not by interaction with its natural enemies as was initially postulated, but by the complicated interaction with its host plant through delayed effects of its own crowding.

In the case of the spruce budworm, very intensive life-table studies were done on its epidemic population for more than 15 years (Morris, 1963). Its eggs, larvae and pupae were all proved to be attacked by various parasites and predators, of which larval parasitoids *Meteorus tachynotus* and *Actic interrupta* contributed the highest percentages of some 40% on average. It i therefore evident that natural enemies are important as mortality factors in this species, too. The result of key-factor analysis, however, showed tha none of these were functioning as key factors for the overall population trends of the budworm from generation to generation, the real key factor being the mortality of old larvae due to unknown causes. Density-dependence was detected in the rate of attack by some of these natural enemies but its degree was so weak both in each species and in total that one canno assume any possibility at all for them to be functioning as a major regulatory factor in the dynamics of the budworm population (Morris, 1963).

The population of the blackheaded budworm studied by Morris (1959 and Miller (1966) is somewhat different from those of the above two specie in that the impact of parasitoids on its dynamics is more conspicuous Namely, its larvae has been proved through their 18-year study to be attacked by many parasitoid species such as *Microgaster penonica, Actia diffidus*, and so on, often at very high rates of more than 90% in total, and yet their overall influence proved to be so variable that it apparently comprised one of the key-factors governing periodical dynamics of the moth population (Miller, 1966; Podoler and Rogers, 1975). Even in this case however, only weak, if any, density-dependence could be detected in the action of these parasitoids, and no indication was obtained after all to justif the assumption that the budworm population is being regulated or effec tively stabilized by these natural enemies.

The general conclusion we can deduce from these intensive studies may thus be as follows. Predation in general is fairly important as a mortality factor for populations of these epidemic insect pests, and may in some special cases also be functioning as one of the key factors causing thei fluctuation. But the role it is playing in their dynamics as a regulatory facto is negligibly small. This conclusion may also apply to many other epidemi insect populations as well, which include those with less regular, sporadi types of outbreaks such as the population of the cinnabar moth studied b Dempster (1975; 1982) (see, Dempster, 1983).

B. Insect Populations – Endemic Species

"Endemic species" here means those species whose populations are kep always at levels far lower than those their habitats may potentially allow, an rarely exhibiting a tendency to outbreaks as in epidemic species describe above. The problem of major interest here is to know whether or not the persistence of these populations at such low levels is the outcome of control by their predators, and then to elucidate, in those cases where the answer is "yes", if that suppression is truly regulatory or not.

Now we take up the three typical cases of such species recently studied in Japan, the rice leafhopper (*Nephotettix cincticeps*) (Hokyo and Kuno, 1977), the thistle lady beetle (*Henosepilachna niponica*) (Ohgushi, 1983; Ohgushi and Sawada, 1985) and the citrus swallowtail butterfly (*Papilio xuthus*) (Hirose *et al.*, 1980). In all these cases intensive life-table data have been collected, adequate to give detailed information for answering the above problem. Reanalysis of data for the three species is made on the common basis, using the technique which was used and proved efficient in simulation experiments of Section IV. The result is shown in Table 7.

The rice leafhopper is a multivoltine species and the data here analysed cover 12 complete generations (two in succession for each year) observed in a paddy field during 1968–1974. It has been well clarified that this species is a typical endemic or non-outbreak type pest whose population is kept stable consistently at levels far lower than the host plant can carry and whose economic importance as a pest comes mainly from the transmission of some virus diseases, not from direct injury (Kuno and Hokyo, 1970).

In this study by Hokyo and Kuno (1970), census data have been obtained for the four successive stages, eggs laid (x_E), larvae that hatched (x_L), adults that emerged (x_P) and adults that escaped parasitism (x_A), giving the four rates of population change, i.e., egg survival rate ($S_E = x_L/x_E$), larval survival rate ($S_L = x_P/x_L$), adult survival rate from parasitism ($S_P = x_A/x_P$) and realized fecundity per healthy adult ($F = x_E/x_A$). Egg mortality in this species is largely attributable to parasitic wasps (*Anagrus* spp.), while larval mortality, to several species of spiders such as *Lycosa pseudoannulata* (e.g., Kiritani *et al.*, 1970; 1972). Parasitism of adults is due to some species of pipunculid flies (*Alloneura* spp.) which lay eggs into host larvae. Thus, the overall importance of these natural enemies as mortality factors of the hopper as evaluated by values of S_E, S_L and S_P proved to be fairly high in this case. Apparently, however, they cannot be regarded as key factors for its population fluctuation, nor as regulatory factors stabilizing the hopper density at low levels, since the values of slope b in the regression of these three rates on the overall reproduction rate $I (= x_A(i + 1)/x_A(i))$ as well as that on the respective population densities are all relatively low and yet the variance of density fluctuation has a clear tendency to decrease from x_A to x_E (see Table 8). Consequently, it is evident from the table that the component functioning as the real key- and regulatory factor in this leafhopper population is the fecundity F which has been proved to be determined mainly by some intraspecies mechanisms leading to density-dependent dispersal of emerging adults (Kuno and Hokyo, 1976).

Table 7

Key-factor analysis for endemic populations of three insect species, the rice leafhopper (RL), the thistle lady beetle (TL) and the citrus swallowtail butterfly (CS). The original data for respective species are from Hokyo and Kuno (1977, Appendix, Table), Ohgushi (1983, Table 3-1) and Hirose *et al.* (1980, Fig. 6).

(a) Mean and variance (in parentheses) of population density and reproductive or survival rate (in log) at each stage.

Species	$\log x_A(i)$	$\log x_E$	$\log x_L$	$\log x_P$	$\log x_A(i+1)$	$\log I$	$\log F$	$\log S_E$	$\log S_L$	$\log S_P$
RL	0·388	1·809	1·658	1·306	1·182	0·794	1·421	−0·151	−0·352	−0·124
	(0·392)	(0·173)	(0·191)	(0·130)	(0·156)	(0·179)	(0·138)	(0·008)	(0·040)	(0·004)
TL	1·899	3·915	3·302	2·498	1·926	0·027	2·016	−0·613	−0·804	−0·572
	(0·328)	(0·035)	(0·049)	(0·098)	(0·328)	(0·621)	(0·190)	(0·026)	(0·039)	(0·086)
CS	−1·045	1·050	0·696	−0·447	−0·920	0·126	2·095	−0·354	−1·142	−0·473
	(0·252)	(0·055)	(0·034)	(0·190)	(0·161)	(0·351)	(0·198)	(0·039)	(0·134)	(0·070)

x_A: adults that emerged (CS), overwintered (TL) or escaped parasitism (RL); x_E: eggs laid; x_L: larvae that hatched; x_P: Pupae (CS)or adults that emerged (RL, TL); $I = x_A(i+1)/x_A(i)$; $F = x_E/x_A(i)$; $S_E = x_L/x_E$; $S_L = x_P/x_L$; $S_P = x_A(i+1)/x_P$.

(b) Regression analysis for detecting key and regulating factors.

Species	Regression analysis for detection of key factor				Regression analysis for detection of regulation				
	b_{FI}	b_{EI}	b_{LI}	b_{PI}	b_{Ix}	b_{Fx}	b_{Ex}	b_{Lx}	b_{Px}
RL	0·803	0·014	0·249	−0·066	−0·529	−0·455	0·027	−0·263	0·080
TL	0·422	0·111	0·198	0·269	−0·959	−0·736	−0·180	0·105	0·698
CS	0·599	−0·005	0·194	0·212	−0·878	−0·785	−0·547	0·339	−0·260

b_{FI}, b_{EI}, b_{LI} and b_{PI}: values of slope b in the regression of $\log F$, $\log S_E$, $\log S_L$ and $\log S_P$ on $\log I$; b_{Ix}, b_{Fx}, b_{Lx} and b_{Px}: values of slope b in the regression of $\log I$, $\log F$, $\log S_E$, $\log S_L$ and $\log S_P$ on $\log x_A(i)$, $\log x_A(i)$, $\log x_E$, $\log x_L$ and $\log x_P$, respectively.

The result of analysis on the thistle lady beetle population is similarly clear though the study period in this case was somewhat shorter (5 years). This beetle is a univoltine species whose population is regarded as one of the most stable ones among phytophagous insects so far studied. Again four stages were surveyed for life-table analysis, x_E, x_L, x_P and x_A, but the latter two indicate here adults that emerged and adults that overwintered, respectively. The egg- and larval mortality in this population has been proved to be caused largely by predators such as dermapterous insects, whereas the mortality of overwintering adults, by physical factors. As seen in Table 7, both S_E and S_L show fairly low values, indicating importance of these predators as mortality factors for the beetle. But they are again not regarded as key factors for the overall population fluctuation, the key factors of primary and secondary importance here being the fecundity F and the adult mortality S_A during hibernation, respectively. It is also clear from both the trend of variance for the densities at different stages and the result of regression analysis for detecting density-dependence that the main factor responsible for such marked regulatory stabilization of the population of this species is not the predation but again the density-dependent changes in realized fecundity, which has proved to be caused by density-dependent adult dispersal (Ohgushi, 1983; Nakamura and Ohgushi, 1981).

The third study by Hirose *et al.* (1980) on the cirus swallowtail butterfly may be of special interest because it is one of the few cases in which the action of predators (egg parasitoid in this case) has been found to be clearly density-dependent (Dempster, 1983). This species has four generations a year. The original life-table data obtained at "grove A" for 11 generations during 1970–1972 were re-analysed here as to the densities in four developmental stages, x_E, x_L, x_P and x_A, where x_P represents here the density of pupae and x_A, that of emerged adults. The egg mortality here is due principally to parasitic wasps, *Trichogramma* spp., the larval mortality, to general predators such as paper wasps (*Polistes* spp.) or birds, and the pupal mortality, to an ichneumonid parasitoid, *Pteromalus puparum*. Generally low values of S_E, S_L and S_P seen in Table 7 indicate that these natural enemies are acting as fairly powerful mortality factors against the butterfly population. But the regression analysis in Table 7 reveals that the key factor for population fluctuation is again the fecundity F, though both larval and pupal survival, S_L and S_P, also make some contribution to the fluctuation of I. The fecundity F is also nominated as the primary factor for population regulation in this species, not only from its most remarkable density-dependence detected by the regression analysis but also from the trend of variance for the density among different stages. It is evident that both egg and pupal parasitoids, especially the former, are surely acting as density-dependent mortality factors as has been shown in their original paper, but the overall evaluation from Table 7 indicates that in effect even here the parasitoids are playing only a minor role, if any, for regulating the natural population of this butterfly.

Thus, it has been shown for all the three typical endemic populations of insects examined here that predators are not playing any significant role in the real regulation of their populations, as was also the case with the epidemic insect populations mentioned above. Evidently, the situation of low and stable density levels coupled with high mortality due to predators does not necessarily mean that the population is under efficient regulation of these predators. In this connection it may be important to note that in all these cases the population regulation has been attained by density-dependent changes in the realized fecundity at adult stage. We have seen in Section IV (see Fig. 13) that, coupled with strong pressures imposed by predators on immature stages of the population, this provides a condition for the "non-regulatory suppression" of the population by predators readily to occur. Apparently all the three populations analysed here may be interpreted as typical examples showing this type of non-regulatory suppression by predators.

Apart from these studies, we have at present a large amount of information from many studies on a variety of endemic field populations of insects, which include intensive and long-term ones such as Richards and Waloff's (1961) on the broom beetle (*Phytodecta olivacea*), Varley and Gradwell's (1968) on the winter moth (*Operophthera brumata*), Klomp's (1965) on the pine looper (*Bupalus piniarius*), and so on. As Dempster's (1983) extensive review of these studies shows, in none of them were predators found to be functioning as effective regulators of the populations under study. It therefore may be safe to conclude here that even endemic insect populations are not subject to regulation by their predators generally, their regulation at low density levels being attained principally by intraspecies mechanisms, with the aid of non-regulatory suppressions by these natural enemies.

C. Populations of Birds and Mammals

Birds and mammals are the two groups of terrestrial animals other than insects on which many population studies have also been concentrated. It therefore may be worthwhile to overview briefly these studies with reference to predator–prey dynamics, though the population interaction with their natural enemies appears to us not so intimate in these higher animals as in insects.

As for bird populations, it may be sufficient for the moment to refer to Lack's (1966) extensive review work. After critical review of intensive population studies on more than 15 species of birds, he concluded:

> In most of the species studied in this book predation was negligible, and though predators took many red grouse, these were chiefly birds without territories

> which would not have bred, while any owners of territories that died were quickly replaced.

And he further writes:

> If predation, disease, and human destruction were unimportant, food shortage was probably the main density-dependent mortality factor, and for this there was positive evidence in several species.

The red grouse (*Lagopus lagopus*) is known as one of the few bird species which have fairly large and cyclic population fluctuations. According to Watson (1971), however, even this cyclicity in population dynamics cannot be attributed to any action of predators.

As for mammals, populations of some species living in the Arctic regions are known to have clear periodical fluctuations with cycles of some 10 years (e.g., Elton, 1924; Elton and Nicholson, 1942). Cyclic population fluctuations of the snowshoe hare (*Lepus americanus*) and the lynx (*Lynx canadensis*) in North America are especially famous. Initially they were occasionally regarded as representing such reciprocal predator–prey oscillations as predicted from the Lotka–Volterra model. But detailed studies on their populations have later revealed that although the population oscillations of the lynx are dependent largely on those of the hare as its prey, those of the latter are fundamentally generated not from the interaction with the predator, but from the periodical depletion of living resources such as food or shelter for the prey by its own outbreaks (e.g., Lack, 1954; Keith, 1983; Keith and Windberg, 1978). Mathematical analysis of their population fluctuations has also supported this (Gilpin, 1973). Fundamentally the same has been accepted as for the cyclic population fluctuations of the lemmings (*Lemmus* spp.) in North America, too (Lack, 1954; Pitelka, 1957).

Small rodents (*Microtus* spp.) are another mammal group which is also known to show periodical population outbreaks occasionally. But again there has been no indication that these oscillations are more or less caused by interaction with their predators, being explicable by some intraspecies mechanisms in relation to either biological or physical conditions of the environment (e.g., Krebs *et al.*, 1973).

Actually, most populations of mammals other than these have much more stable dynamics, rarely showing such outbreaks. Among these "endemic" populations, little evidence has so far been obtained proving that they are regulated by predators. The story of the eruption of the mule deer (*Odocoileus hemionus*) population at the Kaibab North Plateau in North America has long been accepted as one such example, since the outbreak there was observed after intentional removal of many predators such as the puma and the coyote from the habitat (e.g., Allee *et al.*, 1949). But recent reanalysis of the data on this Kaibab deer herd has shown that this explanation about the cause of outbreak may be unreliable. According to Caughley

(1970), not only predators but also many sheep and cattle living there had also been killed before the deer eruption occurred, and he supposes that the deer outbreak was caused directly by the sudden increase of food supply due to the elimination of their competitors, rather than by the elimination of predators.

The population persistence at low density levels that are observed in small mammals such as voles has usually been explained by intraspecies mechanisms such as density-dependent dispersal (e.g., Lidicker, 1961). Erlinge *et al.* (1984), on the other hand, reported a notable (but perhaps rather special) case in Sweden that field populations of voles (*Microtus agrestis* and *Apodemus sylvaticus*) are suppressed by several species of generalist predators at low density levels. In this habitat there also lives the hare, *Oryctolagus cuniculus*, beside the voles as another herbivore more abundant in numbers and these authors inferred with the aid of a simulation model that as an alternative prey this hare population supports and supplies considerable numbers of generalist predators from year to year, which may in turn regulate the coexisting vole populations consistently at low levels through their habit of switching. However, although their inference is of interest in reference to the "switching" model described in Section IV, it still lacks any direct proof for the regulatory response in predators and may need further data for its confirmation.

To conclude, populations of birds and mammals appear to be more free from predators' attack as compared with those of insects. As a result the influence of predators upon the population dynamics of these warm-blooded animals may generally be rather minor even as a simple mortality factor. It thus may follow, as was also the case even with insect populations discussed before, that in most, if not all, cases of bird and mammal populations predation is not playing any major role either in determining or in regulating their dynamics.

VIII. DISCUSSION AND CONCLUSIONS

The main principles of predator–prey interaction that we have deduced from theoretical analyses in Sections II to V may be summarized as follows.

(1) The predator–prey interaction is fundamentally a process which brings about disturbance to the overall system rather than stabilization, having an innate tendency to generate reciprocal oscillations in both populations.

(2) The equilibrium levels and the stability of interacting predator–prey populations are affected in various ways by individual parameters characterizing the respective populations and the habitat concerned (see Table 2). It is noticeable that the conditions that are regarded as advantageous for individuals of either species, such as high carrying capacity of the habitat for

prey and high efficiency of prey searching or reproduction for predators, usually act against the stable persistence of the overall system, often increasing the danger of a crash due to either decline or destabilization of the equilibrium levels. In short, the predator–prey interactive system is fundamentally a delicate entity filled with contradictions, which arise from the fundamental paradox that predators are obliged to exploit and suppress continuously the very population of their prey that are the indispensable resource to assure survival and propagation of the predators themselves.

(3) Theoretically, it is surely possible for predators to achieve by themselves strict regulation of prey populations at low density levels. For this to be realized, however, some specific conditions must be satisfied. They are: (a) very high searching efficiency of the predators; (b) fairly low (but not too low) reproductive capacity of the prey; (c) a stable environment; (d) a habitat which is moderately heterogeneous to provide refuges for the prey or, in case of generalist predators, a habit of switching to currently available prey species. The state of "regulation" here is, therefore, quite fragile, being liable to shift to a state of non-regulatory coexistence of both populations if any one of the above conditions is violated. Under some situations, prey populations can be suppressed at low levels by their predators without any density-dependent process, i.e., the state of "non-regulatory suppression" can occur.

(4) The paradoxes involved in predator–prey interactive systems may inevitably produce a conflict of interests between individual and population in a variety of aspects in their coevolution, which may become serious particularly on the predator's side. Namely, selection on the predator's side is assumed to proceed in such a direction that it raises their efficiency of prey capture and utilization, which might ultimately be maladaptive not only to the prey's but also to their own population, destabilizing the system towards a crash due to overexploitation of the prey. At the same time, however, the individual selection on prey's side is assumed to occur in the reverse direction such that it reduces the predator's capture efficiency, increases the prey's reproductive rate, and thereby reduces the danger of extinction of the system. There is a sound theoretical reason to think that in this predator–prey evolutionary race, the prey in general may have the priority over their predators, and hence the resultant predator–prey systems are likely to be persistent and stable ones in which both predators and prey can successfully coexist. Evidently, the situation that prey populations are strictly regulated by their predators at low density levels contradicts this course of coevolution and is therefore unlikely to be brought about as an outcome of natural selection.

As we have seen in Section VI, some of these principles could be confirmed in the predator–prey dynamics in experimental population systems so far studied.

On the other hand, from the review in Section VII of a wide variety of empirical data so far accumulated on natural populations of insects, birds and mammals, we may draw the following general conclusions as to the characteristics of predator–prey interaction in natural ecosystems.

(1) In natural animal populations, clear-cut reciprocal oscillations due to predator–prey interactions such as are predicted from mathematical models have never been observed. Although some populations such as those of the larch bud moth, the snowshoe hare and the lemming are known to show regular oscillations, the oscillations in all these cases have proved to be caused by interaction, not with their predators, but with the plants as their food.

(2) The role of predators as mortality factors in natural animal populations is fairly important in general, though it of course varies with species, with a general tendency to become less significant in higher animals which usually have developed effective anti-predator defence strategies.

(3) The importance of predators as key factors in population fluctuations is, however, usually minor even among insect populations, and it seems to be virtually negligible in most populations of birds or mammals.

(4) The role of predators in regulating natural animal populations as density-dependent factors is even more trifling. There actually are no natural populations, regardless of whether they are epidemic or endemic, which have been proved to be regulated by their natural enemies, the main regulatory factor almost invariably being the intraspecies competition or interference for restricted food or other related resources.

(5) Thus, animal populations in nature are never in the state of a delicate dynamic balance with their predators as is postulated by the classic Lotka–Volterra model, nor are they subject to the consistent regulation by predators as ecologists have sometimes assumed so far. Instead, they appear to be generally in a state of more loose but fairly stable and robust coexistence with their predators. In other words, most predators in nature are behaving, in effect, as "prudent predators" from the population's viewpoint.

Now we are in a position to discuss the ecological and evolutionary significance of these empirical conclusions in the light of the theoretical ones described before.

First of all it may be necessary to explain why definite oscillations due to predator–prey interaction have never been observed in natural animal populations. There is no doubt now that the tendency to generate reciprocal population oscillations is intrinsic to the behaviour of any predator–prey system. But it is also an evident fact that in nature there exist a number of conditions that have been proved either to damp or to disturb such an innate tendency for oscillation. For example, we know that any prey species in nature is almost always attacked by not a single but several predator species which themselves are very often polyphagous, in turn. We also know that the natural habitats of animal populations are almost always heterogeneous and yet subject to either regular or stochastic environmental disturbances from time to time. Also, the population structure itself may not be homogeneous in either prey or predator species. Accordingly, it turns out to be no wonder that real predator–prey population systems in nature never do show any such clear-cut oscillations but persist robustly with more or less irregular fluctuations. This is to say that the state of reciprocal oscillation in predator–prey systems is fundamentally so fragile that it can be readily masked or broken down by disturbance from the outside.

The fact that most, if not all, animal populations in nature are not subject to the regulatory control of their predators but are being maintained in the state of non-regulatory coexistence with these predators is similarly easy to interpret. In addition to the features of natural ecosystems as described above, it should be really difficult in nature to attain such a high efficiency of prey search by predators and yet to keep such a low fecundity of their prey that are sufficient for the prey population regulation by predators to occur, since these conditions are apparently opposed to the probable direction of their coevolution.

Thus, in view of the theoretical principles summarized before, all these facts consistently indicate the impossibility or difficulty for a situation to occur in nature that enables predators to function as effective regulators of the populations of their prey. Generally speaking, there seems to be no need here to introduce any idea like "group selection" to explain such apparent "prudence" shown by predator populations in nature.

If there is any possibility of true regulation of the prey population being realized in natural predator–prey interactions, it might be only through the attack by powerful generalist predators having the habit of efficient switching, as postulated in Erlinge *et al.* (1984) or in models (27) and (28) in this study. In such a case, increasing search efficiency of predators may not necessarily contradict the natural trend of predator–prey coevolution, since the evolutionary response on the prey's side to such generalist predators would be much less sensitive than to specialist ones. At present, however, we have as yet no definite field evidence to confirm this supposition. The actual role of such generalist predators in animal population regulation still remains unknown and may be a problem of special interest to be elucidated in future.

After severe and long-lasting controversies among ecologists followed by a number of verifying field population studies, it has been accepted now by many people that animal populations in nature are regulated, in the long run, by density-dependent factors or processes within the frame of resources available in their habitats. Although some people such as Dempster (1983) are skeptical in using the term "regulation" in such a wide sense, it seems to be evident now that the regulation of animal populations in nature, in the usual sense as defined by Solomon (1976), is a fact rather than an hypothesis, since it is nothing but a logical corollary of their persistence over a long period (e.g., Royama, 1977).

The question remaining is, then, to what degree predation actually contributes, in comparison with intraspecies competition or interference for limited resources, to such overall regulation as occurs in natural animal populations. The answer is now clear as we have seen above. As Dempster (1983) claims with reference to insect populations, we may conclude generally that the relative role being played by predators as the regulator of animal populations is negligible or subsidiary, at best, in natural ecosystems. In other words, it is not their predators or the animals ranked higher in the food chain, but the organisms ranked lower in the chain as their food resource that fundamentally govern the dynamics of animal populations in nature. This seems to be a fundamental principle that ensures and explains the successful coexistence of various animals of different trophic levels as seen in most ecosystems in the real world, which originally are to be in paradoxical relations to each other.

In reference to the controversy about the role of predators in field populations of insects, Hassell (1985) claimed rightly that one should be careful not to draw hasty conclusions about such critical problems from inefficient statistical analysis of restricted field population data. Our simulations in Section IV, however, clearly showed that the usual regression techniques for key-factor analysis can be efficient enough for practical application to life-table data, as far as the judgment is made comprehensively from various angles. Moreover, even though the population data analysed might remain more or less deficient individually, the overall conclusion could yet be sufficiently persuasive, provided that, as in the present case, all the results consistently point to the same direction.

The situation may be quite different, however, in the cases of biological pest control where intentionally selected new "able" predators are to be released in the habitats which can be also managed so as to become optimal for the predators to work. Although as yet there have been few data to prove this, I maintain along with Hassell (1985) that the true regulation of prey populations by predators at low levels may be possible under some situations, which should of course be the ideal goal for our practical attempts of pest control.

The fundamental condition required for achieving such a situation seems to be quite simple. In short, it is just to raise the attack efficiency of predators to levels as high as possible. Various criteria for predators which have so far been nominated as conditions for successful control, such as prey-specificity, aggregative response to high prey density patches, synchrony of life cycles with their prey, and so forth (e.g., Huffaker and Messenger, 1976; Hassell, 1978) may all be regarded as relevant for attaining this common aim in different ways, even though each of them might not always be indispensable as Murdoch *et al*. (1984) criticize.

It has become evident now that some degree of habitat heterogeneity is a necessary condition for stable persistence of the interactive systems. There may be no need, however, to take this factor into consideration in actual control programmes. Conversely, it may be advisable rather to attempt to reduce the existing habitat heterogeneity, because it often tends to spoil the predator's search efficiency in exchange for stability, and yet after any attempts in the field to reduce the heterogeneity for better performance of predators, the habitat may still remain more than sufficiently heterogeneous to assure system stabilization. Even if the habitat would happen to be made too homogeneous for stable persistence of the system, it would be quite easy to complement it by proper replenishment of either prey or predators from time to time, as we have seen in the relevant models derived in this study.

ACKNOWLEDGEMENTS

I am deeply indebted to Professor A. Macfadyen of the New University of Ulster for critically reading the draft of this paper and improving it with many verbal corrections. Without his generous help and encouragement, the work would not have been accomplished here. My gratitude is also due to Dr Yoshimi Hirose of Kyushu University for permitting me to use the original data on *Papilio xuthus* for analyses in Section VII and Miss Machiko Terabayashi of our laboratory for the help given to me in preparing the manuscript. This study was supported in part by Science Research Fund (No. 58560046) from the Japan Ministry of Education, Science and Culture.

REFERENCES

Allee, W. C., Emerson, A. E., Park, O., Park, T. and Schmidt, K. P. (1949). "Principles of Animal Ecology" W. B. Saunders, Philadelphia and London.

Andrewartha, H. G. and Birch, L. C. (1954). "The Distribution and Abundance of Animals" Chicago University Press, Chicago.

Auer, C. (1968). Erst ergebnisse einfacher stochastischer Modelluntersuchungen über die Ursachen den Populationsbewegung des grauen Lärchenwicklers *Zeiraphera diniana* G. (=*griseana* Hb.) im Oberengadin, 1949/66. *Z. Ang. Ent.* **62**, 202–235.

Bailey, V. A., Nicholson, A. J. and Williams, E. J. (1962). Interactions between hosts and parasites when some host individuals are more difficult to find than others. *J. Theor. Biol.* **3**, 1–18.

Baltensweiler, W. (1984). The role of environment and reproduction in the population dynamics of the larch bud moth, *Zeiraphera diniana* Gn. (Lepidoptera, Tortricidae). *Adv. Invert. Reprod.* **3**, 291–301.

Beddington, J. R. (1975). Mutual interference between parasites or predators and its effect on searching efficiency. *J. Anim. Ecol.* **44**, 331–340.

Beddington, J. R., Free, C. A. and Lawton, J. H. (1978). Characteristics of successful natural enemies in models of biological control of insect pests. *Nature* **273**, 513–519.

Benson, J. F. (1974). Population dynamics of *Bracon hebetor* Say (Hymenoptera: Braconidae) and *Ephestia cautella* (Walker) (Lepidoptera, Phycitidae) in a laboratory ecosystem. *J. Anim. Ecol.* **43**, 71–86.

Benz, G. (1974). Negative Rückkoppelung durch Raum- und Nährungskonkurrenz sowie zyklisch Veränderung der Nährungsgrundlage also Regelprinzip in der Populationsdynamik des grauen Lärchenwicklers, *Zeiraphera diniana* (Guenee) (Lep., Tortricidae). *Z. Ang. Ent.* **76**, 196–228.

Boer, P. J. den (1968). Spreading of risk and stabilization of animal numbers. *Acta Biotheor.* **18**, 165–194.

Burnett, T. (1958). A model of host–parasite interaction. *Proc. Xth Int. Cong. Ent.* (*1956*) **2**, 679–686.

Caswell, H. (1972). A simulation study of a time-lag population model. *J. Theor. Biol.* **34**, 419–439.

Caughley, G. (1970). Eruption of ungulate populations, with emphasis on Himalayan thar in New Zealand. *Ecology* **51**, 53–72.

Dawkins, R. and Krebs, J. R. (1979). Arms races between and within species. *Proc. R. Soc. Lond.* **B205**, 489–511.

DeBach, P., Ed. (1964). "Biological Control of Insect Pests and Weeds" Reinhold, New York.

DeBach, P. and Smith, H. S. (1941). Are population oscillations inherent in the host–parasite relation? *Ecology* **22**, 363–369.

DeBach, P., Rosen, D. and Kennett, C. E. (1971). Biological control of coccids by introduced natural enemies. *In* "Biological Control" (Ed. C. B. Huffaker), Plenum, New York.

Dempster, J. P. (1975). "Animal Population Ecology" Academic Press, London.

Dempster, J. P. (1982). The ecology of the cinnabar moth, *Tyria jacobaeae* L. (Lepidoptera: Arctiidae). *Adv. Ecol. Res.* **12**, 1–36.

Dempster, J. P. (1983). The natural control of populations of butterflies and moths. *Biol. Rev.* **58**, 461–481.

Elton, C. (1924). Periodic fluctuations in the number of animals: their causes and effects. *Brit. J. Exp. Biol.* **2**, 119–163.

Elton, C. S. and Nicholson, M. (1942). The ten-year cycle in numbers of lynx in Canada. *J. Anim. Ecol.* **11**, 215–244.

Embree, D. G. (1965). The population dynamics of the winter moth in Nova Scotia, 1954–1962. *Mem. Ent. Soc. Canada* **46**, 1–57.

Embree, D. G. (1966). The role of introduced parasites in the control of the winter moth in Nova Scotia. *Can. Ent.* **98**, 1159–1168.

Ende, P. van den (1973). Predator–prey interactions in continuous culture. *Science* **181**, 562–564.

Erlinge, S., Goransen, G., Hogstedt, G., Jansson, G., Liberg, O., Loman, J., Nilsson, I. N., von Schantz, T. and Sylven, M. (1984). Can vertebrate predators regulate their prey? *Amer. Nat.* **123**, 125–133.

Fischlin, A. and Baltensweiler, W. (1979). Systems analysis of the larch bud moth system. Part 1: the larch–larch bud moth relationship. *Mitt. Scweiz. Ent. Ges.* **52**, 273–289.

Fulda, J. S. (1981). The logistic equation and population decline. *J. Theor. Biol.* **91**, 255–259.

Gause, G. F. (1934). "The Struggle for Existence" Williams and Wilkins, Baltimore.

Gilpin, M. E. (1972). Enriched predator–prey systems: theoretical stability. *Science* **177**, 902–904.

Gilpin, M. E. (1973). Do hares eat lynx? *Amer. Nat.* **107**, 727–730.

Gilpin, M. E. (1974). A model of the predator–prey interaction. *Theoret. Pop. Biol.* **5**, 333–344.

Gilpin, M. E. (1975). "Group Selection in Predator-Prey Communities" Princeton University Press, Princeton.

Hassell, M. P. (1978). "The Dynamics of Arthropod Predator–Prey Systems" Princeton University Press, Princeton.

Hassell, M. P. (1980). Foraging strategies, population models and biological control: a case study. *J. Anim. Ecol.* **49**, 603–620.

Hassell, M. P. (1985). Insect natural enemies as regulating factors. *J. Anim. Ecol.* **54**, 323–334.

Hassell, M. P. and Commins, H. N. (1978). Sigmoid functional responses and population stability. *Theoret. Pop. Biol.* **14**, 62–67.

Hassell, M. P. and Huffaker, C. B. (1969). Regulatory processes and population cyclicity in laboratory populations of *Anagasta kuhniella* (Zeller) (Lepidoptera: Phycitidae). III. The development of population models. *Res. Popul. Ecol.* **11**, 186–210.

Hassell, M. P. and May, R. M. (1973). Stability in insect host–parasite models. *J. Anim. Ecol.* **42**, 693–726.

Hassell, M. P. and May, R. M. (1974). Aggregation in predators and insect parasites and its effect on stability. *J. Anim. Ecol.* **43**, 563–594.

Hassell, M. P. and Rogers, D. J. (1972). Insect parasite responses in the development of population models. *J. Anim. Ecol.* **41**, 661–676.

Hassell, M. P. and Varley, G. C. (1969). New inductive population model for insect parasites and its bearing on biological control. *Nature* **223**, 1133–1137.

Hassell, M. P. and Waage, J. K. (1984). Host–parasitoid population interactions. *Ann. Rev. Ent.* **29**, 89–114.

Hassell, M. P., Lawton, J. H. and Beddington, J. R. (1977). Sigmoid functional responses by invertebrate predator and parasitoids. *J. Anim. Ecol.* **46** 249–262.

Hassell, M. P., Waage, J. K. and May, R. M. (1983). Variable parasitoid sex ratios and their effect on host–parasitoid dynamics. *J. Anim. Ecol.* **52**, 889–904.

Hirose, Y., Suzuki, Y., Takagi, M., Hiehata, K., Yamasaki, M., Kimoto, H., Yamanaka, M., Iga, M. and Yamaguchi, K. (1980). Population dynamics of the citrus swallowtail, *Papilio xuthus* Linné (Lepidoptera: Papilionidae): mechanisms stabilizing its numbers. *Res. Popul. Ecol.* **21**, 260–285.

Hokyo, N., and Kuno, E. (1977). Life table studies on the paddy field population of the green rice leafhopper, *Nephotettix cincticeps* Uhler (Hemiptera: Cicadellidae), with special reference to the mechanism of population regulation. *Res. Popul. Ecol.* **19**, 107–124.

Holling, C. S. (1959a). The components of predation as revealed by a study of small mammal predation of the European spruce sawfly. *Can. Ent.* **91**, 293–320.

Holling, C. S. (1959b). Some characteristics of simple types of predation and parasitism. *Can. Ent.* **91**, 385–398.

Holling, C. S. (1961). Principles of insect predation. *Ann. Rev. Ent.* **6**, 163–182.

Holling, C. S. (1973). Resilience and stability of ecological systems. *Ann. Rev. Ecol. Syst.* **4**, 1–24.

Huffaker, C. B. (1958). Experimental studies on predation: dispersion factors and predator–prey oscillations. *Hilgardia* **27**, 343–383.

Huffaker, C. B. and Messenger, P. S., Eds (1976). "Theory and Practice of Biological Control" Academic Press, New York.

Hutchinson, G. E. (1948). Circular causal systems in ecology. *Ann. N.Y. Acad. Sci.* **50**, 221–246.

Ives, W. G. H. (1976). The dynamics of larch sawfly (Hymenoptera: Tenthredinidae) populations in southeastern Manitoba. *Can. Ent.* **108**, 701–730.

Ivlev, V. S. (1961). "Experimental Ecology of the Feeding of Fishes" Yale University Press, New Haven.

Keith, L. B. (1983). Role of food in hare population cycles. *Oikos* **40**, 385–395.

Keith, L. B. and Windberg, L. A. (1978). A demographic analysis of the snowshoe hare cycle. *Wildlife Monographs* **58**, 4–70.

Kiritani, K., Hokyo, N., Sasaba, T. and Nakasuji, F. (1970). Studies of population dynamics of the green rice leafhopper, *Nephotettix cincticeps* Uhler: regulatory mechanism of the population density. *Res. Popul. Ecol.* **12**, 137–153.

Kiritani, K., Kawahara, S., Sasaba, T. and Nakasuji, F. (1972). Quantitative evaluation of predation by spiders on the green rice leafhopper, *Nephotettix cincticeps* Uhler, by a sight-count method. *Res. Popul. Ecol.* **13**, 187–200.

Klomp, H. (1965). The dynamics of a field population of the pine looper (*Bupalus piniarius* L.) (Lepidoptera, Geometridae). *Adv. Ecol. Res.* **3**, 207–305.

Krebs, C. J. (1984). "Ecology, the Experimental Analysis of Distribution and Abundance" 3rd Edn, Harper and Row, New York.

Krebs, C. J., Gaines, M. S., Keuer, B. L., Myers, J. H. and Tamarin, R. H. (1973). Population cycles in small rodents. *Science* **179**, 35–41.

Kuno, E. (1973). Statistical characteristics of the density-independent population fluctuations and the evaluation of density-dependence and regulation in animal populations. *Res. Popul. Ecol.* **15**, 99–120.

Kuno, E. (1983). Factors governing dynamical behaviour of insect populations: a theoretical inquiry. *Res. Popul. Ecol. Suppl.* **3**, 27–45.

Kuno, E. and Hokyo, N. (1970). Comparative analysis of the population dynamics of rice leafhoppers, *Nephotettix cincticeps* Uhler and *Nilaparvata lugens* Stål, with special reference to natural regulation of their numbers. *Res. Popul. Ecol.* **12**, 154–184.

Kuno, E. and Hokyo, N. (1976). Population regulation and dispersal of adults in the green rice leafhopper, *Nephotettix cincticeps* Uhler (Hemiptera: Deltocephalidae). *Physiol. Ecol. Japan* **17**, 117–123. (In Japanese with English summary.)

Lack, D. (1954). "The Natural Regulation of Animal Numbers" Oxford University Press, Oxford.

Lack, D. (1966). "Population Studies of Birds" Oxford University Press, Oxford.

Leslie, P. H. (1948). Some further notes on the use of matrices in population mathematics. *Biometrika* **35**, 213–245.

Lidicker, W. Z. Jr. (1961). Emigration as a possible mechanism permitting the regulation of population density below carrying capacity. *Amer. Nat.* **96**, 29–33.

Lotka, A. J. (1925). "Elements of Physical Biology" Williams and Wilkins, Baltimore.

Luckinbill, L. S. (1973). Coexistence in laboratory populations of *Paramecium aurelia* and its predator, *Didinium nasutum*. *Ecology* **54**, 1320–1327.

Luckinbill, L. S. (1974). The effects of space and enrichment on a prey–predator system. *Ecology* **55**, 1142–1147.

May, R. M. (1972). Limit cycles in predator–prey communities. *Science* **177**, 900–902.

May, R. M. (1973). "Stability and Complexity in Model Ecosystems" Princeton University Press, Princeton.
May, R. M. (1978). Host–parasitoid systems in patchy environments: a phenomenological model. *J. Anim. Ecol.* **47**, 833–843.
May, R. M., Hassell, M. P., Anderson, R. M. and Tonkyn, D. W. (1981). Density dependence in host–parasitoid models. *J. Anim. Ecol.* **50**, 855–865.
Maynard Smith, J. (1974). "Models in Ecology" Cambridge University Press, Cambridge.
Maynard Smith, J. and Slatkin, M. (1973). The stability of predator–prey systems. *Ecology* **54**, 384–391.
Miller, C. A. (1966). The black-headed budworm in eastern Canada. *Can. Ent.* **98**, 592–613.
Morris, R. F. (1959). Single factor analysis in population dynamics. *Ecology* **40**, 580–588.
Morris, R. F., Ed. (1963). The dynamics of epidemic spruce budworm populations. *Mem. Ent. Soc. Canada* **31**, 1–332.
Murdoch, W. W. (1969). Switching in general predators: experiments on predator specificity and stability of prey populations. *Ecol. Monog.* **39**, 335–354.
Murdoch, W. W. and Oaten, A. (1975). Predation and population stability. *Adv. Ecol. Res.* **9**, 2–132.
Murdoch, W. W., Chesson, J. and Chesson, P. L. (1985). Biological control in theory and practice. *Amer. Nat.* **125**, 344–366.
Murdoch, W. W., Reeve, J. D., Huffaker, C. B. and Kennett, C. E. (1984). Biological control of olive scale and its relevance to ecological theory. *Amer. Nat.* **123**, 371–392.
Nakamura, K. and Ohgushi, T. (1981). Studies on the population dynamics of a thistle feeding lady beetle, *Henosepilachna pustulosa* (Kono) in a cool temperate climax forest. II. Life tables, key-factor analysis and detection of regulatory mechanisms. *Res. Popul. Ecol.* **23**, 210–231.
Nicholson, A. J. (1954). An outline of the dynamics of animal populations. *Austral. J. Zool.* **2**, 9–65.
Nicholson, A. J., and Bailey, V. A. (1935). The balance of animal populations. Part I. *Proc. Zool. Soc. Lond.* **3**, 551–598.
Ohgushi, T. (1983). Population processes and life history strategy of an herbivorous lady beetle, *Henosepilachna niponica* (Lewis). PhD Thesis, Faculty of Agriculture, Kyoto University. (In Japanese.)
Ohgushi, T., and Sawada, H. (1985). Population equilibrium with respect to available food resource and its behavioural basis in an herbivorous lady beetle, *Henosepilachna niponica. J. Anim. Ecol.* **54**, 781–796.
Olson, D. C. and Pimentel, D. (1974). Evolution of resistance in a host population to an attacking parasite. *Environm. Ent.* **3**, 621–625.
Pielou, E. C. (1977). "Mathematical Ecology" Wiley, New York.
Pimentel, D. (1961). On a genetic feed-back mechanism regulating populations of herbivores, parasites and predators. *Amer. Nat.* **95**, 65–79.
Pimentel, D., Levin, S. A. and Olson, D. C. (1978). Coevolution and the stability of exploiter-victim systems. *Amer. Nat.* **112**, 119–125.
Pitelka, J. P. (1957). Some aspects of population structure in the short-term cycles of the brown lemming in Northern Alaska. *Cold Spring Harbor Symp. Quant. Biol.* **22**, 237–251.
Podoler, H. (1974). Analysis of life tables for a host and parasite (*Plodia-Nemeritus*) ecosystem. *J. Anim. Ecol.* **43**, 653–670.

Podoler, H. and Rogers, D. J. (1975). A new method for the identification of key factors from life-table data. *J. Anim. Ecol.* **44**, 85–114.
Reeve, J. D. and Murdoch, W. W. (1985). Aggregation by parasitoids in the successful control of the California red scale: a test of theory. *J. Anim. Ecol.* **54**, 797–816.
Richards, O. W. and Waloff, N. (1961). A study of a natural population of *Phytodecta olivacea* (Forster) (Coleoptera, Chrisomeloidea). *Phil. Trans. R. Soc.* (*B*) **244**, 205–257.
Rogers, D. (1972). Random search and insect population models. *J. Anim. Ecol.* **41**, 369–383.
Rogers, D. J. and Hassell, M. P. (1974). General models for insect parasite and predator searching behaviour: interference. *J. Anim. Ecol.* **43**, 239–253.
Rosenzweig, M. L. (1971). Paradox of enrichment: destabilization of exploitation ecosystems in ecological time. *Science* **171**, 385–387.
Rosenzweig, M. L. (1973). Evolution of the predator isocline. *Evolution* **27**, 84–94.
Rosenzweig, M. L., and MacArthur, R. H. (1963). Graphical representation and stability conditions of predator–prey interaction. *Amer. Nat.* **97**, 209–223.
Royama, T. (1971). A comparative study of models for predation and parasitism. *Res. Popul. Ecol. Suppl.* **1**, 1–91.
Royama, T. (1977). Population persistence and density dependence. *Ecolog. Monogr.* **47**, 1–35.
Schaffer, W. M. and Rosenzweig, M. L. (1978). Homage to the red queen, I. Coevolution of predators and their victims. *Theor. Pop. Biol.* **14**, 135–157.
Shimazu, Y. (1973). "Systems Ecology" Kyoritsu Shuppan, Tokyo. (In Japanese).
Slatkin, M. and Maynard Smith, J. (1979). Models of coevolution. *Quart. Rev. Biol.* **54**, 233–263.
Solomon, M. E. (1976). "Population Dynamics" 2nd Edn, Edward Arnold, London.
Takahashi, F. (1959). The effect of host finding efficiency of parasite on the cyclic fluctuation of population in the interacting system of *Ephestia* and *Nemeritis*. *Japan. J. Ecol.* **9**, 88–93.
Takahashi, F. (1969). An experimental study on the suppression and regulation of the host population by the action of the parasitic wasp. *Jap. J. Ecol.* **19**, 225–232. (In Japanese with English summary.)
Takafuji, A. (1977). The effect of the role of successful dispersal of a phytoseiid mite, *Phytoseiulus persimilis* Athias-Henriot (Acarina: Phytoseiidae) on the persistence in the interactive system between the predator and its prey. *Res. Popul. Ecol.* **18**, 210–222.
Takafuji, A., Tsuda, Y. and Miki, T. (1983). System behaviour in predator–prey interaction, with special reference to Acarine predator–prey system. *Res. Popul. Ecol. Suppl.* **3**, 75–92.
Taylor, R. J. (1984). "Predation" Chapman and Hall, New York and London.
Thompson, W. R. (1939). Biological control and the theories of the interactions of populations. *Parasitology* **31**, 299–388.
Utida, S. (1953). Population fluctuations in the system of host–parasite interaction. *Res. Popul. Ecol.* **2**, 22–46. (In Japanese with English summary.)
Utida, S. (1957a). Cyclic fluctuations of population density intrinsic to the host–parasite system. *Ecology* **38**, 442–449.
Utida, S. (1957b). Population fluctuation, an experimental and theoretical approach *Cold Spring Harbor Symp. Quant. Biol.* **22**, 139–151.
Varley, G. C. (1947). The natural control of population balance in the knapweed gall-fly (*Urophora jaceana*). *J. Anim. Ecol.* **16**, 139–187.

Varley, G. C. and Gradwell, G. R. (1968). Population models for the winter moth. *In* "Insect Abundance", (Ed. T. R. E. Southwood). Blackwell, Oxford.

Varley, G. C. and Gradwell, G. R. (1971). Recent advances in insect population dynamics. *Ann. Rev. Ent.* **15**, 1–24.

Volterra, V. (1926). Variazioni e fluttuazioni dei numero d'individui in specie animali conviventi. *Mem. Acad. Lincei.* **2**, 31–113. (English translation in: Chapman, R. N. (1931). "Animal Ecology" McGraw-Hill, New York.)

Wangersky, P. J. and Cunningham, W. J. (1957). Time lag in prey–predator population models. *Ecology* **38**, 136–139.

Watanabe, S. (1950). Interaction between a host and its parasite. *Botyu–Kagaku* **15**, 73–79. (In Japanese with English summary.)

Watson, A. (1971). Key factor analysis, density dependence and population limitation in red grouse. *Dyn. Numbers Popul. Proc. Adv. Study Inst.* (1970) 548–559.

Subject Index

A

Acidification *see also* Soil, acidity
  atmospheric, 2, 23–24, 37–40
  experimental, 38
    effects on invertebrates, 214–216
Acids, organic, litter analyses, 10–12
Afforestation, species and H^+ budgets, 32–36
Air pollution,
  effects on soil invertebrates, 214–216
Alkaloids, in plant phloem, 59
Alnus spp
  colonisation, new soil, 37
  and nitrification, 36
  nitrogen fixation, 30, 34
    and soil pH, 30
Aluminium oxides, and sulphate adsorption, 40
Aluminium, soil, 7–8
Ammonia, as fertilizer, 26
Ant–plant interactions, 56–57
  via Homoptera, 66
Ant–plant–Homopteran interactions, 53–73
  abiotic factors, 72
  diagram, 56
  directions for research, 73
  and lepidopteran interactions, 55
Ants
  adaptations to Homoptera, 60–66
  effects on plants, 67, 68–70
  Lasius alienuis, 147
  Myrmica ruginodis, seed dispersal, 147
  seed "predation", 147
  Tetramorium caespitum, seed harvesting, 147
Apodemus sylvaticus, 326
Arctostaphylos uva-ursi
  life cycle, 124
  toxic substances, 109
Arthropods
  cold tolerance, 195
  Critical Thermal Maximum, 199
  freeze susceptibility, 195–198
  heat tolerance, 199–200
  life history, and diapause, 193–195
  osmoregulation, 186–189
  predator–prey systems, 313–317, 319–324
  starvation, 200–206
  temperature control
    behaviour mechanisms, 192–193
    thermoregulation, 191–192

B

Beauveria bassiani, 145
Beetles
  freeze susceptibility, 195–196
  freeze tolerance, 198
  heather beetle (*Lochmaea*), 145
  thistle lady beetle, 321
  water gain/loss patterns, 178–180, 185–186
Betula spp
  and *Calluna* heathlands, 93
  life cycle, and vegetation, 112
  re-establishment, heathland, 110–112
  soil pH, 32
Bud-moths, 319–320

C

Cadmium, effects, invertebrates, 216–233
Calcium cycle, forest soil, 21–23
  inputs/outputs, 27
Carbon cycle, forest soil, 9–12

"Carrying capacity", 254, 266
Calluna vulgaris see also Heathlands
  annual growth and branching, 96
  biomass, locked nutrients, 155–157
  biomass, seasonal changes, 97–99
  competition, mechanisms, 107–110
  and complementary species, 107
  development, 106–107
  dynamics, 90–94
  flow diagrams, dry matter, 102
  grazing and burning, changes (diagram), 151 *see also* Fire
  and *Lochmaea suturalis*, 94
  photosynthate production, 99
    nutrient content, 101
  soil acidification, 110–111
  Sphagnum and regeneration, 93
  succession, following fire, 121–127
Carex pilulifera, seed dispersal, 147
Cervas elephas (red deer), 137–40
Chestnut oak (*Q. prinus*), case study, 40–44
Chromium, effects, invertebrates, 216–233
Collembola
  and acidification, 215
  calcium utilization, 208
  cold tolerance, 196–198
  diapause, 193–195
  feeding, 200–206
    and freezing, 197
  haemolymph osmolalities, 189
  heat tolerance, 199
  heavy metals, concentrations, 217, 219
  life history patterns, 203–206
  niche parameters, and population density, 207
  Pb tolerance, 228
Composting, 214
Conservation issues, heathlands, 147–155
Copper effects, invertebrates, 216–233
Cornus florida
  with conifer needles, litter decomposition, 35

D

Diel activity cycle, arthropods, 192–193
Deschampsia flexuosa
  and *Calluna* heath, 94
  and fire, 121, 123, 126
  and toxic substances, 109, 110
Dorset
  heathlands
    productivity, 100–101
    reduction, 153–154

E

Earthworms
  chlorogenous tissue, function, 224–225
  Cu, toxicity, 230–231
  food preferences, 208–210, 213
  in forest soils, 36
  heavy metals, concentrations, 217–219
  and soil acidification, 215
Empetrum nigrum, 104, 126
  and fire, 126
  two ecotypes, 108–109
Erica spp, heathlands, 104, 108
Eriophorum
  and *Calluna*, burning, 125–126
  heathland changes, 93–94, 150

F

Festuca ovina, replacing *Calluna*, 145
Fires
  forest ecosystems, 29–30, 43
  heathlands
    effects on nutrients, 117–120
    effects on vegetation, 120–126, 149–152
  and herbivory, 128–129
  management
    of forests, 31–32
    of heathlands, 112–117, 117–127, 149–152
  severity, effects on vegetation, 126–127
  weather following, 127
Forest ecosystems
  acid-base status, buffering, 38–40
  acidity, atmospheric deposition, 2, 37–40
  below-ground litter, 11–12
    and fauna, 36–37
  disturbance and regeneration, 29–30
  effects of management, 31–32

Forest ecosystems *continued*
  fires, *see* Fires
  H^+ budgets, 7–22
    buffering, 38–40
    cellular level, 7–9
    chestnut oak forest, 40–45
    and species, 32–37
    summary, 44–45
    time scales, 27–30
  H^+ fluxes (table), 2
  heavy metals, in soil arthropods, 218–220
  management, 31–32
  secondary succession
    time scales, 29–30
Forestry, expansion, on heathland, 149
Fulvic acid, 10 *see also* Humic acid
  and peat aeration, 33

G

Grazing, heathlands, 149–152, 153
Grouse, red, *see* Red grouse

H

Hare, mountain, 140–142
Heathlands *see also* Sand-dunes
  agricultural uses, in Europe, 148
  burning, 117–127, 149–152
    management, 112–117
  Calluna vulgaris, see *Calluna*
  community dynamics, 103
    main phenophases, 104
  definition, 88
  development, to woodland, 110–112
  flowering times, separation, 105–106
  grazing experiments, 149–150, 153
  herbivore dynamics, 129–144
  invertebrates, 144–147
  management and conservation, 89, 147–155
  microclimate, and soil properties, 120
  nutrient availability, 101, 153
    and burning, 117–120
  production of organic matter, 95–103
  reafforestation, 148
  recreation, 152
  seed “predation”, 147
  steady-state, 102–103
  vegetation dynamics, 90–94
Heat tolerance, in Arthropods, 199–200
Heavy metals, effects, invertebrates, 216–233
Herbivore dynamics
  heathlands, 129–147
Homopteran-plant interactions, 57–60, 68–70
Homoptera
  on *Calluna* heathland, 145
  effect on ants, 67, 68
  heathlands, 145–146
Honeydew, Homopteran
  diet for ants, 63–64
    attractants, 65
    plants, effect of, 66
  uses to plants?, 58
Humic acids, 10
  effects, earthworms, 213
  litter analyses, 10–12
  peat aeration, 33

I

Insects *see also* Arthropods; Invertebrates; Parasitism
  endemic species, 320–324
  epidemic species, 319–320
  Malpighian tubules, heavy metals, 225–226
  midgut pH, and heavy metals, 221–223, 225–226
  predator–prey systems, 313–317, 319–324
Invertebrates, effects of pollution, 214–233
  air pollution, 214–216
  heavy metals, 216–233
    behavioural avoidance mechanisms, 220–221
    life history patterns, 227–231
    maximal concentrations, 217–220
    physiological avoidance mechanisms, 221–227
    tolerance mechanisms, 231–233
Invertebrates, water balance, 177–189 *see also* Arthropods; Insects; names of groups
Iron, effects, invertebrates, 216–233

Isopods
  alimentary canal (diagram), 211
  digestion, and pH, 211
  food shortages, 201–202, 206
  food tolerance mechanisms, 210–213
  heavy metals
    concentrations, 217
    Cu tolerance, 231
    Pb/Cu antagonism, 227
    tolerance mechanisms, 231–232
    Zn/Cd, effects, 228, 229
Ixodes ricinus (sheep-tick), associated virus, 150

J

Juniperus communis
  non-survival, burning practice, 152
  susceptibility to grazing, 140
Juniperus virginiana, 34
  soil pH, 34

L

Lagopus lagopus
  populations, 324–325
Lagopus l. scoticus, 129–137
Lead, effects, invertebrates, 216–233
Lepus americanus/*Lynx canadensis*
  oscillations, 325
Lepus timidus, 140–142
Lichens
  colonisation, post-fire, 121
  inhibition of higher plants, 110
Lignin, and tannin, as arthropod foods, 209, 212–213
Lime, application, 31
Lipids, and high temperatures, 200
Lochmaea suturalis
  effects on heath vegetation, 145

M

Markov chain models, 126
Mercury, effects, invertebrates, 217, 228
Microtus agrestis, 325, 326
Mites
  and calcium utilization, 208
  predator–prey systems, 314–315
Molinia
  grazing by red deer, 318–319
  heathland changes, 108, 150
Mosses
  colonization after fire, 121
  moisture retention, heathlands, 115
Mountain hare, 140–142
Mule deer, population eruption, 325–326
Mycorrhizal symbioses, 36–37

N

Nickel, effects, invertebrates, 216–233
Nitric acid vapour, dry deposition, 26
Nitrogen cycle, forest soil, 12–15
  inputs/outputs, 24–26
Nitrogen fixation, 26
  and decline of pH, 30

O

Osmoregulation, in arthropods, 186–189
Oxalate, forest-soil, 37

P

Parasitism
  experimental models, 314–316
  field examples, 316–321
  model, equation, 256
Pheromones, in plant defence, 59
Phosphorus cycle, forest soil, 15–19
Picea glauca
  soil pH, 35
Picea sitchensis
  "check" by *C. vulgaris,* 109
  reafforestation, heathlands, 148
Pinus banksiana, 35
Pinus contorta
  ant–Homopteran associations, 66
  reafforestation, heathlands, 148
  and soil pH, 33
Pinus ponderosa
  resistance to Homoptera, 60
Pinus radiata, 34
Pinus resinosa, 34, 35

Pinus sylvestris
and *Calluna* heathlands, 93, 110
litter analysis, 11–12
nitrate production, 36
susceptibility to grazing, 140
Pinus taeda
fires, and pH, 32
litter decomposition, with hardwood species, 35
Pinus spp
resistance to Homoptera, 58–59
Polytrichum juniperinum
and burning, 121, 127
Populus balsamifera
soil pH, 32
Populus tremuloides
soil pH, profiles, 32
Predator–prey interaction
basic models, 253–256, 262–273, 283
list of parameters, 261
classic models, 252–253, 262, 282
"disc equation", 254
experimental systems, 311–318
field, 316–318
laboratory, 312–316
extended models, 256–261, 274–281, 283–288
list of parameters, 261
natural systems, 318–326
insects, 319–324
mammals and birds, 324–326
predator–prey conflicts, 305–311
predator population
"prudent predators", 328
reproductive efficiency, death rate, 272
selection, 301–302
specialist/generalist, 303–305
summary and conclusions, 326–331
prey population
alternatives, 259–261, 280–281, 285–288
carrying capacity, 266–267
non-regulatory suppression, 296–298
regulation, robustness, 288–296
selection, 299–301
specialist/generalist, 303–305
recruitment effects, 275–280
"regulation"—definition, 282
in extended models, 283–288
Proteins, thermostability, 200
Protozoan predation, 312–313
Psyllids, and *C. vulgaris*, 146
Pteridium aquilinum
control, 153
toxic substances, 109

Q

Quercus prinus (chestnut oak)
case study, 40–44

R

Rabbits, and heathlands, 129
Rain, *see* Acidification, atmospheric
Red deer, 137–140
Red grouse, 129–137
fire management, 113
food supply, 130–131
management, 131–132
population changes, 132–136
populations, 324–325
Root/soil interface
H^+ cycles, 8–9

S

Sand dunes, invertebrates
cold and heat, 190–200
drought, 177–189
mineral shortage, 206–209
starvation, 200–206
toxic compounds, 209–213
"Seed predation", heathlands, 147
Sewage sludge
Eisenia, in, 217
industrial/rural, composition, 214
Sheep, heathlands, 143–144 *see also* Grazing
Soil acidity *see also* Acidification; Forest ecosystems; H^+ budgets
invertebrates, 214–216
litter, analyses, 11–12
nature, 3–7
pH measurements, 5–7, 33
rain and H^+ fluxes, 2
short-term acceleration, 40

species, effects, 32–36
sulphur dioxide deposition, 37–38
variations, 28
Soil, organic matter: 3 types, 10
Spodosols, sulphate adsorption, 40
Springtails, *see* Collembola
Stomatal control, plant groups, 181
Stress factors, and energy physiology, 176
Strophingia ericae, on Calluna, 145–146
Sulphur cycle, forest soil, 19–21
inputs/outputs, 26–27
Sulphur dioxide, and acidification, 37–38
Sulphuric acid
deposition, industrialized regions, 23–24
experimental deposition, 38, 214–215

T

Trees, *see* Forest ecosystems
Trichostrongylus tenuis, 136
Tsuga mertensiana, changes in pH, 29

U

Ultisols, sulphate adsorption, 40
Urea, as fertilizer, 14, 26

V

Vaccinium spp, 104, 105, 107
and *C. vulgaris*
disposal, by grouse, 130, 137
and fire, 126–127
toxic substances, 109

Z

Zinc, effects, invertebrates, 216–233